Modeling of Microbial Communities: Theory and Practice

Modeling of Microbial Communities: Theory and Practice

Editor: Quinn White

www.callistoreference.com

Callisto Reference,
118-35 Queens Blvd., Suite 400,
Forest Hills, NY 11375, USA

Visit us on the World Wide Web at:
www.callistoreference.com

ISBN: 978-1-64116-799-4 (Hardback)

Cataloging-in-Publication Data

Modeling of microbial communities : theory and practice / edited by Quinn White.
  p. cm.
Includes bibliographical references and index.
ISBN 978-1-64116-799-4
1. Microorganisms. 2. Microbiology. 3. Microbial ecology. I. White, Quinn.
QR41.2 .M63 2023
579--dc23

Table of Contents

Preface

I am honored to present to you this unique book which encompasses the most up-to-date data in the field. I was extremely pleased to get this opportunity of editing the work of experts from across the globe. I have also written papers in this field and researched the various aspects revolving around the progress of the discipline. I have tried to unify my knowledge along with that of stalwarts from every corner of the world, to produce a text which not only benefits the readers but also facilitates the growth of the field.

Microbial communities are the groups of microorganisms that live together in a given environment. The microbial populations that make up these communities interact in a variety of ways, such as symbiosis, competition and predation. The diversity of microbial communities encompasses the abundance and quantity distribution of various kinds of organisms in a particular ecosystem. These communities consist of a wide range of biodiversity existing on the Earth. They play an important role in ecosystem functioning, including xenobiotic degradation, decomposition of organic matter and biogeochemical cycling of nutrients. Computational models of interacting microbes are used for understanding the functioning and dynamics of microbial communities. Modeling also plays an integral role in connecting biogeochemical processes to specific microbial metabolic pathways. This book unfolds the innovative aspects of microbial community modeling which will be crucial for the prediction of microbial interactions in the future. It aims to equip students and experts with the advanced topics and upcoming concepts in this area of study.

Finally, I would like to thank all the contributing authors for their valuable time and contributions. This book would not have been possible without their efforts. I would also like to thank my friends and family for their constant support.

Editor

Clustering on Human Microbiome Sequencing Data: A Distance-Based Unsupervised Learning Model

Dongyang Yang [1] and **Wei Xu** [1,2,*]

[1] Division of Biostatistics, Dalla Lana School of Public Health, University of Toronto, Toronto, ON M5T 3M7, Canada; dongyang.yang@mail.utoronto.ca

[2] Department of Biostatistics, Princess Margaret Cancer Centre, Toronto, ON M5G 2M9, Canada

* Correspondence: Wei.Xu@uhnres.utoronto.ca

Abstract: Modeling and analyzing human microbiome allows the assessment of the microbial community and its impacts on human health. Microbiome composition can be quantified using 16S rRNA technology into sequencing data, which are usually skewed and heavy-tailed with excess zeros. Clustering methods are useful in personalized medicine by identifying subgroups for patients stratification. However, there is currently a lack of standardized clustering method for the complex microbiome sequencing data. We propose a clustering algorithm with a specific beta diversity measure that can address the presence-absence bias encountered for sparse count data and effectively measure the sample distances for sample stratification. Our distance measure used for clustering is derived from a parametric based mixture model producing sample-specific distributions conditional on the observed operational taxonomic unit (OTU) counts and estimated mixture weights. The method can provide accurate estimates of the true zero proportions and thus construct a precise beta diversity measure. Extensive simulation studies have been conducted and suggest that the proposed method achieves substantial clustering improvement compared with some widely used distance measures when a large proportion of zeros is presented. The proposed algorithm was implemented to a human gut microbiome study on Parkinson's diseases to identify distinct microbiome states with biological interpretations.

Keywords: clustering; microbiome; unsupervised learning; high-dimension

1. Introduction

Human microbiome carries complicated relationships among species yet profoundly connect with human health. Studies to understand of the effects of the microbiome on human diseases have been conducted recently. For example, evidence linking Parkinson's disease to the gut microbiome is presented in Hill-Burns' work [1]. The microorganisms in the human body consist of over 1000 species of bacteria [2,3]. Modern technology promotes the scope of microbiome research so that the massive microbiome data can be generated by 16S rRNA sequencing or shotgun metagenomic sequencing.

The microbiome data that is generated by the sequencing technology needs to be classified into taxonomic groups. The abundance of a species is quantified based on their similarity to operational taxonomic units (OTUs) and formed into discrete counts of sequence reads. Typical reads have excessive zeros due to either sampling errors or their unique features that only dominant microorganisms are shared among samples. Positive reads are usually skewed with extreme sparse count measures, which are called overdispersion. Traditional statistical methodologies encounter challenges in microbiome studies and result in potential bias [4–6].

An essential research question of microbiome study is to determine whether the microbiota can be stratified into subgroups. If so, how many groups are there, and how to interpret the strata, i.e., does the classification differentiate treatments, diseases, or genetic types. To answer these questions, the measurement of similarity between two microbial communities is desirable. Beta diversity has been proposed to fit diverse purposes, providing various results in assessing the differences between communities. For microbial composition, beta diversity measures the distance among communities based on measurement abundance, either observed counts or relative abundance, calculated based on a dissimilarity or distance measure to quantify the similarity between samples. Many non-parametric statistical methods have been developed to quantify distance measures. For instance, Euclidean and Manhattan distances are most commonly used. Other beta diversity metrics, such as Bray-Curtis Distance (BC) [7], Jensen-Shannon Distance (JS), Jaccard Index, UniFrac distances (unweighted [8], weighted [9], and generalized [10]) are also frequently applied in microbiome studies. Besides the distance metrics, graphical network models have also been introduced in Sparse Inverse Covariance Estimation for Ecological Association Inference (SPIEC-EASI [11,12]). The method applied a centered log-ratio transformation to the OTU data followed by either neighborhood selection or sparse inverse covariance selection to estimate the interaction graph. However, the method encounters difficulty in the underdetermined data regime. For example, the number of OTUs is much larger than the number of samples. In addition, SPIEC-EASI method relies on a single variance-covariance matrix which may not be able to completely recover the underlying OTU network due to the complex structure of the microbial community. In comparison, the mixture models are more flexible from the way of construction, and it may approximate the real distribution of taxa and lead to more accurate estimations of distances that are used in clustering.

The clustering of microbiome samples has been achieved in many studies using a variety of approaches. Clustering algorithms, including distance-based and parametric modeling, have been used to group subjects according to the microbiome samples. Two main types of distance-based approaches are hierarchical clustering [13–15] with different linkage options and discrete clusterings such as k-means [16–19] and Partition Around Medoids (PAM) [20–23]. Discrete clustering requires a pre-specified number of clusters while different linkages for hierarchical clustering such as Ward linkage, complete linkage, simple linkage, and average linkage provide rules to agglomerate. The Dirichlet-multinomial regression model [24] is the most frequently used on microbial metagenomic data for model-based clustering. Extensions such as the Sparse Dirichlet-multinomial regression technique [25] and finite mixtures of the Dirichlet-multinomial model [26,27] have been proposed to improve on different statistical aspects. These regression models investigate the relation between microbiome composition data and environmental or biological factors. However, currently, only a univariate analysis could be performed. On the other hand, distance-based clustering allows us to take multiple OTUs into account simultaneously.

For the clustering algorithms, it is critical to determine the optimal number of clusters K. Therefore, validation measures for clustering have been explored to identify the ideal number of groups K to represent data [28–32]. Validation indices are used to measure the quality of a clustering result in two ways: internal and external. An internal validation index is to use the information from the data only to decide the optimal number of clusters, such as the Silhouette width index [33], prediction strength [34], Calinski-Harabasz index [35], and Laplace approximation. Validation scores can be computed for different K, respectively, and then they identify the optimal K accordingly. An external validation index, on the other hand, uses prior knowledge to compare the predictive results.

There are different combinations of distance measures and algorithms. Koren et al. [23] computed the distances with and without the root square of the JS, BC, weighted and unweighted UniFrac distances and selected the number of clusters with the prediction strength and the silhouette index used in the PAM algorithm. Unlike Koren's approach, Hong et al. [14] applied K-means with Euclidean distance to identify two clusters, in which they believe the number of clusters makes biologically sense in their study. It is noticeable that no standard clustering pipelines are available, and therefore the various approaches to the recognition of subgroups lead to widely different results. This phenomenon is more evident for microbiome datasets due to the features of microbial data—overdispersion and excessive zeros, which will cause more variations in the process of gathering microbiome into groups.

We develop an innovative clustering approach taking a mixture distribution, rather than a beta diversity metric, as the distance measure and applying a clustering algorithm to the microbiome data to characterize sub-populations. The algorithm also involves selecting the optimal number of clusters based on chosen internal indices, and the results are compared between several distance measures and different evaluation methods. The performance of the proposed algorithm is evaluated through comprehensive simulation studies and a real human gut microbiome dataset on Parkinson's diseases.

2. Materials and Methods

A mixture model is a probabilistic model for representing subpopulations within an overall population, which are frequently used in unsupervised learning [36–39]. Simple distributions such as Binomial, Poisson, and Gaussian are occasionally unable to model more complex data. For instance, microbiome data may consist not only one mode (zeros and low counts), high probability mass for larger counts, and smaller probability mass for high counts. In this case, the data is better to be modelled in terms of a mixture of several components, where each component is a simple probabilistic distribution.

To deal with the unique characteristics of microbiome data—sparsity with abundant zeros, we incorporate a mixture model proposed by Shestopaloff [40] to attain the beta diversity measures for partition. The mixture model focuses on the distribution of a single OTU across a population which can address the problem of sparsity between samples. It parametrically models the counts' underlying rate distribution, including low counts OTUs and extremely high counts. For pairwise distances between individual samples, the formulated mixture's probability is used in L_2 norms distances.

By using such a model, the beta diversity measure contains information regarding zero part in the data and distinguishes between the structural and sampling zeros. The proposed mixture models assume the observed counts are from a Poisson distribution with individual-specific rates, and the rates are sampled from some general population distribution, which can be approximated by a set of mixture components. Conditional on the estimated population rate distribution, the subject-specific rate distribution is estimated through individual mixture distribution given the observed sample counts and resolution. After that, beta diversity measure can be calculated by assessing the pairwise differences between samples for a particular OTU using the individual mixture distribution.

2.1. Mixture Model

To introduce some notations for the following section, let n_{ij} be the number of times an OTU was observed from a sample, where i is the subject, $i = 1, \ldots, I$ and j is the OTU, $j = 1, \ldots, J$. Resolution N_i is the sum of the total reads of an individual; thus, it is defined as $N_i = \sum_j n_{ij}$. To connect the general population distribution to the collected data, the rates can be scaled by the average total reads $\bar{N}_j$ for OTU j, $\bar{N}_j = \sum_i N_{ij}/I$. Therefore, the relative resolution t_{ij} is defined as $t_{ij} = N_{ij}/\bar{N}_j$.

The mixture model consists of five components to accommodate the complexity of microbiome data. For individuals who are never disclosed to OTU j, the model assigns a zero point mass $P(n_{ij} = 0) = 1$. For the rates close to zero, a set of adjacent left-skewed distributions with consecutive parameters is used to represent the low rates. For larger rates, the model accommodates a set of Gamma distribution with parameters which are all integers and are derived from the posterior of the Poisson rate λ given an observed count n, that is $\lambda|n \sim \Gamma(n + 1, 1)$. For higher counts that are less dense, the parameters are defined by truncating the interval uniformly after transforming the data range by a log scale. Lastly, for the even higher counts which are too sparse, the model selects a sufficiently large cut-off point and combines all the observations greater than that point into a high point mass $P(n_{ij} > C) = 1$.

The parts other than zeros and extreme high point masses consist of several components with fixed parameters from Gamma distributions. Since each OTU's distribution is estimated independently, we target one OTU per time and drop the subscript j onward. Define the estimated weights for each mixture components described above as $\vec{w} = (w_z, w_1, \ldots, w_M, w_h)$ where w_z is the weight for zero point mass, w_m where $m = 1, \ldots, M$ is the weight for the Gamma components, and w_h is the weight for the high point mass. Weights estimation is utilized to compose the final mixture model and can be calculated by minimizing the squared differences between the observed aggregated counts and the expected ones. For a particular OTU, the observed aggregated counts can be expressed as $y_k = \sum_i \mathcal{I}(n_i = k)$, where k is the number of counts observed in a sample. The expected aggregated counts are the probability of observing k counts from each mixture component in a sample. For the Gamma components, counts are distributed in negative binomial distribution $NB(\alpha_m, \frac{\beta_m}{t_i + \beta_m})$ conditioned on the relative resolution t_i. Define the probability of observing a count k from the mth mixture component conditional on t_i is $p_{kmi} = P_{NB}(K = k|\alpha_m, \beta_m, t_i)$. The expected aggregate counts $\hat{y}_k$ from all mixture components is

$$
\begin{aligned}
\hat{y}_k &= \sum_m \hat{y}_{km} \\
&= \sum_m w_m \cdot p_{km} \cdot I
\end{aligned}
\tag{1}
$$

The estimate of weights $\vec{w}$ is obtained by optimizing the objective function

$$
argmin_{\vec{w}} \sum_{k \in \vec{k}} [y_k - \sum_{w_m \in \vec{w}} w_m \cdot p_{km} \cdot I]^2
\tag{2}
$$

$$
s.t. \sum_m w_m = 1, \; w_m \geq 0, \forall m
$$

and using bootstrap replicates to find an optimal set of models as mixture components. Details of the bootstrap approach can be found in Appendix A.

For each subject i, the probability density function (PDF) of the mixture model is defined as the product of the individual-specific mixture weights and the count probability from mixture components

$$
\begin{aligned}
\mathcal{P}_i &= [\mathcal{P}_i(z), \mathcal{P}_i(0), \ldots, \mathcal{P}_i(C), \mathcal{P}_i(C+)] \\
&= \vec{w}_i^T \cdot [\mathcal{P}(z), \mathcal{P}(0), \ldots, \mathcal{P}(C), \mathcal{P}(C+)],
\end{aligned}
\tag{3}
$$

where

$$
\mathcal{P}(k) = [P_{G_z}(k), P_{G_1}(k), \ldots, P_{G_M}(k), P_{G_h}(k)],
$$

and

$$
\mathcal{P}_i(z) = w_{G_z},
$$

$$
\mathcal{P}_i(h) = 1 - \sum_{k=1}^{C} \mathcal{P}_i(k) - w_{G_z}
$$

2.2. Distance Measures

2.2.1. L_2 Norms Distances

After finalizing the mixture model distribution, distance measures can be calculated through the pairwise distances between samples using probability distribution. Three distance measures based on L_2 norms are considered for comparisons: discrete L_2 PDF norms (L_2-D PDF), discrete L_2 CDF norms (L_2-D CDF), and continuous L_2 CDF norms (L_2-C CDF).

Given a mixture distribution's PDF as $\mathcal{P}_i$ and its cumulative density function (CDF) as $\mathcal{F}_i$ for subject i, the distance of discrete L_2 PDF norms are computed by

$$
\begin{aligned}
\mathcal{D}_{L_2-D,PDF}(i,j) &= \|\mathcal{P}_i - \mathcal{P}_j\|^2 \\
&= [\mathcal{P}(k)]^2 \sum_{q \in \vec{q}} (w_i - w_j)^2
\end{aligned}
\tag{4}
$$

for $i, j = 1, \ldots, I$ and $i \neq j$, where $\vec{q} = (z, 0, 1, \ldots, M, h)'$. Similarly, the distance of discrete L_2 CDF norms can be calculated analogously using the cumulative density function instead of the PDF.

The continuous L_2 CDF norms can be calculated based on the CDF of two individual-specific mixture models. The L_2-C CDF norms can be computed by

$$
\begin{aligned}
\mathcal{D}_{L_2-C,CDF}(F_i, F_j) &= \int_0^C [\mathcal{F}_i(k) - \mathcal{F}_j(k)]^2 dk \\
&= (w_i - w_j)^T G_{q_1,q_2}(w_i - w_j)
\end{aligned}
\tag{5}
$$

where G_{q_1,q_2} is a matrix such that $\int_0^C G_{q_1}(k)G_{q_2}(k)dk$ represents the two components (q_1, q_2) in the mixture model. See details of the derivation in [40].

2.2.2. Other Distances

Other than the distance measures we obtained using the mixture model, some other metrics are selected for comparison, including two standard beta diversity metrics for any ecological distance-based measures, Manhattan and Euclidean distances, and three distance measures specific in microbiome analysis—Bray-Curtis measure, weighted, and generalized UniFrac distances. An unweighted UniFrac distance is not considered in this study since it does not contain taxa abundance information.

Let x_{ij} and x_{ik}, for $i = 1, \ldots, n$, be the observed counts of OTU i in samples j and k, respectively. Let b_i be the length of the branch i in a phylogenetic tree. The Euclidean distance is defined as

$$
\mathcal{D}_E[j, k] = \sqrt{\sum_{i=1}^n (x_{ij} - x_{ik})^2}.
\tag{6}
$$

The Manhattan distance is defined as

$$
\mathcal{D}_M[j, k] = \sum_{i=1}^n |x_{ij} - x_{ik}|.
\tag{7}
$$

The Bray-Curtis distance measure is defined as

$$
\mathcal{D}_{BC}[j, k] = \frac{\sum_{i=1}^n |x_{ij} - x_{ik}|}{\sum_{i=1}^n (x_{ij} + x_{ik})}.
\tag{8}
$$

The weighted UniFrac distance is defined as

$$
\mathcal{D}_w[j, k] = \frac{\sum_{i=1}^n b_i |x_{ij} - x_{ik}|}{\sum_{i=1}^n b_i (x_{ij} + x_{ik})}.
\tag{9}
$$

And the generalized UniFrac distance is defined as

$$
\mathcal{D}_{g(0.5)}[j, k] = \frac{\sum_{i=1}^n b_i \sqrt{x_{ij} + x_{ik}} |\frac{x_{ij} - x_{ik}}{x_{ij} + x_{ik}}|}{\sum_{i=1}^n b_i \sqrt{x_{ij} + x_{ik}}}.
\tag{10}
$$

Note that due to possible zeros in the denominator in Equation (8)–(10), we add a sufficiently small number (1×10^{-8}) in addition to the sum of observed counts. Sensitivity analysis was done and proved that adding a sufficiently small number in the denominator to avoid zeros does not affect the accuracy results.

2.3. Clustering Validation Indices

The clustering assessment utilizes the partition of data by quantifying the results of a clustering algorithm. The indices measure how well the clustering performed regarding both within and between clusters separability. Validation indices can be divided into internal indices and external assessments. When there are no standard labels of the data to evaluate the partition result, internal indices are considered as an assessment of the clustered data itself. Many internal validation indices have been proposed to choose the optimal number of clusters. The number of clusters is data-driven and is usually required to specify in advance by clustering algorithms. Approaches to select the optimal number for partition consider all possible choices that fit the algorithms and then find the best fit of the data after comparing indices. On the other hand, external assessment scores are calculated by directly comparing the partition results with the prior labels, given that the labels are not used in the model-building stage.

2.3.1. Internal Validation Indices

Among the internal indices, similarities are observed in different indices measures. The Dunn index (DI) [41] is a metric for evaluating the separability of within clusters and between clusters. It is the quotient of the minimal distance between points of different clusters and the most substantial within-cluster distance. Let C_k be a cluster of vectors. The diameter of the cluster, which is the largest distance separating two points in cluster C_k is calculated by $\Delta_k = max_{i,j \in R_k, i \neq j}||S_i^k - S_j^k||$. Consider $C_{k'}$ as another cluster other than C_k. Let $\delta(C_k, C_{k'})$ be the inter-cluster distance metric for clusters C_k and $C_{k'}$. It is measured by the distance between their closest points:

$$\delta(C_k, C_{k'}) = min_{i \in R_k, j \in R_{k'}}||S_i^k - S_j^{k'}||, \tag{11}$$

where S_i^k in cluster C_k and $S_j^{k'}$ in $C_{k'}$. The Dunn index is defined as

$$DI_k = \frac{min_{k \neq k'}\delta(C_k, C_{k'})}{max_{1 \leq k \leq K}\Delta_k}. \tag{12}$$

2.3.2. External Validation Assessment Measures

The results from clustering can be quantified by two measures: accuracy and Jaccard index. Both only consider the results obtained by the optimal number of clusters. Accuracy is how close the clustering results compared to the true cluster index. It is defined as the proportion of correctly clustered subjects. Note that when the clustered number of group c is less than the true number of clusters k, the accuracy of c groups are considered. When c is greater than k, a combination of k out of c groups with the highest accuracy is adopted. Jaccard index measures similarity between the clustered results and the original cluster labels which are defined as the ratio of the number of correctly classified subjects (intersection of predicted and real sets) to the number of the total sample size of the two groups (union of two sets):

$$J(C, K) = \frac{|C \cap K|}{|C \cup K|}.$$

2.4. Partitioning Algorithms for Clustering

Partitioning Around Medoids (PAM) algorithm introduced by Kaufman [20], as produced clustering results of this paper, is an adaption of K-means clustering, yet more computational efficient [42] and more robust to the random noises in the data [43]. The aim of clustering for pre-specified K groups is approached in the partitioning algorithm by incorporating two phases: initialization of K medoids and refinement of the initial medoids within clusters. The algorithm takes a greedy search technique in the first step to locate the K medoids in the data with the least computation. Then it uses a swap operation within the neighborhood in the second phase to minimize the objective function

$$\sum_{j \in C_m} d(m, j),$$

where C_m is the cluster containing object m and $d(m, j)$ is the sum of the distances from object j to the closest medoid m.

PAM follows steepest-ascent hill climber algorithm, which can be summarized as follows:

1. Initialize: randomly choose K of the n points in the dataset to be the initial cluster medoids.
2. Assign each data point to the closest medoid based on distance.
3. Refine: for each medoid m and non-medoid data point j, swap j and m and compute the total cost by the new medoid j. Select the best medoid in terms of minimum cost.
4. Repeat steps 2 and 3 until all the medoids are fixed.

A flowchart can be found in Figure 1. PAM selects the points from the original data as medoids, which reduces the difficulty in explaining the cluster. Computation time can be saved by pre-calculating a distance matrix for all the data points. Then in the swap procedure for every iteration, scanning the matrix could be quickly done. In addition, as the refinement phase in step 3 is slow, re-calculation of the distance from only the points that have been moved between clusters to the new medoids in each iteration helps boost efficiency. For all the remaining points, distances can be re-used in objective function calculation.

The Algorithm 1 shows the detailed steps of the proposed clustering procedure using the mixture distribution.

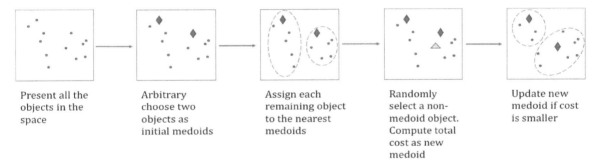

| Present all the objects in the space | Arbitrary choose two objects as initial medoids | Assign each remaining object to the nearest medoids | Randomly select a non-medoid object. Compute total cost as new medoid | Update new medoid if cost is smaller |

Figure 1. A flowchart to illustrate PAM algorithm for two clusters.

Algorithm 1: for clustering based on joint mixture distribution:

1. Specify L models for mixture models with components $G_1, \ldots, G_M$
2. Bootstrap B datasets. For each bootstrap dataset, estimate and select the optimal model based on the distance between observed and expected aggregated counts
3. Calculate weights for the selected model in each replicate and combine with weights for each component to obtain the final weights for the joint mixture model
4. Estimate the subject-specific mixture distribution and calculate the probability of a subject from each component in the mixture model
5. Calculate the distances using L^2 PDF and CDF norms
6. Cluster based on the distances using PAM
7. Compare the scores from internal indices and select the optimal number of clusters as in the final algorithm

---

2.5. Simulation Studies

We conduct extensive simulation studies to evaluate the performance of the proposed algorithm with selected internal validation indices to finalize the optimal number of clusters for the clustering algorithm. To test how well the method performs in clustering, we derive the accuracy and the Jaccard index.

We simulate the data to mimic the OTU counts and their complex structure with class labels. We consider two scenarios with two sub-classes and three sub-classes, and each sub-class contains 200 subjects for total sample sizes of 400 and 600, respectively. All the results are replicated 100 times.

The simulation procedure for each OTU is as follows:

1.   Determine the number of rates r_i in each sub-class using $Beta(\alpha_c, \beta_c)$.
2.   Generate the rate r_i distribution in each sub-class from a mixture distribution with M components including a zero point mass and $M - 1$ Gamma distributions. The number M is randomly chosen between 5 and 15.
3.   Sample the number of rates from $M_{rate} \sim Multinomial(P(M1), P(M2), ..., P(M5), N_c)$ where N_c is the sample size for sub-class c.
4.   Sample t_i for each subject from a $Uniform(2/3, 4/3)$.
5.   Generate the observed count $n_i \sim Poisson(r_i t_i)$.

For each scenario, 25 OTUs are simulated. The simulated count data contains three sets of zero proportions (ZP), first set with 13–27% zeros (low ZP) in each sub-class, second set with 39–61% zeros (medium ZP), and third set with 84–93% zeros (high ZP), to examine clustering performance under different ZP scenarios. ZP in every dataset is controlled by varying α_c in Beta distribution from Step 1. The details of the mixture distribution estimation can be found in Appendix B.

3. Results

3.1. Simulation Results

Figures 2 and 3 show the clustering results of simulated datasets under different scenarios with varying ZPs and number of sub-classes. The distance-based algorithm performance is evaluated through the accuracy and the Jaccard index and presented in boxplots. Specifically, *L2.d.pdf, L2.d.cdf, L2.c.cdf, Manhattan, Euclidean, BC, wUniFrac,* and *gUniFrac* represent the clustering results from a distance calculated by the mixture model using L_2 norms with discrete variable's PDF, discrete variable's CDF, continuous variable's CDF, Manhattan distance, Euclidean distance, Bray-Curtis distance, weighted UniFrac distance, and generalized UniFrac distance, respectively. All the distances were calculated based on the relative abundance data. We conducted additional simulations to calculate the Manhattan, Euclidean, and Bray-Curtis distances after log-transformation. As we explained in methodology, since unweighted UniFrac distances neglect the abundance information and only consider presence/absence of species of branches in a phylogenetic tree, it is not included in the simulation studies. The top three boxplots in Figure 2 illustrate the accuracy of eight comparative distance metrics for high, medium, and low ZP scenarios when the simulated dataset contains two sub-classes. The bottom three boxplots are the accuracy of the 3-subclass simulation scenario. Our proposed distance measures are marked in green as opposed to the other distance metrics in blue. Jaccard index boxplots (Figure 3) are constructed in the same way as in Figure 2. Mean accuracy (MA) and mean Jaccard index (MJI) are shown in Table 1, calculated by averaging the 100 replicates results in each scenario.

We observed that by implementing the proposed distance measures in the clustering algorithm, both accuracy and Jaccard index outperform the results by other distance metrics, especially when the datasets contain a substantial amount of zeros. Clustering using the three proposed L_2 norms in both 2-subclass and 3-subclass scenarios has considerable improvements with the increase of zero proportions in the datasets. The mean accuracy calculated based on 100 replicates achieves around 0.6 for the proposed L_2 norms under high ZP design in 2-subclass scenarios and 0.45 in 3-subclass scenarios. For scenarios with fewer zeros, the L_2-norm distance measures have competitive clustering performance as the competing distance metrics. Among the three L_2 norms, the L_2 discrete PDF distance has better clustering performance across ZP settings. Out of six settings that we investigated in the 3-subclass scenario, the generalized UniFrac distance and the Manhattan distance with log-transformation provide the best partition with a high and low proportion of zeros, respectively. In contract, the L_2 norms show advantages in terms of MA and MJI in the rest of scenarios. Noticeably, the generalized UniFrac distance shows a large variability in the estimation of accuracy. Overall, Manhattan, Euclidean, Bray-Curtis, and weighted UniFrac distance metrics do not distinguish the proportion of zeros in 2-subclass datasets and provide close to the random guess accuracy of 0.5 in 2-subclass scenarios. Jaccard index reveals a similar pattern as accuracy.

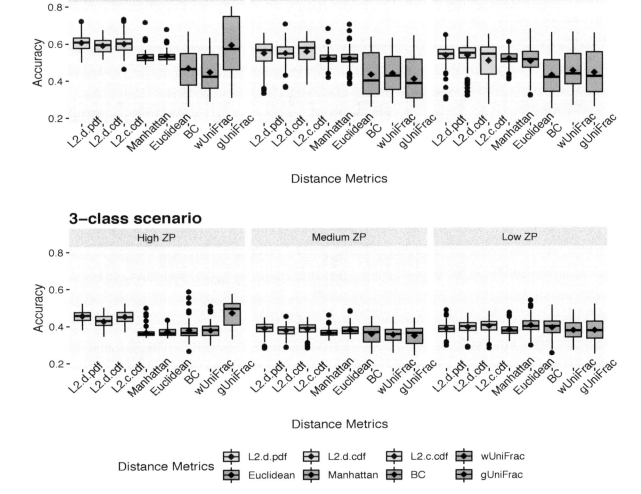

Figure 2. Accuracy boxplots for simulated data. Two-subclass and three-subclass scenarios are considered. Three different cases of proportion of zeros (ZP) are evaluated - high ZP, medium ZP, and low ZP, are presented in left, middle, and right, respectively. For each box of the boxplots, the center line represents the median, the two vertical lines represent the 25th percentiles to the 75th percentiles. The whiskers of the boxplots show 1.5 interquartile range (IQR) below the 25th percentiles and 1.5 IQR above the 75th percentiles. The mean are shown in blue diamond dots.

2–class scenario

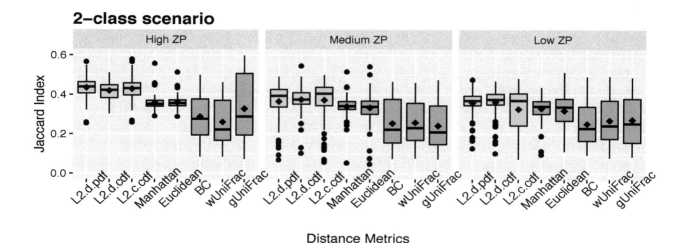

3–class scenario

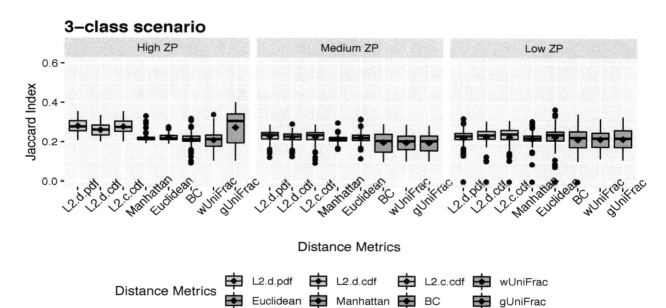

Figure 3. Jaccard index boxplots for simulated data. Two-subclass and three-subclass scenarios are considered. Three different cases of proportion of zeros (ZP) are evaluated - high ZP, medium ZP, and low ZP, are presented in left, middle, and right, respectively. For each box of the boxplots, the center line represents the median, the two vertical lines represent the 25th percentiles to the 75th percentiles. The whiskers of the boxplots show 1.5 interquartile range (IQR) below the 25th percentiles and 1.5 IQR above the 75th percentiles. The mean are shown in blue diamond dots.

The average number of clusters over all the iterations are presented in Table 2. The L_2-D CDF norm predicts the number of clusters the most accurate in 2-subclass scenarios. On the other hand, generalized UniFrac distance and Manhattan distance with log transformation have closer prediction to the actual cluster numbers in the 3-subclass scenarios. However, Bray-Curtis, weighted UniFrac, and generalized UniFrac distances overestimate number of clusters dramatically in the two-subclass situations. Thus, the results in three-subclass scenarios are doubtful to some extent.

Table 1. Mean accuracy (MA) and mean Jaccard index (MJI) estimations. A bold value represents the best cases under each scenario. Distances calculation was conducted on the simulated data inputs with and without log-transformation: Manhattan_log, Euclidean_log, Bray-Curtis_log were models using log-transformation, while L_2-D PDF, L_2-D CDF, L_2-C CDF, Manhattan, Euclidean, Bray-Curtis, weighted UniFrac, and generalized UniFrac were without log-transformation.

Two-Subclass Scenarios						
	High ZP		Medium ZP		Low ZP	
Distance	MA	MJI	MA	MJI	MA	MJI
L_2-D PDF	**0.608**	**0.435**	0.547	0.365	**0.534**	0.359
L_2-D CDF	0.591	0.419	0.547	**0.374**	**0.534**	**0.363**
L_2-C CDF	0.600	0.428	**0.556**	0.371	0.505	0.327
Manhattan	0.530	0.357	0.518	0.340	0.516	0.331
Euclidean	0.530	0.357	0.518	0.341	0.516	0.331
Bray-Curtis	0.467	0.288	0.434	0.258	0.427	0.253
Weighted UniFrac	0.445	0.260	0.437	0.258	0.451	0.270
Generalized UniFrac	0.605	0.420	0.407	0.241	0.441	0.274
Manhattan_log	0.534	0.360	0.520	0.334	0.499	0.317
Euclidean_log	0.534	0.360	0.516	0.333	0.502	0.320
Bray-Curtis_log	0.467	0.287	0.431	0.254	0.427	0.252

Three-Subclass Scenarios						
	High ZP		Medium ZP		Low ZP	
Distance	MA	MJI	MA	MJI	MA	MJI
L_2-D PDF	0.456	**0.281**	**0.386**	**0.230**	0.373	0.223
L_2-D CDF	0.427	0.261	0.375	0.226	0.381	0.228
L_2-C CDF	0.452	0.277	0.383	0.226	0.386	0.228
Manhattan	0.366	0.222	0.364	0.217	0.367	0.217
Euclidean	0.375	0.229	0.364	0.220	0.384	0.234
Bray-Curtis	0.379	0.211	0.348	0.193	0.369	0.207
Weighted UniFrac	0.376	0.210	0.351	0.198	0.374	0.217
Generalized UniFrac	**0.470**	0.274	0.346	0.197	0.374	0.219
Manhattan_log	0.383	0.234	0.379	0.227	**0.404**	**0.243**
Euclidean_log	0.371	0.225	0.377	0.224	0.390	0.228
Bray-Curtis_log	0.378	0.212	0.355	0.199	0.378	0.215

Table 2. Average number of clusters for all simulation scenarios. Optimal number of clusters in each replicate is calculated by Dunn internal indices. A bold value represents the closest estimation of number of clusters to the ground truth.

Distance	Two-Subclass Scenarios			Three-Subclass Scenarios		
	High ZP	Medium ZP	Low ZP	High ZP	Medium ZP	Low ZP
L_2-D PDF	2.47	2.73	2.30	2.48	2.74	2.35
L_2-D CDF	**2.26**	**2.28**	**2.28**	2.16	2.43	2.30
L_2-C CDF	2.40	2.46	2.65	2.26	2.88	2.63
Manhattan	2.87	2.73	2.84	2.48	2.68	2.63
Euclidean	2.86	2.72	2.85	2.93	2.85	2.91
Bray-Curtis	3.08	3.47	3.44	3.31	3.57	3.44
Weighted UniFrac	3.38	3.39	3.23	3.35	3.36	3.29
Generalized UniFrac	2.88	3.39	3.16	**3.01**	3.34	3.20
Manhattan_log	2.72	2.95	3.11	2.82	**2.91**	**2.98**
Euclidean_log	2.76	2.94	3.05	2.46	2.80	2.78
Bray-Curtis_log	3.08	3.50	3.44	2.80	3.48	3.17

4. Real Data Implementation

To demonstrate how well our proposed method works, we analyze the data from Hill-Burns et al. [1] that relates the gut microbiome to Parkinson's disease (PD). The dataset contains stool samples of 197 PD cases and 130 controls. 16S rRNA amplicon sequencing of DNA was extracted for microbial composition performed by Illumina MiSeq. OTUs were picked using a reference of Greengenes 16S rRNA gene sequence database [44] at 97% similarity released in August 2013. The study has shown the association between the dysbiosis of gut microbiome and PD. Besides, the case-only analysis identifies a significant interaction effect between the microbiome and PD medications, including catechol-O-methyl transferase (COMT) inhibitors, anticholinergics, and carbidopa/levodopa.

We apply our algorithm to the PD cases to explore the sub-populations of the PD using gut mirobiome data. The pre-processing step is done by including OTUs on the genus level and excluding ones with the probability of zero relative abundances higher than 80%, resulting in a total of 280 OTUs used for all the samples in our analysis. We compare the proposed L_2 norms with the other three distance metrics with and without log-transformation on the relative abundance data. Various internal indices are applied to distance measures for sensitivity analysis. Selection results of the number of clusters are illustrated in Table 3. The maximum number of clusters is set to ten, meaning that the optimal number is between 2 and 10. Different combinations of distances and internal indices provide moderate variations. For Dunn and Xie-Beni [45] indices, L_2 norms tend to cluster the data into two or three subgroups while in both with and without log-transformation situations, Manhattan, Euclidean, and Bray-Curtis metrics prefer more clusters except for non-transformed Euclidean distance. Wemmert-Gancarski index provides fewer subgroups than the others across the distance measures. No profound trend is found for the Silhouette index.

To illustrate the clustering algorithm, the L_2-D PDF norm is selected as an example for further analysis. We explore OTUs between two clusters for the dataset, and the top 5 significantly different OTUs between clusters are *Akkermansia*, *Anaerotruncus*, *Bacteroides*, *Anaerococcus*, and *Akkermansia*. Among these OTUs, *Akkermansia* [1,46] has previous reported associations. The characteristics of the five OTUs are summarized in Table 4.

Table 3. Mean with standard deviation and median for OTUs that are significantly different ($p < 0.001$) between two clusters using Dunn internal indices and L_2-D PDF distance.

Distance	Dunn	Silhouette Index	Wemmert-Gancarski	Xie-Beni
L_2 discrete PDF	2	7	2	2
L_2 discrete CDF	3	2	2	3
L_2 continuous CDF	2	5	3	3
Manhattan	10	4	4	10
Euclidean	3	3	3	3
Bray-Curtis	7	10	2	10
Manhattan_log	9	2	10	10
Euclidean_log	5	2	5	5
Bray-Curtis_log	9	9	2	10

Table 4. Optimal number of clusters for distance metrics by various internal indices.

OTU	Full Sample (n = 197)	Cluster 1 (n = 166)	Cluster 2 (n = 31)
g_Akkermansia			
Mean (sd)	404.3 (956.8)	479.7 (1025.2)	0.9 (1.4)
Median (Min,Max)	1 (0,5284)	3.5 (0,5284)	1 (0,7)
g_Anaerotruncus			
Mean (sd)	4.5 (11.8)	4.2 (12.5)	6.2 (7.2)
Median (Min,Max)	0 (0,120)	0 (0,120)	3 (0,27)
g_Bacteroides			
Mean (sd)	0.6 (1.5)	0.5 (1.4)	1.3 (1.8)
Median (Min,Max)	0 (0,9)	0 (0,9)	0 (0,5)
g_Anaerococcus			
Mean (sd)	8.8 (40.5)	8.8 (43.6)	8.8 (16.3)
Median (Min,Max)	0 (0,352)	0 (0,352)	2 (0,79)
g_Akkermansia_			
Mean (sd)	198.8 (811.5)	0.7 (1.9)	1259.8 (1709.5)
Median (Min,Max)	0 (0,6278)	0 (0,17)	378 (53,6278)

5. Discussion

We simulate six different scenarios to evaluate the performance of the proposed method thoroughly, using the accuracy and the Jaccard index to reflect clustering results, considering different zero proportions under 2-subclass and 3-subclass settings. Both the accuracy and the Jaccard index are improved or competitive compared to other distance metrics, suggesting better separation among subgroups. Our method performs the best in high and medium zero proportion scenarios, therefore, it is recommended to use our clustering algorithm when a large number of zeros presenting in the data. Under the PAM framework, all distance matrices (Manhattan, Euclidean, Bray-Curtis, and UniFrac) can be used as inputs for clustering. However, as shown in our simulation studies, the pairwise distances calculated by the mixture model perform better than the other distance matrices under a variety of scenarios.

The clustering algorithm involves multiple options, such as the choice of distance measures, the internal indices to specify the number of clusters, and the approach to clustering. Due to a lack of widely accepted standardization, making different choices at each step may lead to various outcomes. Many choices are available regarding the selection of the number of clusters. The decision to make about the optimal number of clusters each time highly relies on data structure, thus case-specific. As we choose to use the Dunn index as internal validation indices for simulation studies, sensitivity analysis is performed using other internal indices. All the considered indices are also compared in real data. Our algorithm classifies subgroups among the PD cases and presented the ability to identify statistically significant distinct OTUs which have association with PD.

The proposed method focuses on distance-based clustering. The next step is to perform partition based on models such as Dirichlet-multinomial and compare with our method. We will also explore the possibility of extending the proposed method to adopt longitudinal trajectories of subjects for deep insights into the dynamic biological mechanisms. The proposed method could be easily extended to high dimensional data with overdispersion. Besides that, we are working on the extension of this proposed method on other microbiome and disease correlation data.

As all clustering methods, one limitation of this algorithm is that suitable internal indices are hard to select for every new data. Thus an optimal and robust number of clusters is difficult to obtain. Besides, for the L_2-norm distances, variable selection is not possible to be developed in clustering. Nevertheless, the proposed algorithm incorporates ad-hoc distance for microbial sequencing data, which provides effective clustering and broader vision to investigate the connection between the microbiome and human health. The introduced clustering algorithm can be seen as a good additional tool for the analysis of microbial data besides the currently used methods.

6. Conclusions

In this article, we propose a distance-based unsupervised machine learning method to cluster subjects based on their microbial structure. We show that our method provide funtional partitions among subjects under various scenarios in simulation studies, and we apply it to a gut microbiome dataset for Parkinson's disease. The distance measures we adopted in the clustering algorithm are capable of capturing the underlying rate distributions of microbial counts, through mixture distributions which take account to zero inflation and overdispersed values. L_2 norms are calculated based on the mixture distributions' PDF and CDF, respectively, and further used in partition around mediod for clustering.

Author Contributions: Conceptualization, D.Y. and W.X.; methodology, data collection and analysis, D.Y.; experimental design and supervision, W.X.; writing-original draft preparation, D.Y.; writing-review and editing, D.Y. and W.X. All authors have read and agreed to the published version of the manuscript.

Funding: This research was funded by Canadian Institutes of Health Research (CIHR Grant 145546) Natural Sciences and Engineering Research Council of Canada (NSERC Grant RGPIN201706672), Crohns and Colitis Canada (CCC Grant CCCGEMIII), CANSSI Ontario Strategic Training for Advanced Genetic Epidemiology (STAGE) Program, and Edwin S.H. Leong Scholarship.

Acknowledgments: We would like to thank Konstantin Shestopaloff and Mei Dong for discussion in this work.

Conflicts of Interest: The authors declare no conflict of interest. The funders had no role in the design of the study; in the collection, analyses, or interpretation of data; in the writing of the manuscript, or in the decision to publish the results.

Abbreviations

The following abbreviations are used in this manuscript:

OTU	operational taxonomic unit
BC	Bray-Curtis distance
JS	Jenson-Shannon distance
PAM	partition around medoids
DI	Dunn index
MA	mean accuracy
MJI	mean Jaccard index
PD	Parkinson's disease

Appendix A. Mixture Model

Appendix A.1. Individual Mixture Distribution Estimation

A non-parametric bootstrap is adopted to estimate a mixture model with a practical consideration to avoid overspecified and overfitted bootstrap data. The selection of the best model among all bootstrap datasets is unnecessary since the final model will be a set of weighted optimal models from all the datasets so that every component can be incorporated in the estimation. Define a set of models $\Psi_l, l = 1, \ldots, L$ with different mixture components combination. Sample B subsets from the original data and estimate weights $\vec{w}_{bl}$ for each model l and subset b. In each bootstrap dataset, the estimated aggregated counts $\vec{\hat{y}}_l = \sum_k \hat{y}_{kl}, k = 0, \ldots, C, C+$ for model l are compared with the observed aggregated counts $\vec{y}$ through the distance $D(\vec{y}, \vec{\hat{y}}_l)$ for $l = 1, \ldots L$ and the best model l_b^* with minimum distance is chosen in subset b. The selected optimal model has weight $v(l) = \sum_b I(l_b^* = l)/B$, and the weight for each component in the mixture distribution is $w_m = \sum_l v(l) w_{ml}$.

Once we have an estimate of the mixture component weights $\vec{w}$ for the final set of mixture components $\mathcal{G} = (G_z, G_1, G_2, \ldots, G_M, G_h)$, the individual-specific mixture probabilities condition on $\vec{w}$

can be obtained for each component. Specifically, the probability of observing counts k from subject i being from each mixture component G_q can be written as

$$P(i \in G_q) = \begin{cases} \mathcal{I}(k=0), & \text{if } q = z \\ w_q \frac{\Gamma(k+\alpha_q)}{\Gamma(k+1)\Gamma(\alpha_q)} \left(\frac{\beta_q}{t_i+\beta_q}\right)^{\alpha_q} \left(1 - \frac{\beta_q}{t_i+\beta_q}\right)^k, & \text{if } q = 1, ..., M \\ \mathcal{I}(k > C), & \text{if } q = h \end{cases} \tag{A1}$$

Denote the individual-specific mixture weights as $\vec{w}_i = (w_{iz}, w_{i1}, \ldots, w_{iM}, w_{ih})$, and weights are calculated as $w_{iq} = P(i \in G_q)/\sum_q P(i \in G_q)$. Since the model components $\mathcal{G}$ remains the same for all the samples, the individual-specific mixture weights become the only variation to compute pairwise distances among samples. Hence distances are calculated based on two parts, individual weights and the probability of an observed count through Poisson-Gamma mixture probabilities $NB(\alpha_q, \frac{\beta_q}{1+\beta_q})$ from mixture component G_q. The probability of observing count k from mixture component G_q, $q = z, 1, \ldots, M, h$ is

$$P_{G_q}(k) = \begin{cases} \mathcal{I}(K=0), & \text{if } q = z \\ P_{NB}(K=k|\alpha_q, \beta_q), & \text{if } q = 1, ...M \\ \mathcal{I}(K > C), & \text{if } q = h \end{cases} \tag{A2}$$

Note that the distribution can estimate structural and non-structural zeros separately, where $P_{G_z}(0) = w_z$ is for structure zeros and $P(k=0) = \sum_q P_{G_q}(0) - w_z$ is for non-structure zeros.

Appendix B. Simulation

Appendix B.1. Mixture Distribution Estimation

The underlying mixture distributionis modeled with pre-set four parts to estimate the simulated data.

- Low rate part: set five Gamma distribution $\Gamma(1,2), \Gamma(1,1), \Gamma(2,1), \Gamma(3,1), \Gamma(4,1)$, and five models, with the first model including all the distributions, the second model including the last 4 distributions, until the fifth model including only the last Gamma distribution;

- Medium rate part: one model with four Gamma distribution $\Gamma(5,1), \Gamma(6,1), \Gamma(7,1), \Gamma(8,1)$;

- Higher rate part: one model with three Gamma distribution $\Gamma(11,1), \Gamma(18,1), \Gamma(19,1)$; each α value in Gamma distribution is chosen by uniformly binned the OTUs on a log-transformed scale from 8% to the 85% quantile;

- High count part: a point mass which combines all the counts greater than the 85% quantile.

The estimation of the joint model is replicated through 300 bootstraps. The estimates of weights in the model are obtained by minimizing the least squares objective function using the Broyden-Fletcher-Goldfarb-Shanno algorithm [47]. We used R packages *NLoptr* [48] for the model estimation, and *cluster* [49] and *clustCrit* [50] for comparison of clustering performance between distance measures.

The appendix is an optional section that can contain details and data supplemental to the main text. For example, explanations of experimental details that would disrupt the flow of the main text, but nonetheless remain crucial to understanding and reproducing the research shown; figures of replicates for experiments of which representative data is shown in the main text can be added here if brief, or as Supplementary data. Mathematical proofs of results not central to the paper can be added as an appendix.

References

1. Hill-Burns, E.M.; Debelius, J.W.; Morton, J.T.; Wissemann, W.T.; Lewis, M.R.; Wallen, Z.D.; Peddada, S.D.; Factor, S.A.; Molho, E.; Zabetian, C.P.; et al. Parkinson's disease and Parkinson's disease medications have distinct signatures of the gut microbiome. *Mov. Disord.* **2017**, *32*, 739–749. [CrossRef] [PubMed]

2. Falony, G.; Joossens, M.; Vieira-Silva, S.; Wang, J.; Darzi, Y.; Faust, K.; Kurilshikov, A.; Bonder, M.J.; Valles-Colomer, M.; Vandeputte, D. Population-level analysis of gut microbiome variation. *Science* **2016**, *352*, 560–564. [CrossRef] [PubMed]

3. Zhernakova, A.; Kurilshikov, A.; Bonder, M.J.; Tigchelaar, E.F.; Schirmer, M.; Vatanen, T.; Mujagic, Z.; Vila, A.V.; Falony, G.; Vieira-Silva, S.; et al. Population-based metagenomics analysis reveals markers for gut microbiome composition and diversity. *Science* **2016**, *352*, 565–569. [CrossRef] [PubMed]

4. Xu, L.; Paterson, A.D.; Turpin, W.; Xu, W. Assessment and selection of competing models for zero-inflated microbiome data. *PLoS ONE* **2015**, *10*, e0129606. [CrossRef]

5. Zhang, X.; Mallick, H.; Tang, Z.; Zhang, L.; Cui, X.; Benson, A.K.; Yi, N. Negative binomial mixed models for analyzing microbiome count data. *BMC Bioinform.* **2017**, *18*, 4. [CrossRef]

6. Fisher, C.K.; Mehta, P. Identifying keystone species in the human gut microbiome from metagenomic timeseries using sparse linear regression. *PLoS ONE* **2014**, *9*, e102451. [CrossRef]

7. Bray, J.R.; Curtis, J.T. An ordination of the upland forest communities of southern Wisconsin. *Ecol. Monogr.* **1957**, *27*, 326–349. [CrossRef]

8. Lozupone, C.; Knight, R. UniFrac: A new phylogenetic method for comparing microbial communities. *Appl. Environ. Microbiol.* **2005**, *71*, 8228–8235. [CrossRef]

9. Lozupone, C.A.; Hamady, M.; Kelley, S.T.; Knight, R. Quantitative and qualitative β diversity measures lead to different insights into factors that structure microbial communities. *Appl. Environ. Microbiol.* **2007**, *73*, 1576–1585. [CrossRef]

10. Chen, J.; Kyle, B.; Emily, S.; Charlson, C.; Hoffmann, J. Associating microbiome composition with environmental covariates using generalized UniFrac distances. *Bioinformatics* **2012**, *28*, 2106–2113. [CrossRef]

11. Zachary, D.; Christian, L.; Emily, R.; Dan, R.; Martin, J. Sparse and compositionally robust inference of microbial ecological networks. *PLoS Comput. Biol.* **2015**, *11*, e1004226.

12. Tsilimigras, M.C.B.; Fodor, A.A. Compositional data analysis of the microbiome: Fundamentals, tools, and challenges. *Ann. Epidemiol.* **2016**, *26*, 330–335. [CrossRef]

13. Forney, L.J.; Gajer, P.; Williams, C.J.; Schneider, G.M.; Koenig, S.S.; McCulle, S.L.; Karlebach, S.; Brotman, R.M.; Davis, C.C.; Ault, K.; et al. Comparison of self-collected and physician-collected vaginal swabs for microbiome analysis. *J. Clin. Microbiol.* **2010**, *48*, 1741–1748. [CrossRef] [PubMed]

14. Hong, B.Y.; Araujo, M.V.F.; Strausbaugh, L.D.; Terzi, E.; Ioannidou, E.; Diaz, P.I. Microbiome profiles in periodontitis in relation to host and disease characteristics. *PLoS ONE* **2015**, *10*, e0127077. [CrossRef]

15. Leake, S.L.; Pagni, M.; Falquet, L.; Taroni, F.; Greub, G. The salivary microbiome for differentiating individuals: Proof of principle. *Microbes Infect.* **2016**, *18*, 399–405. [CrossRef] [PubMed]

16. Neyman, J. *Proceedings of the Third Berkeley Symposium on Mathematical Statistics and Probability: Held at the Statistical Laboratory, University of California, 21 June–18 July 1970, 9–12 April, 16–21 June, 19–22 July 1971;* University of California Press: Berkeley, CA, USA, 1972.

17. Gury-BenAri, M.; Thaiss, C.A.; Serafini, N.; Winter, D.R.; Giladi, A.; Lara-Astiaso, D.; Levy, M.; Salame, T.M.; Weiner, A.; David, E.; et al. The spectrum and regulatory landscape of intestinal innate lymphoid cells are shaped by the microbiome. *Cell* **2016**, *166*, 1231–1246. [CrossRef] [PubMed]

18. Poole, A.C.; Goodrich, J.K.; Youngblut, N.D.; Luque, G.G.; Ruaud, A.; Sutter, J.L.; Waters, J.L.; Shi, Q.; El-Hadidi, M.; Johnson, L.M.; et al. Human salivary amylase gene copy number impacts oral and gut microbiomes. *Cell Host Microbe* **2019**, *25*, 553–564. [CrossRef] [PubMed]

19. Maia, M.C.; Poroyko, V.; Won, H.; Almeida, L.; Bergerot, P.G.; Dizman, N.; Hsu, J.; Jones, J.; Salgia, R.; Pal, S.K. Association of Microbiome and Plasma Cytokine Dynamics to Nivolumab Response in Metastatic Renal Cell Carcinoma (mRCC). *J. Clin. Oncol.* **2018**, *36*, 656–656. [CrossRef]

20. Kaufman, L.; Rousseeuw, P.J. Partitioning around medoids (program pam). *Find. Groups Data Introd. Clust. Anal.* **1990**, *344*, 68–125.

21. Arumugam, M.; Raes, J.; Pelletier, E.; Le Paslier, D.; Yamada, T.; Mende, D.R.; Fernandes, G.R.; Tap, J.; Bruls, T.; Batto, J.M.; et al. Enterotypes of the human gut microbiome. *Nature* **2011**, *473*, 174–180. [CrossRef]

22. McMurdie, P.J.; Holmes, S. Waste not, want not: Why rarefying microbiome data is inadmissible. *PLoS Comput. Biol.* **2014**, *10*, e1003531. [CrossRef] [PubMed]

23. Koren, O.; Knights, D.; Gonzalez, A.; Waldron, L.; Segata, N.; Knight, R.; Huttenhower, C.; Ley, R.E. A guide to enterotypes across the human body: Meta-analysis of microbial community structures in human microbiome datasets. *PLoS Comput. Biol.* **2013**, *9*, e1002863. [CrossRef] [PubMed]

24. Wu, G.D.; Chen, J.; Hoffmann, C.; Bittinger, K.; Chen, Y.Y.; Keilbaugh, S.A.; Bewtra, M.; Knights, D.; Walters, W.A.; Knight, R.; et al. Linking long-term dietary patterns with gut microbial enterotypes. *Science* **2011**, *334*, 105–108. [CrossRef] [PubMed]

25. Chen, J.; Li, H. Variable selection for sparse Dirichlet-multinomial regression with an application to microbiome data analysis. *Ann. Appl. Stat.* **2013**, *7*, 418–442. [CrossRef]

26. Holmes, I.; Harris, K.; Quince, C. Dirichlet multinomial mixtures: Generative models for microbial metagenomics. *PLoS ONE* **2012**, *7*, e30126. [CrossRef]

27. Feng, Z.; Subedi, S.; Neish, D.; Bak, S. Cluster Analysis of Microbiome Data via Mixtures of Dirichlet-Multinomial Regression Models. *J. R. Stat. Soc. Ser. C (Appl. Stat.)* **2015**, *69*, *1163–1187*.

28. Calinski, T.; Harabasz, J. A Dendrite Method for Cluster Analysis. *Comm. Stat. Simulat. Comp.* **1974**, *3*, 1–27. [CrossRef]

29. Davies, D.L.; Bouldin, D.W. A Cluster Separation Measure. *IEEE Trans. Pattern Anal. Mach. Intell.* **1979**, *PAMI-1*, 224–227. [CrossRef]

30. Strehl, A.; Ghosh, J. Cluster Ensembles—A Knowledge Reuse Framework for Combining Multiple Partitions. *J. Mach. Learn. Res.* **2003**, *3*, 583–617.

31. Zhao, Q.; Franti, P. WB-index: A sum-of-squares based index for cluster validity. *Data Knowl. Eng.* **2014**, *92*, 77–89. [CrossRef]

32. Joonas, H.; Susanne, J.; Tommi, K. Comparison of Internal Clustering Validation Indices for Prototype-Based Clustering. *Algorithms* **2017**, *10*, 105–105.

33. Rousseeuw, P.J. Silhouettes: A graphical aid to the interpretation and validation of cluster analysis. *J. Comput. Appl. Math.* **1987**, *20*, 53–65. [CrossRef]

34. Tibshirani, R.; Walther, G. Cluster validation by prediction strength. *J. Comput. Graph. Stat.* **2005**, *14*, 511–528. [CrossRef]

35. Hennig, C.; Liao, T.F. *Comparing Latent Class and Dissimilarity Based Clustering for Mixed Type Variables with Application to Social Stratification*; Research Report No. 308; Department of Statistical Science, University College London: London, UK, 2010.

36. Figueiredo, M.A.T.; Jain, A.K. Unsupervised learning of finite mixture models. *IEEE Trans. Pattern Anal. Mach. Intell.* **2002**, *24*, 381–396. [CrossRef]

37. Bouguila, N.; Ziou, D.; Vaillancourt, J. Unsupervised learning of a finite mixture model based on the Dirichlet distribution and its application. *IEEE Trans. Image Process. A Publ. IEEE Signal Process. Soc.* **2004**, *13*, 1533–1543. [CrossRef]

38. Xu, P.; Peng, H.; Huang, T. Unsupervised Learning of Mixture Regression Models for Longitudinal Data. *Comput. Stats Data Anal.* **2018**, *125*, 44–56. [CrossRef]

39. Mohamed, M.B.I.; Frigui, H. Unsupervised clustering and feature weighting based on Generalized Dirichlet mixture modeling. *Inf. Sci.* **2014**, *274*, 35–54.

40. Shestopaloff, K.; Escobar, M.D.; Xu, W. Analyzing differences between microbiome communities using mixture distributions. *Stat. Med.* **2018**, *37*, 4036–4053. [CrossRef]

41. Dunn, J.C. A Fuzzy Relative of the ISODATA Process and Its Use in Detecting Compact Well-Separated Clusters. *J. Cybern.* **1973**, *3*, 32–57. [CrossRef]

42. García-Jiménez, B.; Wilkinson, M.D. Robust and automatic definition of microbiome states. *PeerJ* **2019**, *7*, e6657. [CrossRef]

43. Struyf, A.; Hubert, M.; Rousseeuw, P.J. Integrating robust clustering techniques in S-PLUS. *Comput. Stat. Data Anal.* **1997**, *26*, 17–37. [CrossRef]

44. McDonald, D.; Price, M.N.; Goodrich, J.; Nawrocki, E.P.; DeSantis, T.Z.; Probst, A.; Andersen, G.L.; Knight, R.; Hugenholtz, P. An improved Greengenes taxonomy with explicit ranks for ecological and evolutionary analyses of bacteria and archaea. *ISME J.* **2012**, *6*, 610–618. [CrossRef] [PubMed]

45. Xie, X.L.; Beni, G. A validity measure for fuzzy clustering. *IEEE Trans. Pattern Anal. Mach. Intell.* **1991**, *13*, 841–847. [CrossRef]

46.   Keshavarzian, A.; Green, S.J.; Engen, P.A.; Voigt, R.M.; Naqib, A.; Forsyth, C.B.; Mutlu, E.; Shannon, K.M. Colonic bacterial composition in Parkinson's disease. *Mov. Disord.* **2015**, *30*, 1351–1360. [CrossRef]

47.   Nocedal, J. Updating quasi-Newton matrices with limited storage. *Math. Comput.* **1980**, *35*, 773–782. [CrossRef]

48.   Ypma, J. Introduction to Nloptr: An R Interface to NLopt. R Package. 2 August 2014. Available online: https://docplayer.net/39407286-Introduction-to-nloptr-an-r-interface-to-nlopt.html (accessed on 20 October 2020).

49.   Maechler, M.; Rousseeuw, P.; Struyf, A.; Hubert, M.; Hornik, K. Cluster: Cluster Analysis Basics and Extensions. R Package Version 2.0.1. 2015. Available online: https://www.scirp.org/(S(lz5mqp453edsnp55rrgjct55))/reference/ReferencesPapers.aspx?ReferenceID=2062247 (accessed on 20 October 2020).

50.   Desgraupes, B. Clustering indices. *Univ. Paris Ouest-Lab Modal X* **2013**, *1*, 34.

The Expanding Computational Toolbox for Engineering Microbial Phenotypes at the Genome Scale

Daniel Craig Zielinski [1,†], **Arjun Patel** [1,†] ⓘ and **Bernhard O. Palsson** [1,2,*]

[1] Department of Bioengineering, University of California, San Diego, San Diego, CA 92093, USA; dczielin@ucsd.edu (D.C.Z.); arpatel@eng.ucsd.edu (A.P.)

[2] Novo Nordisk Foundation Center for Biosustainability, Technical University of Denmark, 2800 Lyngby, Denmark

* Correspondence: palsson@ucsd.edu

† These authors contributed equally to this work.

Abstract: Microbial strains are being engineered for an increasingly diverse array of applications, from chemical production to human health. While traditional engineering disciplines are driven by predictive design tools, these tools have been difficult to build for biological design due to the complexity of biological systems and many unknowns of their quantitative behavior. However, due to many recent advances, the gap between design in biology and other engineering fields is closing. In this work, we discuss promising areas of development of computational tools for engineering microbial strains. We define five frontiers of active research: (1) Constraint-based modeling and metabolic network reconstruction, (2) Kinetics and thermodynamic modeling, (3) Protein structure analysis, (4) Genome sequence analysis, and (5) Regulatory network analysis. Experimental and machine learning drivers have enabled these methods to improve by leaps and bounds in both scope and accuracy. Modern strain design projects will require these tools to be comprehensively applied to the entire cell and efficiently integrated within a single workflow. We expect that these frontiers, enabled by the ongoing revolution of big data science, will drive forward more advanced and powerful strain engineering strategies.

Keywords: synthetic biology; metabolic modeling; machine learning; metabolic engineering

1. Introduction

Microbes have been engineered for a broad number of applications. As cell factories, cells have been designed to convert low-value substrates into valuable chemical products, including biofuels [1], commodity chemicals [2], bioactive compounds [3], and foods [4]. To benefit the environment, microbes have been engineered for bioremediation [5] and biosensing [6] of toxic compounds and pollutants. As engineered tools, microbes have been programmed using cell circuits to exhibit elaborate behaviors, from synchronized fluorescence [7] to hunting down tumors to deliver chemotherapeutics [8]. Finally, as cellular products, microbes themselves are increasingly of interest for probiotic and nutritional supplements [9].

The experimental workflow to engineer a new microbial strain has a number of common steps, although the order may vary (Figure 1A) [10]. First, a background organism and strain is chosen for the application of interest. Genes may be knocked out, introduced, knocked down, or overexpressed for a variety of purposes, such as control of transcriptional regulation, redirection of metabolic flux to desired pathways, or removal of unwanted or wasteful processes. Bioprocess conditions can be optimized through control of various factors including media, feed rate, growth rate, pH, and temperature. Specific sequence variants can be introduced through rational design or selected through screens and adaptive laboratory evolution to control expression, alter enzyme activity, or remove regulatory sites from proteins. The typical strain design workflow thus requires a large number of decisions on how to improve strain behavior. Left to a trial and error approach, the complexity of biological systems makes efficient engineering of strains a daunting task.

To aid strain design efforts, computational tools have been integrated from various fields into the strain design workflow [11]. These tools offer the promise of restricting the experimental search space by either identifying modifications that are more likely to improve strain performance or proposing entirely new designs through mathematical modeling of cell behavior. However, many steps in the strain design process are still driven by rational approaches, rules of thumb, and extensive experimental screening and trial and error. Workflows driven purely by predictive tools would have the advantage of efficiency of execution through fewer experimental steps, reduced time, and ultimately improved performance through careful guidance toward an optimal desired phenotype. We describe two approaches that show promise as systematic tools for cell design: genetic circuits and genome-scale modeling.

One strategy for constructing synthetic strains has been to engineer desired behaviors through the use of genetic circuits [12]. The key concept is to carefully characterize and often mathematically model the behavior of a 'circuit', typically a small transcriptional regulatory network, to control a cell phenotype. As greater numbers of these small circuits are characterized, they begin to comprise a 'parts list' of available phenotypes from which an engineer can choose or can be assembled automatically by an algorithm [13]. Larger and larger circuits can then be constructed of well-characterized smaller circuits to engineer more complex phenotypes. This strategy has been employed for a number of promising applications [14,15].

Another successful paradigm for computational design of cells is genome-scale network modeling [16]. While genetic circuits approaches utilize highly controllable systems of limited scope, genome-scale models seek to predict cell phenotype by comprehensively modeling all known functions of the cell. As part of the Constraint-based Reconstruction and Analysis (COBRA) framework, genome-scale models of metabolism utilize a metabolic network reconstruction to predict metabolic phenotypes and analyze genome-scale datasets [17]. These models deal with the large scope of the system by utilizing the constraint-based modeling framework, which requires few parameters to generate predictions. The challenge of managing these large-scale models is achieved through community enforcement of rigid requirements, testing, and data standards [18,19]. Although these models were originally developed for metabolism, they have recently been extended to include transcription and translation machinery [20–22] and even further to whole-cell kinetic simulations [23].

Although computational methods have undoubtedly augmented rational strain design efforts, there are a number of challenges in a strain design workflow that still cannot be effectively addressed by existing computational tools [24] (Figure 1B). For example: (1) Organisms are often chosen for a strain design project due to historical knowledge and convenience, rather than fundamental benefits provided by the organism that could be calculated computationally *a priori*, (2) Gaps in gene annotation make choosing non-model organisms a risk, (3) The difficulty in accounting for enzyme kinetics makes the understanding of metabolic and allosteric regulation a challenge, (4) A lack of understanding of regulatory networks impedes the understanding and control of gene expression, and (5) Insufficient annotation of the organism genome makes it difficult to interpret the functional implications of sequence variation. Challenges such as these present major barriers to interpreting data and predicting strain phenotype.

There are many methods currently being developed that may directly meet these challenges to enable fully predictive strain design workflows (Figure 1C). For example, advances in metabolic modeling could enable the optimization of bioprocess conditions or the identification of optimal expression levels of pathway genes [25,26]. However, these models are still in development and have not been shown to enable accurate predictions at scale. In this perspective, we describe five frontiers consisting of promising developments in computational strain design that may pave the way toward achieving comprehensive and integrated strain design workflows (Figure 2).

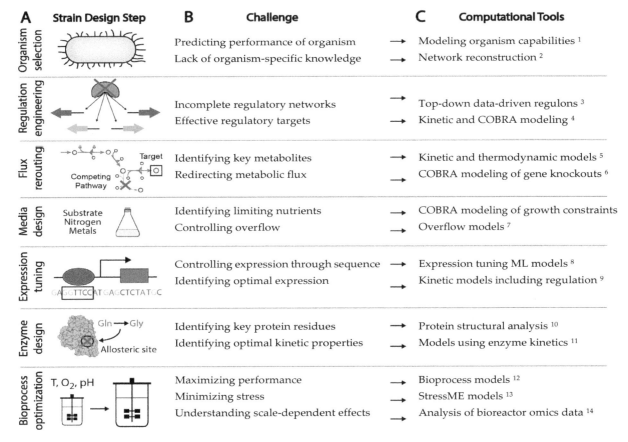

Figure 1. Challenges and computational solutions for a typical strain design workflow. (**A**) Typical experimental steps in the development of a new strain design. (**B**) Common challenges encountered at each strain design step. (**C**) Computational tools that may be used to meet the strain design challenges. Note that the design steps, challenges, and computational tools highlighted here are intended to be exemplative rather than comprehensive. [1], Modeling organism capabilities [27]; [2], Network reconstruction [18]; [3], Top-down data-driven regulons [28]; [4], Kinetic and COBRA modeling [26]; [5], Kinetic and thermodynamic models [29]; [6], COBRA modeling of gene knockouts [30]; [7], Overflow models [31]; [8], Expression tuning ML models [32]; [9], Kinetic models including regulation [33]; [10], Protein structural analysis [34]; [11], Models using enzyme kinetics [35]; [12], Bioprocess models [36]; [13], StressME models [37–39]; [14], Analysis of bioreactor omics data [40].

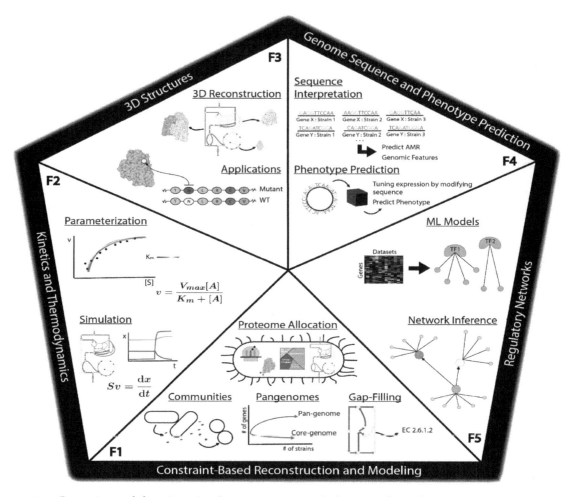

Figure 2. Overview of frontiers in the computational design of synthetic organisms. Frontier 1: Constraint-based Reconstruction and Modeling, consisting of tools for analyzing pan-genomes, microbial communities, gap-filling metabolic networks, and modeling proteome allocation. Frontier 2: Kinetics and Thermodynamics, consisting of tools for parameterizing and simulating kinetic and thermodynamic models. Parameterization can utilize the Michaelis-Menten equation where [A] is the substrate concentration, whereas simulation uses dynamic mass balance equations where S is the stoichiometric matrix. Frontier 3: 3D Structures, consisting of methods for the reconstruction of 3D metabolic networks with protein structural information and subsequent applications of these 3D reconstructions. Frontier 4: Genome Sequence and Phenotype Prediction, consisting of workflows for analyzing strain variations in genome sequence as well as building machine learning models based on genome sequence to predict strain phenotype. Frontier 5: Regulatory Networks, consisting of methods for the determination of transcriptional regulatory networks and subsequence models of gene expression and strain phenotype utilizing regulatory network information.

2. Frontier 1: Constraint-Based Reconstruction and Modeling

COBRA methods and tools have been used and refined for over 15 years in the field of systems biology [17,41–43]. Many current applications of constraint-based models answer biologically meaningful questions such as max growth rate, metabolite production, limiting nutrients, and gene essentiality. COBRA methods are also often used when little parameter and/or metabolite information is available given the size of the networks. These methods have predominantly been limited by annotation gaps in the underlying reconstructed networks [44]. COBRA models have also been expanded to include other cellular processes, such as transcription and translation, in order to increase the predictive capabilities of the models. Previous models did not have the capacity to accurately model suboptimal states, such as thermal or oxidative stress, nor could they properly describe how microbes interact within a community [24]. In the following sections, we describe recent computational advances that have benefitted COBRA models and methods.

2.1. Proteome Allocation Models

One of the main challenges for omics integration with M-models is that they only indirectly relate expression to metabolic fluxes. Metabolism and macromolecular expression (ME) models were developed to be multiscale models that explicitly include multiple cellular processes such as transcription, translation, and post-translational modifications. Early ME-models were able to compute optimal proteome compositions of a growing cell but suffered from significant model sizes and complexity, which has been addressed with the recent development of the COBRAme platform [20]. ME-models have since been produced for a variety of organisms and can now accurately predict overflow metabolism and cofactor/metal usage given media composition [45,46]. Additionally, ME-models have led to increased confidence in the selection of strain designs with robust growth-coupled production, in addition to having better byproduct secretion predictions than M-models [47,48]. Time course ME simulations are also possible and can predict substrate utilization hierarchy on mixed carbon source medium. These simulations compute distinct proteome compositions over time [49]. Prediction capabilities for ME-models have also expanded to include suboptimal states such as stress and mitigation responses. These 'StressME' models are able to more accurately model the cell's unused proteome that acts as a hedging mechanism for preparedness functions under low pH, thermal, and oxidative stress [37–39].

Unfortunately, developing a ME-model requires a deep understanding of the organism's cellular machinery. For organisms that lack the required annotations, less complex alternatives exist. MOMENT, or Metabolic Modeling with Enzyme Kinetics, was first developed to better predict metabolic fluxes and growth rates by using enzyme turnover rates and molecular weights [50]. GEMs can also be integrated with enzyme constraints using kinetic and omics data, which is also known as GECKO. Since GECKO does not require detailed knowledge about every step in protein synthesis its been applied to other organisms such as the eukaryal *Saccharomyces cerevisiae* [51,52]. Unlike GECKO, MOMENT does not allow for the direct incorporation of measured enzyme concentrations. Short MOMENT or sMOMENT was recently developed to allow for this type of integration while also greatly reducing the size and complexity of the original MOMENT model [53].

2.2. Communities

Modeling communities or co-cultures are important for healthcare and biotech applications since many times strains are paired based on metabolic coupling. For example, pairing phototrophs with heterotrophs is a promising prospect for sustainable biotechnology, since it enables heterotrophs like *Escherichia. coli* to grow in minimal media devoid of organic carbon sources. Community metabolic models (CM-models) can be constructed for co-cultures to aid in this strain selection. By modeling and simulating various synthetic microbial co-cultures, researchers are able to identify optimal pairs that produce the most active community [54]. In some cases, dynamic community metabolic models can be generated for co-cultures using each organism's genome-scale metabolic network. These dynamic simulations are able to predict metabolite concentration profiles for the community as well metabolic exchange flux profiles for individual organisms [55]. Along a similar vein, communities are able to partition metabolic functions among community members like in cases of auxotrophy. These specialized pairings are capable of improving product yield and accomplishing more complex tasks as compared to a single strain [56]. The OptAux algorithm was created to aid in designing auxotrophic strains that need a metabolite cross-feeding co-culture [57].

2.3. Pangenomes and Multistrain Models

Reconstructing GEMs for multiple strains across a single species has enabled a systems-level approach to study and characterize the pan-metabolic capabilities of the species. Pangenomic studies have been accomplished for a wide range of species from *Staphylococcus aureus* to *Klebsiella pneumoniae* [58–63]. Integration of genomics, phenomics, transcriptomics, and genome-scale modeling for seven commonly used *E. coli* strains linked molecular features to strain-specific phenotypes. The integrated models showed that certain strains are better suited to produce specific compounds or phenotypes, which has implications for strain selection when choosing a platform strain for microbial engineering [27].

2.4. Gap filling, Discovery, and Annotation

In order to more accurately predict an organism's phenotype, there needs to be a more complete genome annotation to better understand the organism's capabilities. Even in *E. coli*, one of the best-studied model organisms, 35% of its genome lacks functional annotations, with many of these genes being experimentally linked to phenotypes [64]. Recently, there has been a multitude of computational tools that have taken strides in elucidating the possible functions of these genes.

Transposable elements (TEs) are of high interest when engineering a strain due to the deleterious effects they can have if uncontrolled. In platform strains, TEs or insertion sequences (ISs) are often removed in order to preserve the intended genomic content. Issues arise if the organism of interest is poorly characterized and the TEs are not known, however, new machine learning algorithms that utilize genome sequence are able to identify the TEs and ISs in both eukaryotic and prokaryotic species [65,66].

Gap-filling has been commonly used for reconstructing genome-scale metabolic models but over the past few years has quickly advanced in coordination with the recent developments in machine learning. Current methods such as DeepEC are able to predict enzyme commission numbers based on genome sequence with high accuracy [67]. Additionally, machine learning algorithms that use genome sequence can predict systems-wide enzyme promiscuity or candidate genes/enzymes for orphan reactions [68,69]. By filling these annotation gaps and identifying possible cases of promiscuity in microbial networks, researchers will be able to achieve more accurate and comprehensive predictions.

3. Frontier 2: Kinetics and Thermodynamics

Kinetic modeling of metabolism is a field with a long history dating back to the original work understanding enzyme kinetics at the beginning of the 20th century. These ODE-based models offer the promise of mechanistically accounting for the saturation and regulatory state of every enzyme, providing a direct representation of the mechanisms underlying cellular homeostasis [35]. Metabolic control analysis, rooted in steady-state analysis of a kinetic model, paved the way for quantitative analysis of metabolic networks in the early days of metabolic engineering [70]. Sharing much of the same underlying theory, constraint-based thermodynamic models calculate the energetic driving forces underlying metabolic fluxes and can be used to determine physiological constraints on reaction reversibility in the metabolic network [29,71–74]. Multi-scale models accounting for kinetics and thermodynamics have begun to emerge as well [75–77], as discussed in the earlier section on Proteome Allocation Models. The development of kinetic and thermodynamic models has been hampered by the difficulty in acquiring the necessary parameters and validating model behavior at large-scale [78]. Furthermore, the complexity of these models, need for accounting for parameter uncertainty, and additional confounding factors such as numerical instability in kinetic models or constraint infeasibilities in thermodynamic models, substantially increase computational requirements for large-scale modeling. However, recent advances in parameterization and simulation of kinetic and thermodynamic models promise to greatly expand the scope and accuracy of these models.

3.1. Parameterization

The critical step in the construction of a kinetic or thermodynamic model is specifying the values of the necessary parameters. This effort is complicated by the lack of required data which leaves these models largely underdetermined. Algorithms must be developed to fit parameters to available data and account for parameter uncertainty.

For parameterizing kinetic models, a number of approaches are now available: (1) Systems-level fitting, where all parameters of the model are fit simultaneously to systems-level data such as metabolomics and fluxomics [79,80], (2) estimation of kinetic parameters directly from in vivo data without the need of a model [81,82], (3) machine learning to estimate kinetic parameters [83], and (4) bottom-up reconstruction of kinetics on an enzyme by enzyme basis [84]. Similarly, methods to account for parameter uncertainty have advanced through powerful algorithms [85]. Thus, the 'kinetome', a genome-scale collection of the kinetic properties of metabolic enzymes, may soon be within reach [86].

In thermodynamic models, the majority of work has focused on estimating the standard Gibbs free energy of reaction, dG^0_r, which can be readily converted to the reaction equilibrium constant K_{eq}. Experimentally, reaction equilibria are directly measured under a variety of biologically relevant experimental conditions, such as pH, T, IS, and magnesium concentration. To estimate the equilibrium properties of reactions lacking experimental data, the most popular approach has been the group-contribution family of methods [87,88], which has led to software such as eQuilibrator [89]. The ability to correct these estimates accurately for pH [90] and temperature [91] have since been added. However, there are a number of inherent flaws in group-contribution as an estimator, including fundamental limitations of the underlying additivity assumptions [92]. Methods for estimating compound Gibbs energies based on direct quantum chemistry predictions are a promising alternative [93].

3.2. Simulation

Kinetic models also present substantial computational challenges in simulation and analysis. Numerous issues including model stiffness due to poor conditioning, dynamic instability, and complex dynamic properties require sophisticated tools to manage dealing with kinetic models effectively. A number of software packages have emerged to meet this challenge [94–96]. Additionally, specialized methods for dealing with large-scale kinetic models have emerged by necessity [23,97,98].

Simulation of the thermodynamic properties of a metabolic network faces a distinct set of challenges. Thermodynamic simulation at the genome-scale, through constraint-based methods such as thermodynamic flux balance analysis (tFBA) [99] and network-embedded thermodynamics (NET) [29], must carefully account for uncertainty in thermodynamic parameters and metabolic concentration constraints. Thermodynamic optimization algorithms, such as tFBA, often involve a mix of integer and linear constraints, and MILP, MIQP, and MINLP algorithms may compute slowly at the genome-scale without efficient solving approaches. Integration of thermodynamic constraints with other biophysical constraints presents further complications due to the non-convexity of the resulting space [100].

4. Frontier 3: 3D Structures

Currently, structures are commonly used for analyzing observed sequence variants or identifying functional sites of a protein for targeted engineering. While structure-guided enzyme design remains a promising application, it is incredibly complex and often coupled to large experimental screens [101–103]. Software has been developed to aid in enzyme design, such as the Iterative Protein Redesign and Optimization (IPRO) method, but incorporating enzyme design into a strain design workflow remains difficult due to the required expertise and experimental validation required [104]. On the other hand, integrating protein structures into systems biology has shown promise as a more accessible addition to strain design workflows, but remains a challenge due to differences between the fields causing a steep learning curve. Here, we highlight some of the advances that have lowered the learning curve for using and integrating structures data with systems biology approaches.

4.1. 3D Reconstruction

GEMs have now been expanded to include protein structural information, which has enabled comparative structural proteome analysis between strains and organisms. These GEMs with protein structures (GEM-PROs) allow for a direct mapping of gene to protein structure to phenotype [34,105]. Software has been developed to aid with the construction of high-quality GEM-PROs and to visualize/annotate structures. The pipeline and software, ssbio, is available for use on GitHub (http://github.com/SBRG/ssbio) and is implemented in Python [106]. For cases where a protein is poorly characterized and the necessary mapping fails to exist, homology modeling or tools such as I-TASSER can be used to predict 3D structures and structure-based functional annotations [107]. AlphaFold was also recently announced as the best protein folding solution at the Critical Assessment of protein Structure Prediction (CASP)-14 competition [108]. CASP was founded in 1994 with the goal to establish the state of the art in protein structure prediction based, and AlphaFold 2 has recently achieved predictions competitive with results obtained from experimental methods, something that was once thought to be impossible.

4.2. Applications

Structural information is often used for 3D mutational mapping or visualization but has more capabilities when used in an integrated workflow. For example, GEM-PROs for multiple strains of the same species enables the comparison of sequence variants among conserved genes [105]. Additionally, combining machine learning approaches and 3D structural mutational mapping has identified genetic signatures of antimicrobial resistance evolution to multiple antibiotics in *M. tuberculosis* [109]. Brunk et al. developed a multiscale workflow to better understand the roles and mechanisms of protein post-translational modifications (PTMs) due to the challenges they present in engineering organisms and their interference with drug action [110]. This workflow incorporates genome-scale modeling, genome editing, and molecular enzyme assays to identify specific roles of PTMs and how they regulate cell phenotype [110]. Well-established software incorporating three-dimensional structural information also exists today. Amber is a package suite of computer programs and has been in development for over 40 years. Amber simulates molecular dynamics for proteins and other biomolecules using structural information and molecular mechanical force fields [111,112].

5. Frontier 4: Genome Sequence and Phenotype Prediction

The genome sequence lies at the heart of a strain design workflow. Heterologous genes must either be added via plasmids with established expression behavior or integrated into the chromosome, and the behavior of these genes depends on the sequence of both coding and non-coding regions of the genes. Further, mutations occur in any mutagenesis or adaptive laboratory evolution strain that control phenotype. Strain design projects may choose between different strains of a species, and the sequence variations between these strains may lead to diverse differences in behavior. Quantifying the sum of sequence factors to predict strain phenotype such as gene expression is an active area of research. Two related strain design tasks utilize the genome sequence: (1) analysis of observed sequence variation data and (2) prediction of phenotype based on genome sequence.

5.1. Sequence Interpretation

A combination of natural, selected, or randomized sequence variants of different genes will typically be observed or generated throughout a typical strain design project. Interpreting the effect of these mutations presents a significant challenge. Natural variants occurring across strains of a species have been analyzed to understand phenotypes related to antimicrobial resistance [109,113–115]. Machine learning models can be trained on a phenotype of interest, and key variables extracted to understand which genetic features are important for determining the phenotype [116]. A well-established example of this type of analysis is the identification of protein-DNA binding motifs, for example by the

MEME suite [117]. Sequence variants occurring in the course of adaptive laboratory evolution have also been collected [118], and analyzed [119]. This type of sequence analysis is possible due to the recent development of data structures for contextualizing multiple data types within the genome [120]. Further follow up is required to understand the mechanistic basis establishing the relationship between these genetic features and the phenotype.

5.2. Phenotype Prediction

In addition to analysis of observed mutations, the genome sequence can also be used to directly predict strain phenotypes. The genome sequence directly affects the function of the protein product, the mRNA transcription rate, and the protein translation rate of a transcript, termed the translation efficiency. Not surprisingly, models capturing these effects are of great interest. While classically, the assignment of function to a new gene has been done primarily through sequence homology, more sophisticated methods based on machine learning are being developed that have the potential to better capture the sequence-function relationship [67]. Similarly, a number of machine learning and mechanistic models of methods predicting mRNA expression [121] and translation efficiency [122] from promoter and transcript sequence, respectively, have been developed. Other methods predict protein expression directly from sequence, integrating both transcription and translation effects [123]. These methods often are trained on large assays based on synthetic libraries of sequence variants. A key challenge is to achieve general applicability of these models in new experimental conditions and strain backgrounds.

6. Frontier 5: Regulatory Networks

Transcriptional regulatory networks (TRNs) for multiple organisms have been studied rigorously and have made significant advancements with the expansion of ChIP-seq data. Even with this deluge of experimental data, many questions still remain, even for *E. coli* with one of the best characterized TRNs. Computational approaches combining experimental datasets with machine learning methods have shown promise in further elucidating the underlying regulatory network, even from old low-resolution datasets. These new approaches and insights have opened the door for more accurately integrating metabolic modeling with regulation, a long time goal in the field of computational systems biology. The following will discuss these methods and their implications.

6.1. Regulatory Network Machine Learning Models

A few methods have been developed for further elucidating an organism's TRN using machine learning. Applying independent component analysis (ICA), an unsupervised machine learning algorithm, to diverse transcriptomics compendia reveal statistically independent signals that modulate the expression of genes. In *E. coli*, 66% of the 92 identified signals represent the effects of transcriptional regulators, whereas 27% represent biological or genetic explanations. ICA decomposition has also proven effective for other organisms, such as *B. subtilis, and S. aureus* [28,59,124]. Models for these organisms and more information on the method can be found in the iModulon database (iModulonDB.org) [28,125]. Another method, Transcriptional Regulatory Network Analysis, or TReNA, combines TF/target gene correlations and TF DNA binding information to create gene regulatory predictions. The resulting predictions approach mechanistic accuracy when using high-resolution data collection techniques like single-cell RNA-seq and ChIP-seq, but bulk mRNA data and low-resolution techniques still provide a coarse-grained result. TReNA thus enables researchers to work along this spectrum [126]. While these processes utilize transcriptomics to predict regulators, algorithms also exist to predict gene expression as a function of transcription factor expression levels [127]. Probabilistic regulation of metabolism, or PROM, is a method for integrating TRNs and metabolic models. PROM uses conditional probabilities to represent gene states and gene-TF interactions, as well as FBA for modeling the metabolic network. The method was one of the first to enable automated integration of transcriptional and metabolic networks using high-throughput datasets [128].

6.2. Network Inference

Being able to predict network structure and identify the roles of specific vs. global regulators is necessary for a more complete understanding of how gene expression changes under varying sample conditions. Rustad et al. created a transcription factor overexpression (TFOE) library by cloning and overexpressing 206 TFs in *M. tuberculosis*. The resulting strains were used to identify sets of genes affected by TFOE and assembled into a global transcriptional map. The TFOE regulatory map identifies potential regulators of gene sets and was used by Rustad et al. to predict and validate the phenotype of a regulator affecting the susceptibility of isoniazid, an antibiotic used to treat tuberculosis [129]. Kochanowski et al. developed computational methods to identify global vs. specific transcriptional regulation based on promoter activity and to discern metabolites that serve as potential transcriptional regulators. The simple mathematical models use high-throughput measurements of metabolite concentrations and promoter activity. A model of *E. coli* central carbon metabolism showed that 90% of expression changes are due to a few regulatory metabolites (F1P, FBP, and cAMP) and global TFs (Crp and Cra) [130].

7. Drivers of Advances in Computational Tools for Strain Design

The improvements in both scope and accuracy of the computational tools discussed here have been driven by a combination of experimental and machine learning advances. The availability of large-scale datasets that enable comprehensive modeling has exploded. To list a few examples: (1) Continued determination of experimentally-determined protein 3D structures has empowered the rapid development of homology modeling of protein structures, (2) Falling sequencing costs and improved sequencing quality have resulted in a rapid increase of the number, assembly quality, and annotation quality of available microbial genomes, with similar increases to other sequence-based omics types, such as RNA-seq and ribo-seq, and (3) Improved quantitative proteomics datasets have enabled the large-scale estimation of in vivo enzyme turnover rates [81], greatly empowering kinetic and proteome-allocation modeling approaches. Alongside this increased availability of data, machine learning methods to effectively utilize this experimental data to parameterize biological models have greatly improved in recent years. In addition to machine learning algorithms, any expert in machine learning knows the importance of data curation and data cleaning to successful model development. For this reason, the growth and maintenance of highly curated organism-specific knowledgebases, such as RegulonDB and EcoCyc for *E. coli*, has been critical to providing a source of reliable data for many computational tools [131,132]. Thus, as these companion fields continue to evolve, it is expected that the accuracy of the computational tools built upon these data and algorithms will continue to rapidly improve.

8. Outlook for Synthetic Genome Design

Although there has been substantial progress in each of the individual fields discussed above, there are additional challenges with integrating these tools into an effective strain design workflow. Workflow: While we discussed many tools as they relate to individual strain design tasks, these tasks must be synthesized into a coherent end-to-end design workflow. The decisions of the order of operations in the development of a strain could greatly benefit from computational predictions, but much work is yet to be done to identify a strain design workflow that maximizes efficiency and minimizes cost and risk. Expertise: Any workflow that integrates many different computational tools will require domain expertise in each tool to decide details of implementation, from parameters to valid use cases. Thus, strain designers will be required to have broad computational skillsets that exceed what is taught by most current training programs. Software: The practical difficulty of implementing many separate computational tools can become a substantial burden, spanning various details from licensing issues to file formats. However, the number of software packages enabling these workflows continues to increase, and we mention many examples in this work (Figure 3). Thanks to these efforts, finding compatible tools for easily integrated workflows is becoming easier. Validation: Tools must be validated to clearly established accuracy metrics under physiological conditions. Validation of tools on individual datasets, for example on a single wild type strain background, is likely to be insufficient as the strain is engineered further from the wild type. To meet these challenges, it is critical to take a systematic approach that includes dedicated training, effective documentation of tools, and extensive validation of tools in real applications. There will be a significant challenge reaching a standard where strain design researchers can effectively conduct analyses and understand results from multiple tools across a typical workflow.

	Tool	Platform	Description	Link
F1	COBRApy [1]	Python	COBRA modeling in Python	github.com/opencobra/cobrapy
	COBRA Toolbox [2]	MATLAB	COBRA modeling in MATLAB	github.com/opencobra/cobratoolbox
	COBRAme [3]	Python	COBRApy extension for ME-modeling	github.com/SBRG/cobrame
	GECKO [4]	Python	Package for enzyme constrained modeling	github.com/SysBioChalmers/GECKO
	CAMEO [5]	Python	Computer aided metabolic engineering optimization	github.com/biosustain/cameo
F2	pyTFA [6]	Python	Thermodynamics based flux analysis in Python	github.com/EPFL-LCSB/pytfa
	COPASI [7]	Any OS	Software for simulating and analyzing biochemical networks	copasi.org
	MASSpy [8]	Python	Mass action stoichiometric simulation package in python	masspy.readthedocs.io/en/latest
	eQuilibrator [9]	Python	Interface for thermodynamic analysis	gitlab.com/equilibrator/equilibrator
F3	SSBIO [10]	Python	Framework for structural systems biology	github.com/SBRG/ssbio
	Amber [11]	Python	Framework for structural systems biology	ambermd.org
	I-TASSER [12]	Linux	Molecular mechanics software for simulating protein structures	zhanglab.ccmb.med.umich.edu/I-TASSER
F4	Bitome [13]	Python	Software suite for structure prediction and function annotation	github.com/SBRG/bitome
	MEME [14]	Any OS	Genomic data organization	meme-suite.org
F5	iModulonDB [15]	Website	Suite of tools for sequence motif analysis	imodulondb.org
	PRECISE [16]	Python	Knowledgebase of microbial TRNs	github.com/SBRG/precise-db

Figure 3. A selection of actively maintained software for computational design and analysis of microbial phenotypes. We focus on Python tools due to the popularity of the language as well as potential for integration in a single strain design workflow, but also include important packages in other languages and standalone applications. Frontier 1: Packages for constraint-based reconstruction and modeling, proteome allocation modeling, and strain design optimization. Frontier 2: Kinetics and thermodynamics packages for model parameterization, simulation, and thermodynamics constrained modeling. Frontier 3: Software for annotating and visualizing structures as well as integrating 3D structural information with systems biology approaches. Frontier 4: Python package for storing, organizing, and analyzing genome sequences. Frontier 5: Online knowledgebase and software for determining transcriptional regulatory networks using ICA decomposition methods. [1] COBRApy [42]; [2] COBRA Toolbox [17]; [3] COBRAme [20]; [4] GECKO [52]; [5] CAMEO [43]; [6] pyTFA [95]; [7] COPASI [96]; [8] MASSpy [94]; [9] eQuilibrator [89]; [10] SSBIO [106]; [11] Amber [111]; [12] I-TASSER [107]; [13] Bitome [120]; [14] MEME [117]; [15] iModulonDB [125]; [16] PRECISE [28].

The field is nearing an important milestone in synthetic biology, that of the comprehensive and computationally-driven strain design workflow. We may soon enter an era of 'computational genome design', where rational approaches finally give way to biological design algorithms dominated by computational predictions. Thus, one of the early promises of the field of systems biology may finally be nearing its realization. The practical applications of such a cell design workflow are endless, from the chemical industry to the environment to human health.

Author Contributions: Conceptualization, D.C.Z., B.O.P.; writing—original draft preparation, D.C.Z., A.P., B.O.P.; writing—review and editing, D.C.Z., A.P., B.O.P.; visualization, D.C.Z., A.P.; funding acquisition, B.O.P. All authors have read and agreed to the published version of the manuscript.

Acknowledgments: We would like to thank Patrick Phaneuf and Tobias Alter for helpful comments on the manuscript.

References

1. Liao, J.C.; Mi, L.; Pontrelli, S.; Luo, S. Fuelling the Future: Microbial Engineering for the Production of Sustainable Biofuels. *Nat. Rev. Microbiol.* **2016**, *14*, 288–304. [CrossRef] [PubMed]
2. Lee, S.Y.; Kim, H.U.; Chae, T.U.; Cho, J.S.; Kim, J.W.; Shin, J.H.; Kim, D.I.; Ko, Y.-S.; Jang, W.D.; Jang, Y.-S. A Comprehensive Metabolic Map for Production of Bio-Based Chemicals. *Nat. Catal.* **2019**, *2*, 18–33. [CrossRef]
3. Kalia, V.C.; Saini, A.K. (Eds.) *Metabolic Engineering for Bioactive Compounds: Strategies and Processes*; Springer: Singapore, 2017.
4. Matassa, S.; Boon, N.; Pikaar, I.; Verstraete, W. Microbial Protein: Future Sustainable Food Supply Route with Low Environmental Footprint. *Microb. Biotechnol.* **2016**, *9*, 568–575. [CrossRef] [PubMed]
5. Das, S.; Dash, H.R. 1—Microbial Bioremediation: A Potential Tool for Restoration of Contaminated Areas. In *Microbial Biodegradation and Bioremediation*; Das, S., Ed.; Elsevier: Oxford, UK, 2014; pp. 1–21.
6. Bereza-Malcolm, L.T.; Mann, G.; Franks, A.E. Environmental Sensing of Heavy Metals Through Whole Cell Microbial Biosensors: A Synthetic Biology Approach. *ACS Synth. Biol.* **2015**, *4*, 535–546. [CrossRef] [PubMed]
7. Danino, T.; Mondragón-Palomino, O.; Tsimring, L.; Hasty, J. A Synchronized Quorum of Genetic Clocks. *Nature* **2010**, *463*, 326–330. [CrossRef] [PubMed]
8. Din, M.O.; Danino, T.; Prindle, A.; Skalak, M.; Selimkhanov, J.; Allen, K.; Julio, E.; Atolia, E.; Tsimring, L.S.; Bhatia, S.N.; et al. Synchronized Cycles of Bacterial Lysis for in Vivo Delivery. *Nature* **2016**, *536*, 81–85. [CrossRef]
9. Yadav, R.; Singh, P.K.; Shukla, P. Metabolic Engineering for Probiotics and Their Genome-Wide Expression Profiling. *Curr. Protein Pept. Sci.* **2018**, *19*, 68–74. [CrossRef]
10. Lee, S.Y.; Kim, H.U. Systems Strategies for Developing Industrial Microbial Strains. *Nat. Biotechnol.* **2015**, *33*, 1061–1072. [CrossRef]
11. St. John, P.C.; Bomble, Y.J. Approaches to Computational Strain Design in the Multiomics Era. *Front. Microbiol.* **2019**, *10*. [CrossRef]
12. Brophy, J.A.N.; Voigt, C.A. Principles of Genetic Circuit Design. *Nat. Methods* **2014**, *11*, 508–520. [CrossRef]
13. Nielsen, A.A.K.; Der, B.S.; Shin, J.; Vaidyanathan, P.; Paralanov, V.; Strychalski, E.A.; Ross, D.; Densmore, D.; Voigt, C.A. Genetic Circuit Design Automation. *Science* **2016**, *352*, aac7341. [CrossRef] [PubMed]
14. Sedlmayer, F.; Aubel, D.; Fussenegger, M. Synthetic Gene Circuits for the Detection, Elimination and Prevention of Disease. *Nat. Biomed. Eng.* **2018**, *2*, 399–415. [CrossRef] [PubMed]
15. Khalil, A.S.; Collins, J.J. Synthetic Biology: Applications Come of Age. *Nat. Rev. Genet.* **2010**, *11*, 367–379. [CrossRef] [PubMed]
16. Kim, W.J.; Kim, H.U.; Lee, S.Y. Current State and Applications of Microbial Genome-Scale Metabolic Models. *Curr. Opin. Syst. Biol.* **2017**, *2*, 10–18. [CrossRef]
17. Heirendt, L.; Arreckx, S.; Pfau, T.; Mendoza, S.N.; Richelle, A.; Heinken, A.; Haraldsdóttir, H.S.; Wachowiak, J.; Keating, S.M.; Vlasov, V.; et al. Creation and Analysis of Biochemical Constraint-Based Models Using the COBRA Toolbox v.3.0. *Nat. Protoc.* **2019**, *14*, 639–702. [CrossRef]
18. Mendoza, S.N.; Olivier, B.G.; Molenaar, D.; Teusink, B. A Systematic Assessment of Current Genome-Scale Metabolic Reconstruction Tools. *Genome Biol.* **2019**, *20*, 158. [CrossRef]

19. Lieven, C.; Beber, M.E.; Olivier, B.G.; Bergmann, F.T.; Ataman, M.; Babaei, P.; Bartell, J.A.; Blank, L.M.; Chauhan, S.; Correia, K.; et al. MEMOTE for Standardized Genome-Scale Metabolic Model Testing. *Nat. Biotechnol.* **2020**, *38*, 272–276. [CrossRef]
20. Lloyd, C.J.; Ebrahim, A.; Yang, L.; King, Z.A.; Catoiu, E.; O'Brien, E.J.; Liu, J.K.; Palsson, B.O. COBRAme: A Computational Framework for Genome-Scale Models of Metabolism and Gene Expression. *PLoS Comput. Biol.* **2018**, *14*, e1006302. [CrossRef]
21. O'Brien, E.J.; Monk, J.M.; Palsson, B.O. Using Genome-Scale Models to Predict Biological Capabilities. *Cell* **2015**, *161*, 971–987. [CrossRef]
22. O'Brien, E.J.; Lerman, J.A.; Chang, R.L.; Hyduke, D.R.; Palsson, B.Ø. Genome-Scale Models of Metabolism and Gene Expression Extend and Refine Growth Phenotype Prediction. *Mol. Syst. Biol.* **2013**, *9*, 693. [CrossRef]
23. Karr, J.R.; Sanghvi, J.C.; Macklin, D.N.; Gutschow, M.V.; Jacobs, J.M.; Bolival, B., Jr.; Assad-Garcia, N.; Glass, J.I.; Covert, M.W. A Whole-Cell Computational Model Predicts Phenotype from Genotype. *Cell* **2012**, *150*, 389–401. [CrossRef] [PubMed]
24. McCloskey, D.; Palsson, B.Ø.; Feist, A.M. Basic and Applied Uses of Genome-Scale Metabolic Network Reconstructions of Escherichia Coli. *Mol. Syst. Biol.* **2013**, *9*, 661. [CrossRef] [PubMed]
25. Richelle, A.; David, B.; Demaegd, D.; Dewerchin, M.; Kinet, R.; Morreale, A.; Portela, R.; Zune, Q.; von Stosch, M. Towards a Widespread Adoption of Metabolic Modeling Tools in Biopharmaceutical Industry: A Process Systems Biology Engineering Perspective. *NPJ Syst. Biol. Appl.* **2020**, *6*, 6. [CrossRef] [PubMed]
26. Andreozzi, S.; Chakrabarti, A.; Soh, K.C.; Burgard, A.; Yang, T.H.; Van Dien, S.; Miskovic, L.; Hatzimanikatis, V. Identification of Metabolic Engineering Targets for the Enhancement of 1,4-Butanediol Production in Recombinant E. Coli Using Large-Scale Kinetic Models. *Metab. Eng.* **2016**, *35*, 148–159. [CrossRef] [PubMed]
27. Monk, J.M.; Koza, A.; Campodonico, M.A.; Machado, D.; Seoane, J.M.; Palsson, B.O.; Herrgård, M.J.; Feist, A.M. Multi-Omics Quantification of Species Variation of Escherichia Coli Links Molecular Features with Strain Phenotypes. *Cell Syst.* **2016**, *3*, 238–251.e12. [CrossRef]
28. Sastry, A.V.; Gao, Y.; Szubin, R.; Hefner, Y.; Xu, S.; Kim, D.; Choudhary, K.S.; Yang, L.; King, Z.A.; Palsson, B.O. The Escherichia Coli Transcriptome Mostly Consists of Independently Regulated Modules. *Nat. Commun.* **2019**, *10*, 5536. [CrossRef]
29. Kümmel, A.; Panke, S.; Heinemann, M. Putative Regulatory Sites Unraveled by Network-Embedded Thermodynamic Analysis of Metabolome Data. *Mol. Syst. Biol.* **2006**, *2*, 2006.0034. [CrossRef]
30. Burgard, A.P.; Pharkya, P.; Maranas, C.D. Optknock: A Bilevel Programming Framework for Identifying Gene Knockout Strategies for Microbial Strain Optimization. *Biotechnol. Bioeng.* **2003**, *84*, 647–657. [CrossRef]
31. De Groot, D.H.; Lischke, J.; Muolo, R.; Planqué, R.; Bruggeman, F.J.; Teusink, B. The Common Message of Constraint-Based Optimization Approaches: Overflow Metabolism Is Caused by Two Growth-Limiting Constraints. *Cell. Mol. Life Sci.* **2020**, *77*, 441–453. [CrossRef]
32. Zrimec, J.; Börlin, C.S.; Buric, F.; Muhammad, A.S.; Chen, R.; Siewers, V.; Verendel, V.; Nielsen, J.; Töpel, M.; Zelezniak, A. Deep Learning Suggests That Gene Expression Is Encoded in All Parts of a Co-Evolving Interacting Gene Regulatory Structure. *Nat. Commun.* **2020**, *11*, 6141. [CrossRef]
33. Kotte, O.; Zaugg, J.B.; Heinemann, M. Bacterial Adaptation through Distributed Sensing of Metabolic Fluxes. *Mol. Syst. Biol.* **2010**, *6*, 355. [CrossRef] [PubMed]
34. Brunk, E.; Mih, N.; Monk, J.; Zhang, Z.; O'Brien, E.J.; Bliven, S.E.; Chen, K.; Chang, R.L.; Bourne, P.E.; Palsson, B.O. Systems Biology of the Structural Proteome. *BMC Syst. Biol.* **2016**, *10*, 1–6. [CrossRef] [PubMed]
35. Kim, O.D.; Rocha, M.; Maia, P. A Review of Dynamic Modeling Approaches and Their Application in Computational Strain Optimization for Metabolic Engineering. *Front. Microbiol.* **2018**, *9*, 1690. [CrossRef] [PubMed]
36. Jabarivelisdeh, B.; Waldherr, S. Optimization of Bioprocess Productivity Based on Metabolic-Genetic Network Models with Bilevel Dynamic Programming. *Biotechnol. Bioeng.* **2018**, *115*, 1829–1841. [CrossRef]
37. Chen, K.; Gao, Y.; Mih, N.; O'Brien, E.J.; Yang, L.; Palsson, B.O. Thermosensitivity of Growth Is Determined by Chaperone-Mediated Proteome Reallocation. *Proc. Natl. Acad. Sci. USA* **2017**, *114*, 11548–11553. [CrossRef]
38. Du, B.; Yang, L.; Lloyd, C.J.; Fang, X.; Palsson, B.O. Genome-Scale Model of Metabolism and Gene Expression Provides a Multi-Scale Description of Acid Stress Responses in Escherichia Coli. *PLoS Comput. Biol.* **2019**, *15*, e1007525. [CrossRef]

39. Yang, L.; Mih, N.; Anand, A.; Park, J.H.; Tan, J.; Yurkovich, J.T.; Monk, J.M.; Lloyd, C.J.; Sandberg, T.E.; Seo, S.W.; et al. Cellular Responses to Reactive Oxygen Species Are Predicted from Molecular Mechanisms. *Proc. Natl. Acad. Sci. USA* **2019**, *116*, 14368–14373. [CrossRef]

40. Wang, G.; Haringa, C.; Tang, W.; Noorman, H.; Chu, J.; Zhuang, Y.; Zhang, S. Coupled Metabolic-Hydrodynamic Modeling Enabling Rational Scale-up of Industrial Bioprocesses. *Biotechnol. Bioeng.* **2020**, *117*, 844–867. [CrossRef]

41. Monk, J.; Nogales, J.; Palsson, B.O. Optimizing Genome-Scale Network Reconstructions. *Nat. Biotechnol.* **2014**, *32*, 447–452. [CrossRef]

42. Ebrahim, A.; Lerman, J.A.; Palsson, B.O.; Hyduke, D.R. COBRApy: COnstraints-Based Reconstruction and Analysis for Python. *BMC Syst. Biol.* **2013**, *7*, 74. [CrossRef]

43. Cardoso, J.G.R.; Jensen, K.; Lieven, C.; Lærke Hansen, A.S.; Galkina, S.; Beber, M.; Özdemir, E.; Herrgård, M.J.; Redestig, H.; Sonnenschein, N. Cameo: A Python Library for Computer Aided Metabolic Engineering and Optimization of Cell Factories. *ACS Synth. Biol.* **2018**, *7*, 1163–1166. [CrossRef] [PubMed]

44. Bordbar, A.; Monk, J.M.; King, Z.A.; Palsson, B.O. Constraint-Based Models Predict Metabolic and Associated Cellular Functions. *Nat. Rev. Genet.* **2014**, *15*, 107–120. [CrossRef]

45. Lerman, J.A.; Hyduke, D.R.; Latif, H.; Portnoy, V.A.; Lewis, N.E.; Orth, J.D.; Schrimpe-Rutledge, A.C.; Smith, R.D.; Adkins, J.N.; Zengler, K.; et al. In Silico Method for Modelling Metabolism and Gene Product Expression at Genome Scale. *Nat. Commun.* **2012**, *3*, 1–10. [CrossRef]

46. Liu, J.K.; Lloyd, C.; Al-Bassam, M.M.; Ebrahim, A.; Kim, J.-N.; Olson, C.; Aksenov, A.; Dorrestein, P.; Zengler, K. Predicting Proteome Allocation, Overflow Metabolism, and Metal Requirements in a Model Acetogen. *PLoS Comput. Biol.* **2019**, *15*, e1006848. [CrossRef] [PubMed]

47. Dinh, H.V.; King, Z.A.; Palsson, B.O.; Feist, A.M. Identification of Growth-Coupled Production Strains Considering Protein Costs and Kinetic Variability. *Metab. Eng. Commun.* **2018**, *7*, e00080. [CrossRef] [PubMed]

48. King, Z.A.; O'Brien, E.J.; Feist, A.M.; Palsson, B.O. Literature Mining Supports a next-Generation Modeling Approach to Predict Cellular Byproduct Secretion. *Metab. Eng.* **2017**, *39*, 220–227. [CrossRef] [PubMed]

49. Yang, L.; Ebrahim, A.; Lloyd, C.J.; Saunders, M.A.; Palsson, B.O. DynamicME: Dynamic Simulation and Refinement of Integrated Models of Metabolism and Protein Expression. *BMC Syst. Biol.* **2019**, *13*, 2. [CrossRef]

50. Adadi, R.; Volkmer, B.; Milo, R.; Heinemann, M.; Shlomi, T. Prediction of Microbial Growth Rate versus Biomass Yield by a Metabolic Network with Kinetic Parameters. *PLoS Comput. Biol.* **2012**, *8*, e1002575. [CrossRef]

51. Massaiu, I.; Pasotti, L.; Sonnenschein, N.; Rama, E.; Cavaletti, M.; Magni, P.; Calvio, C.; Herrgård, M.J. Integration of Enzymatic Data in Bacillus Subtilis Genome-Scale Metabolic Model Improves Phenotype Predictions and Enables in Silico Design of Poly-γ-Glutamic Acid Production Strains. *Microb. Cell Fact.* **2019**, *18*, 3. [CrossRef]

52. Sánchez, B.J.; Zhang, C.; Nilsson, A.; Lahtvee, P.-J.; Kerkhoven, E.J.; Nielsen, J. Improving the Phenotype Predictions of a Yeast Genome-Scale Metabolic Model by Incorporating Enzymatic Constraints. *Mol. Syst. Biol.* **2017**, *13*, 935. [CrossRef]

53. Bekiaris, P.S.; Klamt, S. Automatic Construction of Metabolic Models with Enzyme Constraints. *BMC Bioinformatics* **2020**, *21*, 19. [CrossRef] [PubMed]

54. Zuñiga, C.; Li, T.; Guarnieri, M.T.; Jenkins, J.P.; Li, C.-T.; Bingol, K.; Kim, Y.-M.; Betenbaugh, M.J.; Zengler, K. Synthetic Microbial Communities of Heterotrophs and Phototrophs Facilitate Sustainable Growth. *Nat. Commun.* **2020**, *11*, 3803. [CrossRef] [PubMed]

55. Özcan, E.; Seven, M.; Şirin, B.; Çakır, T.; Nikerel, E.; Teusink, B.; Toksoy Öner, E. Dynamic Co-Culture Metabolic Models Reveal the Fermentation Dynamics, Metabolic Capacities and Interplays of Cheese Starter Cultures. *Biotechnol. Bioeng.* **2020**. [CrossRef] [PubMed]

56. Zhou, K.; Qiao, K.; Edgar, S.; Stephanopoulos, G. Distributing a Metabolic Pathway among a Microbial Consortium Enhances Production of Natural Products. *Nat. Biotechnol.* **2015**, *33*, 377–383. [CrossRef] [PubMed]

57. Lloyd, C.J.; King, Z.; Sandberg, T.; Hefner, Y.; Feist, A. Model-Driven Design and Evolution of Non-Trivial Synthetic Syntrophic Pairs. *BioRxiv* **2018**. [CrossRef]

58. Monk, J.M.; Charusanti, P.; Aziz, R.K.; Lerman, J.A.; Premyodhin, N.; Orth, J.D.; Feist, A.M.; Palsson, B.Ø. Genome-Scale Metabolic Reconstructions of Multiple Escherichia Coli Strains Highlight Strain-Specific Adaptations to Nutritional Environments. *Proc. Natl. Acad. Sci. USA* **2013**, *110*, 20338–20343. [CrossRef]

59. Poudel, S.; Tsunemoto, H.; Seif, Y.; Sastry, A.V.; Szubin, R.; Xu, S.; Machado, H.; Olson, C.A.; Anand, A.; Pogliano, J.; et al. Revealing 29 Sets of Independently Modulated Genes in Staphylococcus Aureus, Their Regulators, and Role in Key Physiological Response. *Proc. Natl. Acad. Sci. USA* **2020**, *117*, 17228–17239. [CrossRef]

60. Norsigian, C.J.; Attia, H.; Szubin, R.; Yassin, A.S.; Palsson, B.Ø.; Aziz, R.K.; Monk, J.M. Comparative Genome-Scale Metabolic Modeling of Metallo-Beta-Lactamase-Producing Multidrug-Resistant Klebsiella Pneumoniae Clinical Isolates. *Front. Cell. Infect. Microbiol.* **2019**, *9*. [CrossRef]

61. Seif, Y.; Kavvas, E.; Lachance, J.-C.; Yurkovich, J.T.; Nuccio, S.-P.; Fang, X.; Catoiu, E.; Raffatellu, M.; Palsson, B.O.; Monk, J.M. Genome-Scale Metabolic Reconstructions of Multiple Salmonella Strains Reveal Serovar-Specific Metabolic Traits. *Nat. Commun.* **2018**, *9*, 3771. [CrossRef]

62. Prigent, S.; Nielsen, J.C.; Frisvad, J.C.; Nielsen, J. Reconstruction of 24 Penicillium Genome-Scale Metabolic Models Shows Diversity Based on Their Secondary Metabolism. *Biotechnol. Bioeng.* **2018**, *115*, 2604–2612. [CrossRef]

63. Fouts, D.E.; Matthias, M.A.; Adhikarla, H.; Adler, B.; Amorim-Santos, L.; Berg, D.E.; Bulach, D.; Buschiazzo, A.; Chang, Y.-F.; Galloway, R.L.; et al. What Makes a Bacterial Species Pathogenic?: Comparative Genomic Analysis of the Genus Leptospira. *PLoS Negl. Trop. Dis.* **2016**, *10*, e0004403. [CrossRef] [PubMed]

64. Ghatak, S.; King, Z.A.; Sastry, A.; Palsson, B.O. The Y-Ome Defines the 35% of Escherichia Coli Genes That Lack Experimental Evidence of Function. *Nucleic Acids Res.* **2019**, *47*, 2446–2454. [CrossRef] [PubMed]

65. Xie, Z.; Tang, H. ISEScan: Automated Identification of Insertion Sequence Elements in Prokaryotic Genomes. *Bioinformatics* **2017**, *33*, 3340–3347. [CrossRef] [PubMed]

66. Tarailo-Graovac, M.; Chen, N. Using RepeatMasker to Identify Repetitive Elements in Genomic Sequences. *Curr. Protoc. Bioinform.* **2009**, *25*, 1–4. [CrossRef] [PubMed]

67. Ryu, J.Y.; Kim, H.U.; Lee, S.Y. Deep Learning Enables High-Quality and High-Throughput Prediction of Enzyme Commission Numbers. *Proc. Natl. Acad. Sci. USA* **2019**, *116*, 13996–14001. [CrossRef]

68. Hadadi, N.; MohammadiPeyhani, H.; Miskovic, L.; Seijo, M.; Hatzimanikatis, V. Enzyme Annotation for Orphan and Novel Reactions Using Knowledge of Substrate Reactive Sites. *Proc. Natl. Acad. Sci. USA* **2019**, *116*, 7298–7307. [CrossRef]

69. Oberhardt, M.A.; Zarecki, R.; Reshef, L.; Xia, F.; Duran-Frigola, M.; Schreiber, R.; Henry, C.S.; Ben-Tal, N.; Dwyer, D.J.; Gophna, U.; et al. Systems-Wide Prediction of Enzyme Promiscuity Reveals a New Underground Alternative Route for Pyridoxal 5'-Phosphate Production in E. Coli. *PLoS Comput. Biol.* **2016**, *12*, e1004705. [CrossRef]

70. Moreno-Sánchez, R.; Saavedra, E.; Rodríguez-Enríquez, S.; Olín-Sandoval, V. Metabolic Control Analysis: A Tool for Designing Strategies to Manipulate Metabolic Pathways. *J. Biomed. Biotechnol.* **2008**, *2008*. [CrossRef]

71. Noor, E.; Flamholz, A.; Liebermeister, W.; Bar-Even, A.; Milo, R. A Note on the Kinetics of Enzyme Action: A Decomposition That Highlights Thermodynamic Effects. *FEBS Lett.* **2013**, *587*, 2772–2777. [CrossRef]

72. Henry, C.S.; Broadbelt, L.J.; Hatzimanikatis, V. Thermodynamics-Based Metabolic Flux Analysis. *Biophys. J.* **2007**, *92*, 1792–1805. [CrossRef]

73. Hamilton, J.J.; Dwivedi, V.; Reed, J.L. Quantitative Assessment of Thermodynamic Constraints on the Solution Space of Genome-Scale Metabolic Models. *Biophys. J.* **2013**, *105*, 512–522. [CrossRef] [PubMed]

74. Du, B.; Zielinski, D.C.; Monk, J.M.; Palsson, B.O. Thermodynamic Favorability and Pathway Yield as Evolutionary Tradeoffs in Biosynthetic Pathway Choice. *Proc. Natl. Acad. Sci. USA* **2018**, *115*, 11339–11344. [CrossRef] [PubMed]

75. Salvy, P.; Hatzimanikatis, V. ETFL: A Formulation for Flux Balance Models Accounting for Expression, Thermodynamics, and Resource Allocation Constraints. *bioRxiv* **2019**. [CrossRef]

76. Salvy, P.; Hatzimanikatis, V. Emergence of Diauxie as an Optimal Growth Strategy under Resource Allocation Constraints in Cellular Metabolism. *bioRxiv* **2020**. [CrossRef]

77. Pandey, V.; Hadadi, N.; Hatzimanikatis, V. Enhanced Flux Prediction by Integrating Relative Expression and Relative Metabolite Abundance into Thermodynamically Consistent Metabolic Models. *PLoS Comput. Biol.* **2019**, *15*, e1007036. [CrossRef]

78. Saa, P.A.; Nielsen, L.K. Formulation, Construction and Analysis of Kinetic Models of Metabolism: A Review of Modelling Frameworks. *Biotechnol. Adv.* **2017**, *35*, 981–1003. [CrossRef]

79. Gopalakrishnan, S.; Dash, S.; Maranas, C. K-FIT: An Accelerated Kinetic Parameterization Algorithm Using Steady-State Fluxomic Data. *Metab. Eng.* **2020**, *61*, 197–205. [CrossRef]

80.   Saa, P.A.; Nielsen, L.K. Construction of Feasible and Accurate Kinetic Models of Metabolism: A Bayesian Approach. *Sci. Rep.* **2016**, *6*, 1–13. [CrossRef]
81.   Davidi, D.; Noor, E.; Liebermeister, W.; Bar-Even, A.; Flamholz, A.; Tummler, K.; Barenholz, U.; Goldenfeld, M.; Shlomi, T.; Milo, R. Global Characterization of in Vivo Enzyme Catalytic Rates and Their Correspondence to in Vitro Kcat Measurements. *Proc. Natl. Acad. Sci. USA* **2016**, *113*, 3401–3406. [CrossRef]
82.   Heckmann, D.; Campeau, A.; Lloyd, C.J.; Phaneuf, P.V.; Hefner, Y.; Carrillo-Terrazas, M.; Feist, A.M.; Gonzalez, D.J.; Palsson, B.O. Kinetic Profiling of Metabolic Specialists Demonstrates Stability and Consistency of in Vivo Enzyme Turnover Numbers. *Proc. Natl. Acad. Sci. USA* **2020**. [CrossRef]
83.   Heckmann, D.; Lloyd, C.J.; Mih, N.; Ha, Y.; Zielinski, D.C.; Haiman, Z.B.; Desouki, A.A.; Lercher, M.J.; Palsson, B.O. Machine Learning Applied to Enzyme Turnover Numbers Reveals Protein Structural Correlates and Improves Metabolic Models. *Nat. Commun.* **2018**, *9*, 5252. [CrossRef] [PubMed]
84.   Beard, D.A.; Vinnakota, K.C.; Wu, F. Detailed Enzyme Kinetics in Terms of Biochemical Species: Study of Citrate Synthase. *PLoS ONE* **2008**, *3*, e1825. [CrossRef]
85.   Andreozzi, S.; Miskovic, L.; Hatzimanikatis, V. iSCHRUNK—In Silico Approach to Characterization and Reduction of Uncertainty in the Kinetic Models of Genome-Scale Metabolic Networks. *Metab. Eng.* **2016**, *33*, 158–168. [CrossRef] [PubMed]
86.   Nilsson, A.; Nielsen, J.; Palsson, B.O. Metabolic Models of Protein Allocation Call for the Kinetome. *Cell Syst.* **2017**, *5*, 538–541. [CrossRef] [PubMed]
87.   Jankowski, M.D.; Henry, C.S.; Broadbelt, L.J.; Hatzimanikatis, V. Group Contribution Method for Thermodynamic Analysis of Complex Metabolic Networks. *Biophys. J.* **2008**, *95*, 1487–1499. [CrossRef] [PubMed]
88.   Noor, E.; Haraldsdóttir, H.S.; Milo, R.; Fleming, R.M.T. Consistent Estimation of Gibbs Energy Using Component Contributions. *PLoS Comput. Biol.* **2013**, *9*, e1003098. [CrossRef]
89.   Flamholz, A.; Noor, E.; Bar-Even, A.; Milo, R. eQuilibrator—the Biochemical Thermodynamics Calculator. *Nucleic Acids Res.* **2012**, *40*, D770–D775. [CrossRef]
90.   Noor, E.; Bar Even, A.; Flamholz, A.; Lubling, Y.; Davidi, D.; Milo, R. An Integrated Open Framework for Thermodynamics of Reactions That Combines Accuracy and Coverage. *Bioinformatics* **2012**, *28*, 2037–2044. [CrossRef]
91.   Du, B.; Zhang, Z.; Grubner, S.; Yurkovich, J.T.; Palsson, B.O.; Zielinski, D.C. Temperature-Dependent Estimation of Gibbs Energies Using an Updated Group-Contribution Method. *Biophys. J.* **2018**, *114*, 2691–2702. [CrossRef]
92.   Du, B.; Zielinski, D.C.; Palsson, B.O. Estimating Metabolic Equilibrium Constants: Progress and Future Challenges. *Trends Biochem. Sci.* **2018**, *43*, 960–969. [CrossRef]
93.   Jinich, A.; Rappoport, D.; Dunn, I.; Sanchez-Lengeling, B.; Olivares-Amaya, R.; Noor, E.; Even, A.B.; Aspuru-Guzik, A. Quantum Chemical Approach to Estimating the Thermodynamics of Metabolic Reactions. *Sci. Rep.* **2014**, *4*, 1–6. [CrossRef] [PubMed]
94.   Haiman, Z.B.; Zielinski, D.C.; Koike, Y.; Yurkovich, J.T.; Palsson, B.O. MASSpy: Building, Simulating, and Visualizing Dynamic Biological Models in Python Using Mass Action Kinetics. *bioRxiv* **2020**. [CrossRef]
95.   Salvy, P.; Fengos, G.; Ataman, M.; Pathier, T.; Soh, K.C.; Hatzimanikatis, V. pyTFA and matTFA: A Python Package and a Matlab Toolbox for Thermodynamics-Based Flux Analysis. *Bioinformatics* **2019**, *35*, 167–169. [CrossRef] [PubMed]
96.   Hoops, S.; Sahle, S.; Gauges, R.; Lee, C.; Pahle, J.; Simus, N.; Singhal, M.; Xu, L.; Mendes, P.; Kummer, U. COPASI—A COmplex PAthway SImulator. *Bioinformatics* **2006**, *22*, 3067–3074. [CrossRef]
97.   Khodayari, A.; Zomorrodi, A.R.; Liao, J.C.; Maranas, C.D. A Kinetic Model of Escherichia Coli Core Metabolism Satisfying Multiple Sets of Mutant Flux Data. *Metab. Eng.* **2014**, *25*, 50–62. [CrossRef]
98.   Tokic, M.; Hatzimanikatis, V.; Miskovic, L. Large-Scale Kinetic Metabolic Models of Pseudomonas Putida KT2440 for Consistent Design of Metabolic Engineering Strategies. *Biotechnol. Biofuels* **2020**, *13*, 1–9. [CrossRef]
99.   Soh, K.C.; Hatzimanikatis, V. Constraining the Flux Space Using Thermodynamics and Integration of Metabolomics Data. *Methods Mol. Biol.* **2014**, *1191*, 49–63.
100. Akbari, A.; Palsson, B.O. Scalable Computation of Intracellular Metabolite Concentrations. *arXiv* **2020**. [CrossRef]
101. Chowdhury, R.; Ren, T.; Shankla, M.; Decker, K.; Grisewood, M.; Prabhakar, J.; Baker, C.; Golbeck, J.H.; Aksimentiev, A.; Kumar, M.; et al. PoreDesigner for Tuning Solute Selectivity in a Robust and Highly Permeable Outer Membrane Pore. *Nat. Commun.* **2018**, *9*, 3661. [CrossRef]

102. Huang, P.-S.; Boyken, S.E.; Baker, D. The Coming of Age of de Novo Protein Design. *Nature* **2016**, *537*, 320–327. [CrossRef]

103. Arnold, F.H. Directed Evolution: Bringing New Chemistry to Life. *Angew. Chem. Int. Ed. Engl.* **2018**, *57*, 4143–4148. [CrossRef] [PubMed]

104. Pantazes, R.J.; Grisewood, M.J.; Li, T.; Gifford, N.P.; Maranas, C.D. The Iterative Protein Redesign and Optimization (IPRO) Suite of Programs. *J. Comput. Chem.* **2015**, *36*, 251–263. [CrossRef] [PubMed]

105. Monk, J.M.; Lloyd, C.J.; Brunk, E.; Mih, N.; Sastry, A.; King, Z.; Takeuchi, R.; Nomura, W.; Zhang, Z.; Mori, H.; et al. iML1515, a Knowledgebase That Computes Escherichia Coli Traits. *Nat. Biotechnol.* **2017**, *35*, 904–908. [CrossRef]

106. Mih, N.; Brunk, E.; Chen, K.; Catoiu, E.; Sastry, A.; Kavvas, E.; Monk, J.M.; Zhang, Z.; Palsson, B.O. Ssbio: A Python Framework for Structural Systems Biology. *Bioinformatics* **2018**, *34*, 2155–2157. [CrossRef] [PubMed]

107. Yang, J.; Yan, R.; Roy, A.; Xu, D.; Poisson, J.; Zhang, Y. The I-TASSER Suite: Protein Structure and Function Prediction. *Nat. Methods* **2015**, *12*, 7–8. [CrossRef] [PubMed]

108. Senior, A.W.; Evans, R.; Jumper, J.; Kirkpatrick, J.; Sifre, L.; Green, T.; Qin, C.; Žídek, A.; Nelson, A.W.R.; Bridgland, A.; et al. Improved Protein Structure Prediction Using Potentials from Deep Learning. *Nature* **2020**, *577*, 706 710. [CrossRef] [PubMed]

109. Kavvas, E.S.; Catoiu, E.; Mih, N.; Yurkovich, J.T.; Seif, Y.; Dillon, N.; Heckmann, D.; Anand, A.; Yang, L.; Nizet, V.; et al. Machine Learning and Structural Analysis of Mycobacterium Tuberculosis Pan-Genome Identifies Genetic Signatures of Antibiotic Resistance. *Nat. Commun.* **2018**, *9*, 4306. [CrossRef]

110. Brunk, E.; Chang, R.L.; Xia, J.; Hefzi, H.; Yurkovich, J.T.; Kim, D.; Buckmiller, E.; Wang, H.H.; Cho, B.-K.; Yang, C.; et al. Characterizing Posttranslational Modifications in Prokaryotic Metabolism Using a Multiscale Workflow. *Proc. Natl. Acad. Sci. USA* **2018**, *115*, 11096–11101. [CrossRef]

111. Case, D.A.; Belfon, K.; Ben-Shalom, I.Y.; Brozell, S.R.; Cerutti, D.S.; Cheatham, T.E., III; Cruzeiro, V.W.D.; Darden, T.A.; Duke, R.E.; Giambasu, G.; et al. *AMBER 2020*; University of California: San Francisco, CA, USA, 2020.

112. Salomon-Ferrer, R.; Case, D.A.; Walker, R.C. An Overview of the Amber Biomolecular Simulation Package. *WIREs Comput. Mol. Sci.* **2013**, *3*, 198–210. [CrossRef]

113. Kavvas, E.S.; Yang, L.; Monk, J.M.; Heckmann, D.; Palsson, B.O. A Biochemically-Interpretable Machine Learning Classifier for Microbial GWAS. *Nat. Commun.* **2020**, *11*, 2580. [CrossRef]

114. Davis, J.J.; Boisvert, S.; Brettin, T.; Kenyon, R.W.; Mao, C.; Olson, R.; Overbeek, R.; Santerre, J.; Shukla, M.; Wattam, A.R.; et al. Antimicrobial Resistance Prediction in PATRIC and RAST. *Sci. Rep.* **2016**, *6*. [CrossRef] [PubMed]

115. Arango-Argoty, G.; Garner, E.; Pruden, A.; Heath, L.S.; Vikesland, P.; Zhang, L. DeepARG: A Deep Learning Approach for Predicting Antibiotic Resistance Genes from Metagenomic Data. *Microbiome* **2018**, *6*, 23. [CrossRef] [PubMed]

116. Haugen, S.P.; Ross, W.; Gourse, R.L. Advances in Bacterial Promoter Recognition and Its Control by Factors That Do Not Bind DNA. *Nat. Rev. Microbiol.* **2008**, *6*, 507–519. [CrossRef] [PubMed]

117. Bailey, T.L.; Boden, M.; Buske, F.A.; Frith, M.; Grant, C.E.; Clementi, L.; Ren, J.; Li, W.W.; Noble, W.S. MEME SUITE: Tools for Motif Discovery and Searching. *Nucleic Acids Res.* **2009**, *37*, W202–W208. [CrossRef] [PubMed]

118. Phaneuf, P.V.; Gosting, D.; Palsson, B.O.; Feist, A.M. ALEdb 1.0: A Database of Mutations from Adaptive Laboratory Evolution Experimentation. *Nucleic Acids Res.* **2019**, *47*, D1164–D1171. [CrossRef] [PubMed]

119. Phaneuf, P.V.; Yurkovich, J.T.; Heckmann, D.; Wu, M.; Sandberg, T.E.; King, Z.A.; Tan, J.; Palsson, B.O.; Feist, A.M. Causal Mutations from Adaptive Laboratory Evolution Are Outlined by Multiple Scales of Genome Annotations and Condition-Specificity. *BMC Genom.* **2020**, *21*, 514. [CrossRef]

120. Lamoureux, C.R.; Choudhary, K.S.; King, Z.A.; Sandberg, T.E.; Gao, Y.; Sastry, A.V.; Phaneuf, P.V.; Choe, D.; Cho, B.-K.; Palsson, B.O. The Bitome: Digitized Genomic Features Reveal Fundamental Genome Organization. *Nucleic Acids Res.* **2020**, *48*, 10157–10163. [CrossRef]

121. Einav, T.; Phillips, R. How the Avidity of Polymerase Binding to the −35/−10 Promoter Sites Affects Gene Expression. *Proc. Natl. Acad. Sci. USA* **2019**, *116*, 13340–13345. [CrossRef]

122. Tuller, T.; Waldman, Y.Y.; Kupiec, M.; Ruppin, E. Translation Efficiency Is Determined by Both Codon Bias and Folding Energy. *Proc. Natl. Acad. Sci. USA* **2010**, *107*, 3645–3650. [CrossRef]

123. Bonde, M.T.; Pedersen, M.; Klausen, M.S.; Jensen, S.I.; Wulff, T.; Harrison, S.; Nielsen, A.T.; Herrgård, M.J.; Sommer, M.O.A. Predictable Tuning of Protein Expression in Bacteria. *Nat. Methods* **2016**, *13*, 233–236. [CrossRef]

124. Rychel, K.; Sastry, A.V.; Palsson, B.O. Machine Learning Uncovers Independently Regulated Modules in the Bacillus Subtilis Transcriptome. *bioRxiv* **2020**. [CrossRef]

125. Rychel, K.; Decker, K.; Sastry, A.V.; Phaneuf, P.V.; Poudel, S.; Palsson, B.O. iModulonDB: A Knowledgebase of Microbial Transcriptional Regulation Derived from Machine Learning. *Nucleic Acids Res.* **2020**. [CrossRef] [PubMed]

126. Ament, S.; Shannon, P.; Richards, M. *TReNa: Fit. Transcriptional Regulatory Networks Using Gene Expression, Priors, Machine Learning*; Bioconductor: Washington, DC, USA, 2017. [CrossRef]

127. Fang, X.; Sastry, A.; Mih, N.; Kim, D.; Tan, J.; Yurkovich, J.T.; Lloyd, C.J.; Gao, Y.; Yang, L.; Palsson, B.O. Global Transcriptional Regulatory Network for Escherichia Coli Robustly Connects Gene Expression to Transcription Factor Activities. *Proc. Natl. Acad. Sci. USA* **2017**, *114*, 10286–10291. [CrossRef] [PubMed]

128. Chandrasekaran, S.; Price, N.D. Probabilistic Integrative Modeling of Genome-Scale Metabolic and Regulatory Networks in Escherichia Coli and Mycobacterium Tuberculosis. *Proc. Natl. Acad. Sci. USA* **2010**, *107*, 17845–17850. [CrossRef]

129. Rustad, T.R.; Minch, K.J.; Ma, S.; Winkler, J.K.; Hobbs, S.; Hickey, M.; Brabant, W.; Turkarslan, S.; Price, N.D.; Baliga, N.S.; et al. Mapping and Manipulating the Mycobacterium Tuberculosis Transcriptome Using a Transcription Factor Overexpression-Derived Regulatory Network. *Genom. Biol.* **2014**, *15*, 502. [CrossRef]

130. Kochanowski, K.; Gerosa, L.; Brunner, S.F.; Christodoulou, D.; Nikolaev, Y.V.; Sauer, U. Few Regulatory Metabolites Coordinate Expression of Central Metabolic Genes in Escherichia Coli. *Mol. Syst. Biol.* **2017**, *13*, 903. [CrossRef]

131. Santos-Zavaleta, A.; Salgado, H.; Gama-Castro, S.; Sánchez-Pérez, M.; Gómez-Romero, L.; Ledezma-Tejeida, D.; García-Sotelo, J.S.; Alquicira-Hernández, K.; Muñiz-Rascado, L.J.; Peña-Loredo, P.; et al. RegulonDB v 10.5: Tackling Challenges to Unify Classic and High Throughput Knowledge of Gene Regulation in E. Coli K-12. *Nucleic Acids Res.* **2019**, *47*, D212–D220. [CrossRef]

132. Keseler, I.M.; Mackie, A.; Santos-Zavaleta, A.; Billington, R.; Bonavides-Martínez, C.; Caspi, R.; Fulcher, C.; Gama-Castro, S.; Kothari, A.; Krummenacker, M.; et al. The EcoCyc Database: Reflecting New Knowledge about Escherichia Coli K-12. *Nucleic Acids Res.* **2017**, *45*, D543–D550. [CrossRef]

Photorespiration and Rate Synchronization in a Phototroph-Heterotroph Microbial Consortium

Fadoua El Moustaid [1], Ross P. Carlson [2], Federica Villa [3] and Isaac Klapper [4],*

[1] Department of Biological Sciences, Virginia Tech University, Blacksburg, VA 24061, USA; fadoua@vt.edu
[2] Center for Biofilm Engineering, Montana State University, Bozeman, MT 59717, USA; rossc@montana.edu
[3] Department of Food, Environmental and Nutritional Sciences, Università degli Studi di Milano, 20133 Milano, Italy; federica.villa@unimi.it
[4] Department of Mathematics, Temple University, Philadelphia, PA 19122, USA
* Correspondence: klapper@temple.edu

Academic Editor: Hyun-Seob Song

Abstract: The process of oxygenic photosynthesis is robust and ubiquitous, relying centrally on input of light, carbon dioxide, and water, which in many environments are all abundantly available, and from which are produced, principally, oxygen and reduced organic carbon. However, photosynthetic machinery can be conflicted by the simultaneous presence of carbon dioxide and oxygen through a process sometimes called photorespiration. We present here a model of phototrophy, including competition for RuBisCO binding sites between oxygen and carbon dioxide, in a chemostat-based microbial population. The model connects to the idea of metabolic pathways to track carbon and degree of reduction through the system. We find decomposition of kinetics into elementary flux modes a mathematically natural way to study synchronization of mismatched rates of photon input and chemostat turnover. In the single species case, though total biomass is reduced by photorespiration, protection from excess light exposures and its consequences (oxidative and redox stress) may result. We also find the possibility that a consortium of phototrophs with heterotrophs can recycle photorespiration byproduct into increased biomass at the cost of increase in oxidative product (here, oxygen).

Keywords: photosynthesis; photorespiration; chemostat model; phototroph-heterotroph consortium

1. Introduction

Life on earth, in large part, has oxygenic photosynthesis at its foundation, and much of that photosynthesis occurs in microbes. Oxygenic phototrophic microorganisms such as cyanobacteria are common in reliably lit environments, where impinging photons provide, often, a more than sufficient energy source even at low intensity, and carbon dioxide (or related chemical species) provides a reliable and abundant carbon source. When the other fundamental component of photosynthesis, water, is also available, then phototrophic based life is likely. In many cyanobacteria, nitrogen fixation can even be supported due to the abundance of photon energy. It is perhaps surprising, then, that the process of photosynthetic fixation of carbon dioxide into reduced carbon suitable for biosynthesis has, seemingly, a significant inefficiency due to the competition by oxygen for inorganic carbon binding sites, here denoted as photorespiration.

Thus, we focus on processing of inorganic carbon, i.e., carbon fixation, a central component of oxygenic phototrophy, and on its principle byproduct, molecular oxygen. Oxygenic phototrophy uses photon energy to extract electrons from water and eventually apply those electrons to fix inorganic carbon, while, in the process, oxygen is produced: electron source (effectively here, water and light) feeds electron sink (inorganic carbon) while producing oxidative byproduct (molecular oxygen).

Implicit to this assembly line is the need for extracellular, macroscale transfer of inorganic carbon and of oxygen. Rates of macroscale transport (advective and/or diffusive) are largely beyond the control of individual cells and thus oxygen concentration may serve as a signal of transport limitation, triggering photorespiration. High photon flux can be an even further aggravating factor if transport of inorganic carbon into the photosynthesizing machinery cannot keep pace.

Nevertheless, the oxygenic phototrophic "business model" is generally robust and capable of being largely self sufficient. Strikingly, however, phototrophic organisms are often found in multispecies consortia together with heterotrophs. It is not immediately clear why this should be the case, as competition for resources, e.g., space or nutrients, is possible, and it seems that oxygenic phototrophs might be expected to be able to outcompete heterotrophic neighbors for those resources. Even so, multispecies communities are observed including in environments where heterotrophs might not be able to persist on their own [1]. Further, there are at least some examples of communities where resident phototrophs lack critical anabolic capabilities and must instead rely on nearby organisms to supply them [2]. Here, we explore the possible utility of interaction via organic/inorganic carbon exchange. Note there are other possible advantages in adding a heterotroph to the autotroph community. For example, heterotroph-induced oxygen usage or moderation of variation in redox potential may mitigate transport limitation.

The models presented here, both for single species (an oxygenic phototroph we call cyanobacteria) as well as for a combined two species system (cyanobacteria plus a generic heterotroph) are based in a chemostat platform. The chemostat serves as a simple and convenient way to mimic an environment where, over long times, nutrient inflow and byproduct outflow occur at rates determined by external environmental factors. From this viewpoint, a chemostat is a natural choice here due to its simplicity and also the steady oligotrophic environment it models, and thus hopefully is a reasonable bridge between abstract modeling and empirical observations.

In fact, comparison of population models with population scale observations has a well established methodology in microbial ecology. Of late, however, rapidly increasing use of molecular level technology (e.g., high throughput sequencing) has dramatically changed the nature and scale of these observations. As a result, in principle and increasingly also in practice, detailed data describing microbial capability and function is available. This information can and should potentially be used to understand how microbes exploit and alter their environment. There is a substantial gap, however, between molecular behavior at the cellular microscale and emergent community function at the population macroscale. Intermediate between the two, progress is being made in translating genomics information into models of cell dynamics [3]. Annotation of gene sequences into so-called wiring diagrams is becoming increasingly reliable and automatable. These diagrams encode cell physiology along with regulatory machinery and are accompanied by an intimidating list of unknown rate constants. However, gene encoded functions relevant to metabolic processes are naturally organized into gene pathways [4–6], and then, under the often reasonable assumption of steady state, balance of influx and outflux through these pathways makes choice of individual reaction rates within any particular pathway unnecessary, replaceable instead by a single flux through that entire path. Regulatory function can be characterized as a management of resource allocation between different paths and then modeled by imposing optimality criteria on that allocation [7,8]. The result is an enormous simplification: cell function is now characterized by only a limited number of rates of cellular inflow and outflow of substrates and byproducts together with an optimization principle to divide them between available metabolic pathways.

Still, there are two significant though not unrelated requirements for use of such analyses. First, despite the reduction, there remain, generally, many available and redundant metabolic flux pathways encoded by any one genome and so, as mentioned, some principle is necessary in order to decide how flux is to be distributed between those pathways. Second, also as mentioned, rates of substrate flow into the cell and byproduct flow out of the cell need to be characterized. The first of these issues couples to the second which then couples to the environment in which the cell and its community find themselves [9]. Conversely, though quantities on the large environmental scale are oftened characterized by concentrations, from this point of view it seems, rather, that fluxes are natural quantities at the cell scale. Thus, beyond the immediate aim of studying photorespiration, a further goal of this study is to suggest ways to match models at cell and population levels.

2. Materials and Methods

2.1. Model Description

We study productivity of two interacting species, one a photoautotroph with cell density $\widehat{P}_1(t)$ and the other a heterotroph with cell density $\widehat{P}_2(t)$, both of which are growing in a well mixed chemostat with dilution rate D [10], exposed to photon influx, see Figure 1. For simplicity, we neglect transport across the chemostat-air interface or suppose that such an interface does not exist, and let external inflow and outflow of dissolved quantities be governed by the chemostat dilution rate D. Conversely, since our aim is to study possible mutualistic or commensal effects over long times, we do not include diel light variation effects, considering them, for such purposes, to be relatively short time phenomena that can also be averaged out.

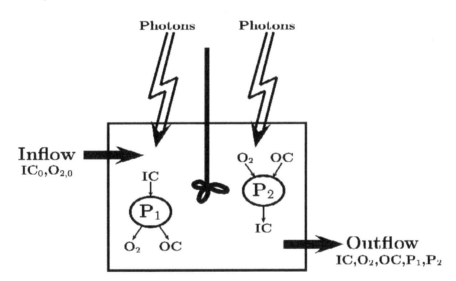

Figure 1. Chemostat diagram: photoautotrophic (P_1) and heterotrophic (P_2) microbial communities interact in a well mixed tank, exposed to light, with constant and equal inflow and outflow. Dissolved inorganic carbon (IC, inflow concentration IC_0), organic carbon (OC, inflow concentration zero), and oxygen (O_2, inflow concentration $O_{2,0}$) are also mixed throughout the tank, and in the inflow. Transport across any fluid-air interface is neglected for simplicity.

A central element of the model is the tracking of carbon flow through a microbial communtiy. As such it is convenient to measure all carbon carrying quantities in terms of carbon moles (Cmoles), e.g., to measure phototroph and heterotroph populations by the total moles of carbon they incorporate. We assume here, for convenience only, that cell sizes and densities are similar, i.e., that the total carbon moles per microbial cell, denoted c, is a constant and is the same constant for both phototrophs and heterotrophs. To convert populations from units of cells/volume to units of Cmoles/volume, we change to $P_1(t) = c\widehat{P}_1(t)$ and $P_2(t) = c\widehat{P}_2(t)$, both with units Cmoles/volume. In addition, we measure both dissolved component densities $IC(t)$ (pooled inorganic carbon, Cmoles/volume), $OC(t)$ (organic carbon, Cmoles/volume) in Cmoles, and $O_2(t)$ (oxygen, molecular oxygen moles/volume) in moles of molecular oxygen. In computations, we use liters as volume units. It is assumed that oxygen concentrations always remain sufficiently low so that oxygen remains in solution and a gas phase does not occur. Note that we use the notation O_2 both to denote molecular oxygen and its concentration. Inorganic carbon in solution, $IC(t)$, consists of aqueous CO_2

and related dissolved forms, notably aqueous bicarbonate HCO_3^-; we do not distinguish between the forms here, though phototrophs generally do. Organic carbon here is supposed for specificity to consist of glycolic acid $C_2H_4O_3$, a byproduct of photorespiration. Note that the term photorespiration has been used in the literature to designate a number of different mechanisms that, effectively, oxidize photosynthetically fixed carbon [11]. We consider here only one of those mechanisms, namely oxygenic activity of RuBisCO (ribulose bisphosphate carboxylase) secretion from the cell of partially oxidized organic carbon in the form of glycolic acid. For brevity, however, we will use the umbrella label photorespiration for this single type.

For purposes of tracking carbon, we could, as is commonly done in microbial population models, also include a microbial species $Q_1(t)$ (Cmoles/volume) consisting of inactive photoautotroph biomaterial damaged (or killed/lysed) due to oxidative stress or in some other manner, as well as a similar heterotroph damage species $Q_2(t)$ (Cmoles/volume). For simplicity and clarity, however, we include oxidative damage only through its direct effect on photosynthetic machinery. Note, though, that as a result, importance of oxidative damage and its amelioration are, if anything, likely underestimated in the later results.

The general form of the equations used here for a chemostat with photon flux ν are

$$\frac{d}{dt}P_1 = (\eta g_1(\text{IC}, O_2; \nu) - D)P_1, \tag{1}$$

$$\frac{d}{dt}P_2 = (g_2(\text{OC}, O_2) - D)P_2, \tag{2}$$

$$\frac{d}{dt}\text{IC} = -Y_{P_1,\text{IC}}^{-1} g_1(\text{IC}, O_2; \nu)P_1 + y_{P_2,\text{IC}}^{-1} g_2(\text{OC}, O_2)P_2 + D(\text{IC}_0 - \text{IC}), \tag{3}$$

$$\frac{d}{dt}\text{OC} = Y_{\text{IC},\text{OC}}^{-1} (1-\eta)g_1(\text{IC}, O_2; \nu)P_1 - y_{P_2,\text{OC}}^{-1} g_2(\text{OC}, O_2)P_2 - D\,\text{OC}, \tag{4}$$

$$\frac{d}{dt}O_2 = ((Y_{\text{IC},O_2}^{-1} - Y_{\text{OC},O_2}^{-1}(1-\eta))g_1(\text{IC}, O_2; \nu))P_1$$
$$\quad -y_{P_2,O_2}^{-1} g_2(\text{OC}, O_2)P_2 + D(O_{2,0} - O_2), \tag{5}$$

where the various subscripted $Y_{\alpha,\beta}$'s (associated with P_1) and $y_{\alpha,\beta}$'s (associated with P_2) are yield coefficients, all of which are fixed by stoichiometry, with units of Cmoles of α per Cmoles of β or moles of O_2. The parameters k_1 and k_2 indicate specific rates of deactivation of active biomaterial and could be functions of O_2. The function $\eta = \eta(\text{IC}, O_2)$ is related to photorespiration, and will be defined later. Terms containing rate g_1 are involved in the photobiosynthesis and/or photorespiration pathways and terms containing rate g_2 are involved in the heterotrophic biosynthesis pathway. All internal metabolic rates are fixed by the three pathway (phototroph biosynthesis, photorespiration, and heterotroph biosynthesis) rates so that they need not be parameterized in detail except through the single rate functions g_1 and g_2 together with branching parameter η: this is a consequence of the powerful assumption of short timescale equilibration of metabolic pathways [6]. For easy reference, see Tables 1 and 2. Details for individual terms in (1)–(5) will be provided below.

Table 1. State variables (left) and key environmental parameters (right) for system (1)–(5).

STATE QUANTITIES			KEY ENVIRONMENTAL PARAMETERS		
Symbol	Description	Units	Symbol	Description	Units
P_1	Phototroph Concentration	Cmol·L^{-1}	D	Dilution Rate	s^{-1}
P_2	Heterotroph Concentration	Cmol·L^{-1}	ν	Photon Flux	$\mu\text{E m}^{-2}\text{·s}^{-1}$
IC	Inorganic Carbon Concentration	Cmol·L^{-1}	IC_0	Inflow IC Conc.	Cmol·L^{-1}
OC	Organic Carbon Concentration	Cmol·L^{-1}			
O_2	Oxygen Concentration	Omol·L^{-1}	$O_{2,0}$	Inflow O_2 Conc.	Omol·L^{-1}

Table 2. State variables (left) and key environmental parameters (right) for system (1)–(5).

RATE FUNCTIONS			YIELD PARAMETERS		
Symbol	Description	Units	Symbol	Description	Reference
η	Photorespiration Branching Function	–			
g_1	Photosynthesis Rate Photobiosynthesis Rate $= \eta g_1$ Photorespiration rate $= (1-\eta)g_1$	s^{-1}	$Y_{\alpha,\beta}$	Phototroph Yields α per β	see Section 2.3.1
g_2	Heterotroph Biosynthesis Rate	s^{-1}	$y_{\alpha,\beta}$	Heterotroph Yields α per β	see Section 2.4

The system energy is supplied through the photon flux ν which serves to drive carbon reduction through photosynthesis. We assume that microbe populations are sufficiently sparse in the chemostat so that no significant shading occurs, though one could introduce a shaded photon flux $\nu_{shade} = \nu_{shade}(\nu, P_1, P_2, Q_1, Q_2)$ (note that even in the case of shading, all microbes effectively receive the same photon flux over time due to the well mixed assumption). The other environmental conditions included are the dissolved concentrations IC, OC, and O_2. Inorganic carbon and oxygen flow into the chemostat at concentrations IC_0 and $O_{2,0}$, respectively. Exchange of O_2 and CO_2 with the atmosphere is neglected but inclusion would not be expected to change qualitative conclusions. The inflow is assumed to be free of organic carbon. Note that non-negative initial conditions are required for all quantities but have only transient influence, on a D^{-1} time scale, except/unless $P_1(0) = 0$ (in which case $P_1(t) = 0$ for all t) or $P_2(0) = 0$ (in which case $P_2(t) = 0$ for all t). Hence, later, we will ignore transients and study steady states.

Photosynthesis drives ecology through conversion of photons to chemical energy (photons power ADP $\rightarrow$ ATP, say) but also, and possibly more importantly, through production of reducing power, referred to here as electrons. In fact, we will not consider energy production and, rather, implicitly track electrons through degree of reduction (see Appendix A) as the more important quantity. A key step in oxygenic photosynthesis is the splitting of H_2O into, for our purposes, a combination of O_2 and reducing power. Oxygen's importance goes beyond its role as reactant; it also is an important contributor to degree of reduction balance of the entire oxygenic photosynthesizing system. In fact, in the model presented here, oxygen is the only explicit quantity with negative degree of reduction and hence, by proxy, its concentration is central to community redox state and hence to community function.

2.2. Metabolic Pathways

From an engineering point of view, organism metabolics operate somewhat like chemical processing networks so that they and implicitly resulting ecological interactions, are conveniently represented in terms of what are called metabolic pathways, chains of reactions that convert external substrates into external byproducts (though cycles of internal reactions might also be considered as pathways). Organisms themselves might be viewed as collections of such reaction chains, interacting with each other while producing fluxes at rates which must be consistent with external flux constraints. For example, in the case of a simple chemostat, external inflow and outflow fluxes are set by dilution rate D. While we look here to adopt the point of view of organism metabolisms as collections of pathways, at the same time we want a simple system able to illustrate basic principles of a phototroph-heterotroph interactions. Thus, while detailed metabolic models exist including for cyanobacteria [12–15] as well as for communities [16,17], we reduce system metabolics to the interaction of three particular pathways: photosynthesis-driven biosynthesis, photorespiration (in a restricted sense as previously noted) in the phototrophs, and aerobic respiration-driven heterotrophic biosynthesis. Community function is determined by the rates at which these pathways operate; the environment, through chemostat inflow and outflow, constrains community function by constraining these rates, though the community, specifically here the photoautotrophic cyanobacteria, have some freedom to choose them.

There are two specific rate functions in system (1)–(5): rate of carbon fixation $g_1(\text{IC}, O_2; \nu)$ of the photoautotrophs via the photosynthesis/photorespiration pathways and rate of growth $g_2(\text{IC}, O_2)$ of heterotrophs via a pathway for catabolysis of available organic carbon. In addition, there is a branching parameter η which determines percent of fixed carbon going to photoautotroph growth versus photorespiration. Between the three g_1, g_2, and η, the rates of the three pathways are determined. In balance, there are essentially two types of constraints: (1) cellular inflow rates of photons and inorganic carbon, with photosynthesis determined by the minimum rate of the two, as well as (2) system inflows and outflows determined by the chemostat dilution rate D. Cells must be able to synchronize system rates, dilution rate here, to direct pathway inputs, photon and inorganic carbon in the case of photosynthesis, and oxygen and organic carbon in the case of heterotrophic anabolysis. In principle, cellular outflow rates for pathway products may also constrain, but we suppose here, for the particular pathways studied, that these rates are essentially free.

Relation to Pathway Analysis

Metabolic network analysis of a system of m metabolites with internal concentrations c_i, $1 \leq i \leq m$, and n reactions with rates $v_j(c_1, c_2, \ldots, c_m)$, $1 \leq j \leq n$, starts from a metabolic map that can be represented by a set of equations of the form

$$\frac{dc_i}{dt} = \sum_j N_{ij} v_j$$

where N_{ij} is a stoichiometric coefficient, possibly negative, for production of metabolite i via reaction j. A rate v_j can be determined as a function of the concentrations c_i and is parameterized by rate constants. These rate constants are often unknown, but if steady state is assumed then the problem reduces to characterization of the null space

$$N\mathbf{v} = \mathbf{0} \tag{6}$$

of the $m \times n$ stoichiometric matrix N in a useful way by somehow identifying important pathway vectors $\mathbf{v}$ from this null space. (Precisely, a pathway consists of the reactions corresponding to non-zero entries in a pathway vector $\mathbf{v}$; a pathway vector encodes the flux through each of those reactions.) Note that knowledge of rate constants is unnecessary to solve the steady state Equation (6) and thus also unnecessary to determine pathway vectors, though steady state internal concentrations c_i cannot be computed without these rate constants.

One objective here is to proceed a further step by connecting internal metabolic activity, as encoded by those distinguished pathways, to community dynamics, e.g., connecting information extracted from (6) to the community model, as stated in (1)–(5). To do so, we use the (significantly) reduced metabolic maps as shown in Figure 2, explained in detail later. The interiors of the dashed domains in Figure 2 correspond to the interiors of the circled objects P_1 and P_2 of Figure 1. Circled objects in Figure 2 are generalized metabolites and arrowed curves are generalized reactions. Metabolites that are associated with reactions exiting a dashed domain are "seen" by the environment and hence explicitly tracked in the model (1)–(5); other metabolites are internal (in this case, only electrons e) so not directly observed in the environment and thus not explicit in the model, i.e., do not have tracking equations. Interior dynamics are assumed to be at quasi-steady state, that is, are able to quickly equilibrate to time on the community interaction time scale. Later, we will suppose a third, longer time scale on which the community also reaches steady state.

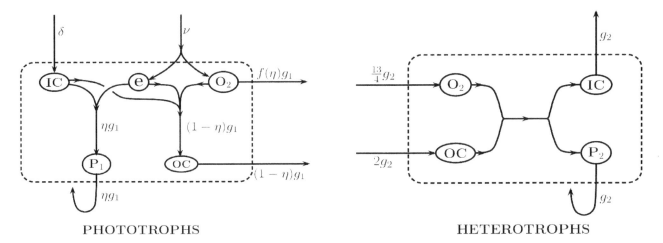

PHOTOTROPHS   HETEROTROPHS

Figure 2. Reduced metabolic maps for phototrophic population (P_1, left) and heterotrophic population (P_2, right) corresponding to the model used in Equations (1)–(5). The dashed curves represent boundaries between cell insides and outsides. Circled quantities correspond to metabolites (in a generalized sense): IC = inorganic carbon, O_2 = oxygen, P_1 = phototrophic biomass, P_2 = heterotrophic biomass, OC = organic carbon, and label e stands for electrons, see Section 2.3.1. Arrows correspond to reactions (in a generalized sense). Some of the reaction rates are labeled. Note that quantities P_1 and P_2 are turned into new cells. Also note that this representation of phototrophs can be viewed as the "insides" of P_1 in Figure 1 and, similarly, the representation of heterotrophs can be viewed as the "insides" of P_2 in the same figure.

We do not construct here the stoichiometric matrix corresponding to the metabolic network in Figure 2, but rather proceed directly to its elementary flux modes [6] which mathematically are, where reversible reactions are not present as is the case here, non-negative solutions of (6) for which no other non-negative solution containing a proper subset of non-zero entries exists. That is, an elementary flux mode is, roughly, a realizable pathway through the metabolic network that does not contain within itself any smaller realizable pathways. Non-negative linear combinations of the complete list of all elementary flux modes of a given network generate all allowable solutions of (6). For the system in Figure 2, there are three elementary modes, see Figure 3, corresponding to (1) biomass production and (2) photorespiration in the phototrophs and to (3) biomass production in the heterotrophs. All realizable steady states of the system can be uniquely written as positive combinations of these three elementary modes.

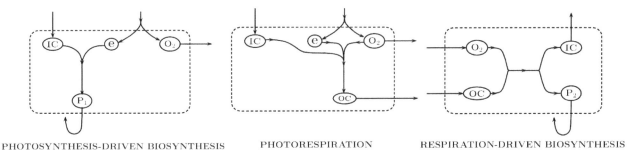

PHOTOSYNTHESIS-DRIVEN BIOSYNTHESIS   PHOTORESPIRATION   RESPIRATION-DRIVEN BIOSYNTHESIS

Figure 3. Elementary flux modes for metabolic maps in Figure 2. Left and center modes are phototrophic biosynthesis and photorespiration, right mode is heterotrophic biosynthesis. The phototrophic biosynthesis mode takes as inputs extracellular inorganic carbon and photons and produces as outputs extracellular oxygen and new biomass. The photorespiration mode also takes as inputs extracellular inorganic carbon and photons but produces extracellular oxygen and organic carbon. The heterotrophic biosynthesis mode takes as inputs extracellular oxygen and organic carbon and produces as outputs extracellular inorganic carbon and new biomass.

The metabolic state of the two population of microbes is thus described by the rates of the three elementary modes. These rates are determined by the flux rates into cells of mode inputs and out of cells of mode outputs, all of which generally depend on external concentrations of those inputs and outputs. Note that, for a given mode, setting any one of the input or output rates determines all reaction rates through the entire mode due to stoichiometric constraints and the steady state assumption. Hence, though many individual reactions are involved, each with its own rate constant, the three modes can be described by only three rates in total (one for each).

It should be noted that, for the sake of simplicity, we in fact compromise flux mode balance in one respect: flux balance through the phototroph electron compartment (circled e in Figure 2. left) is not explicitly enforced in the carbon limited case when the supply of inorganic carbon is insufficient to match the electron supply. Instead, in that particular case, some excess electrons are removed from the system. However, we keep track of their flux implicitly through photoinhibition – those excess electrons effectively combine with biomass and oxygen to remove some biomass. To maintain explicit flux balance, we would need to track them say to reaction with reactive oxygen species or wherever else they may go. As we implicitly suppose that such products leave the system, explicit tracking would complicate the model without advantage, as particular mechanisms of excess electron removal and damage are not the principle focus here.

The two phototroph modes operate in parallel and hence compete directly, in a sense, for inputs (inorganic carbon and photons). They operate in series with the single heterotroph mode and interact with it indirectly through external concentrations of dissolved quantities. In the case of a chemostat with dilution rate D, transport in and out of the chemostat of all quantities also proceeds at rate D. It must thus necessarily be the case that biosynthesis modes also operate at rate D (or else at rate 0) placing two constraints on mode rates. Hence we have one remaining condition determined, that of photorespiration. From there, concentrations of external quantities are determined by consistency with mode rates. These external concentrations effectively determine steady state biomass concentrations.

2.3. Photosynthesis

Microbial oxygenic photosynthesis can be divided into two steps, the *light* reaction followed by the *dark* reaction (also called the Calvin cycle), so named because photons are involved only in the first step [11]. The entire process uses energy from incoming photons to split H_2O producing O_2 and electrons, which, in the form NADPH, are used to fix CO_2. The light reaction, which is the oxygenic step, can be summarized for our purposes by

$$2H_2O + 8 \text{ photons} \longrightarrow 4H^+ + 4 \text{ electrons} + O_2 \tag{7}$$

and the dark reaction, the carbon fixation step, can be summarized, again for our purposes, by

$$CO_2 + \frac{1}{2}(1 - \eta)O_2 + \omega \text{ electrons} \longrightarrow \eta CH_{1.7}O_{0.5}N_{0.2} + (1 - \eta)CH_2O_{1.5}, \tag{8}$$

with both formulas balanced for carbon and degree of reduction (and only carbon and degree of reduction, for simplicity). Note that the dark reaction also consumes energy in the form of ATP, which may also be of importance for cellular energy balances, but ATP cycling is not considered here. Model parameters associated with (7)-(8) are described in Table 3.

The division into two steps as formulated by (7) and (8) has important consequences. In particular, at least within the formulation of the model, the light reaction is governed by photon supply (H_2O being assumed to be abundant) whereas the dark reaction rate is determined by both output rate of the light reaction (in electrons) as well in inflow rate of inorganic carbon. Hence, effectively, the carbon fixation rate function g_1 is determined by the rate at which the dark reaction proceeds, which may be slower than that of the light reaction in the case of limiting CO_2 (or, for us, IC). If so, the excess electrons effectively recombine with oxygen or some other oxidant, but may cause inhibition of photosynthesis in the process.

A second point of note is that photosynthesis as formulated by (7) and (8) is divided into two separate pathways weighted by the branching function η, with $0 \leq \eta \leq 1$. The valuation $\eta = 1$ corresponds to all fixed carbon being used for a growth pathway, while $\eta < 1$ indicates that some of the fixed carbon is instead allocated to a pathway that results in secretion of organic carbon from the cell (with $\eta = 0$ corresponding to all fixed carbon going to the secretion pathway, though this outcome would not allow a viable population). Within the model presented here, there are seemingly two apparent advantages to $\eta < 1$ over $\eta = 1$: first, $\eta < 1$ results in consumption of oxygen, see (8), which may alleviate effects of oxygen oversupply, and second, $\eta < 1$ results in secreted organic carbon which can be used to supply a population of heterotrophs which in turn produce inorganic carbon as a byproduct. These points are discussed in more detail below, but note that we suppose here that the value of η is subject to control by cyanobacterial cells themselves. Hence it may be that η is in some way regulated through an optimization process.

2.3.1. Carbon Fixation

The righthand side of (8) consists of two types of fixed organic carbon, each produced as a consequence of Calvin cycle reactions, a carboxylase reaction and an oxygenase reaction. That is, formula (8) actually combines contributions from two pathways: production of $CH_{1.7}O_{0.5}N_{0.2}$ via biosynthesis, weighted by η, and production of $CH_2O_{1.5}$ via photorespiration, weighted by $1 - \eta$.

New biomass is approximated as $CH_{1.7}O_{0.5}N_{0.2}$ [18]. In actuality, the dark reaction only produces a precursor (glyceraldehyde 3-phosphate) and biosynthesis is completed elsewhere, but for purposes of electron balance, it is convenient to use the biomass proxy formula $CH_{1.7}O_{0.5}N_{0.2}$ as the ultimate biomass output, see the source term in (1). $CH_2O_{1.5}$ (carbon-normalized glycolate) is a soluble byproduct of photorespiration and is assumed to be excreted from the cell; we track its concentration in the chemostat as the quantity $OC(t)$, see (4). Recall that, for purposes of representation in terms of carbon moles, all carbon compounds are normalized so that the number of carbon atoms is one.

Given the branching function η, described and parameterized below, then

$$\omega = \omega(\eta) = 4.7\eta + 5(1 - \eta) = 5 - 0.3\eta \tag{9}$$

is the electron demand (emole/Cmole), the number of moles of electrons needed to fix a mole of inorganic carbon (ω is related to carbon-oxygen-demand, a quantity sometimes used in engineering applications). We use degree of reduction 4.7 for $CH_{1.7}O_{0.5}N_{0.2}$ (autotroph) and degree of reduction 3.0 for $CH_2O_{1.5}$, see Appendix A. The coefficient in (9) of $(1 - \eta)$ is 5 rather than 3, though 3 is the degree of reduction of $CH_2O_{1.5}$, because the left-hand side term $(1/2)(1 - \eta)O_2$ in (8), with degree of reduction $-2(1 - \eta)$, effectively transfers to the righthand side for purposes of computing electron demand. For reference, note that the yield coefficients of moles of molecular oxygen per fixed Cmole are computed from (7) and (8) to be

$$Y_{IC,O_2}^{-1} = \frac{\omega(\eta)}{4},$$

$$Y_{OC,O_2}^{-1} = \frac{1}{2},$$

with rates proportional to g_1P_1, see the righthand side of (5). The coefficient of the first term in the righthand side of (5),

$$f(\eta) = Y_{IC,O_2}^{-1} - Y_{OC,O_2}^{-1}(1 - \eta) = \frac{3}{4} + \frac{17}{40}\eta, \tag{10}$$

indicates the net yield of oxygen moles per mole of photosynthetically fixed carbon and is important in the results presented here. Note that by varying η between 0 (all fixed carbon goes to excreted, soluble carbon) and 1 (all fixed carbon goes to new biomass), organisms vary oxygen production by about

60%. Of course, this should be understood as at best a rough estimate, since the presented model greatly simplifies the true biochemistry, but nevertheless the variation is potentially significant. At the same time, electron demand is much less sensitive, varying only by about 6% as η varies between 0 and 1, so it might seem that photorespiration provides a means to reduce oxygen production without significantly reducing capacity to process the photosynthetically driven electron stream.

Electrons are not normally free in solution but, rather, are transported by carrier compounds, particularly $NADP^+/NADPH$, and passed along through redox reactions [19]. As a convenience, though, we account for electrons directly rather than track $NADP^+$ and $NADPH$. Note that in each reaction Cmoles and degree of reduction (see Appendix A) are balanced as the key governing quantities. Other reaction components, e.g., N, are considered of secondary importance and are not balanced. Doing so would require introduction of more reactions, obscuring the main points. For example, protons are in excess and can be assumed to be buffered by the aqueous environment through mechanisms not of direct importance to the modeling aims here. Note in passing, though, that in some instances growth is limited by availability of quantities other than those tracked here, e.g., by limitation in fixed nitrogen. We suppose that this is not the case here.

Reactions (7) and (8) together comprise the photosynthesis and photorespiration pathways (which branch from each other in the dark reaction step), with $CH_{1.7}O_{0.5}N_{0.2}$ being the output of the photobiosynthesis pathway and $CH_2O_{1.5}$ the output of the photorespiration pathway. Excess molecular oxygen is also an output of both. An emphasis on rate rather than concentration is key and all internal reaction rates are effectively slaved to rates of inflow and outflow to/from the cell. The only other needed parameters are the stoichiometric ones, which are known from the pathway descriptions, in this case (7) and (8). Hence it is important to characterize governing rates, particularly those that have limiting or other important roles. The principle inputs of interest to photosynthesis are photons and CO_2 (water is plentiful at least in a chemostat) and growth rate is limited by the lesser availability of the two. We assume that the principle bottleneck for CO_2 inflow is transport (more specifically, here, transport of inorganic carbon – recall that we do not distinguish between inorganic carbon species) from outside the cell to the photosynthetic machinery inside. As is commonly practice, e.g., [20], we approximate this transport rate by a Michaelis-Menten function of the form

$$\delta(\text{IC}) = r_{\text{IC}} \frac{\text{IC}}{K_{\text{IC}} + \text{IC}} \qquad (\text{time}^{-1}) \qquad (11)$$

where r_{IC} and K_{IC} are, respectively, maximal transport rate and half-saturation of cross-membrane transport, see Appendix B. There are a number of mechanisms cells can use to influence transport, notably carbon capturing and active transporters [11,21]. From our point of view, carbon capturing effects can, roughly, be replaced by decrease in the inverse specificity factor γ_1 defined below, and active transporters can influence parameters r_{IC} and K_{IC}. Ultimately transport rate of inorganic carbon into cells over time is limited by concentrations outside of the cell and, more particularly, transport rates of inorganic carbon into and out of the local environment.

Rate δ then needs to be compared to the rate at which the dark reaction (8) can use the electrons to match with the inflowing inorganic carbon. The light reaction (7) provides that electron supply. Photons flow through the chemostat with constant flux ν (photons/area·time) set externally as a parameter, but in actuality enter photosynthesis machinery at an effective rate $\nu_{\text{eff}} = A\alpha\nu$ (photons/cell·time) where A is cell cross-section (area/cell) and $0 \leq \alpha \leq 1$ is an efficiency factor (unitless), see Appendix B. The parameter alpha accounts for photons that impact the cell but do not result in oxygen and electron production, either because they do not enter the photosynthetic process at all or because their end impact is shunted to non-photochemical quenching precesses such as Mehler reactions. These latter mechanisms may have other outcomes such as ATP generation which are supposed here to be non-limiting (though can have negative impacts at high enough levels) so are not considered. Note that alpha can be scaled into nu, so changing efficiency is equivalent, in the model, to changing light intensity. Note that α in fact measures of the efficiency of the process of electrons impacting the cell all the way to production of electrons and reductant, and in principle may be a function of conditions such as photon flux, oxygen concentration, etc., though we do not try to model these effects here. The light reaction component of photosynthesis then produces electrons at rate $Y_\nu \nu_{\text{eff}}$ (electrons/cell·time), with yield $Y_\nu = 1/2$ electron/photon. Electron production rate by the light reaction, per Cmole of biomass, is thus

$$e = \frac{Y_\nu \nu_{\text{eff}}}{c} = \frac{A\alpha\nu}{2c} \qquad \text{(emole/Cmole} \cdot \text{time)} \qquad (12)$$

(recall that c is the number of Cmoles per cell). The dark reaction then processes these electrons together with inorganic carbon. Note however, from (8) and (9), that this processing depends on the division of output between biomass and soluble inorganic production. Hence the rate at which electrons are actually consumed by the dark reaction (at least if inorganic carbon supply is not limited) is

$$\epsilon = \frac{e}{\omega} \qquad \text{(time}^{-1}), \qquad (13)$$

which can be understood as the maximum rate at which photon flux can drive carbon fixation through combined photobiosynthesis and photorespiration.

In fact, reduction of inorganic carbon proceeds by matching, in a sense, incoming inorganic carbon with incoming electrons, with rates set by δ(IC) and ϵ, respectively. Since δ(IC) $\neq \epsilon$ in general, then in fact reduction can proceed at best at rate $\min(\delta, \epsilon)$. In the spirit of rate-based modeling, we suppose this minimum to largely govern the actual reduction rate, so that photosynthesis rate, more particularly, the dark reaction rate, is

$$g_1(\text{IC}, O_2; \nu) = \begin{cases} \epsilon & \epsilon - \delta \leq 0 \quad \text{(light limited)} \\ \delta I & \epsilon - \delta > 0 \quad \text{(carbon limited)} \end{cases} \qquad (14)$$

where $I = I(\epsilon - \delta, O_2)$ is a photoinhibition function, defined below, of excess electrons should there be any. Note that O_2 dependence in g_1 arises from O_2 dependence in ω and I.

Table 3. Key photosynthesis-related functions and parameters.

Symbol	Description	Units	Definition
A	Average cell cross-sectional area	μm^2	Appendix B
c	Carbon moles per cell	Cmole/cell	Appendix B
e	Electron production rate by the light reaction	emole/Cmole·s	Equation (12)
f	Net oxygen per photosynthetically fixed carbon	Omole/Cmole	Equation (10)
I	Photoinhibition function	−	Equation (15)
α	photosynthesis efficiency factor	−	Equation (12)
γ_1	inverse specificity factor	Cmole/Omole	Equation (16)
γ_2	excess electron capacity	s/Omole	Equation (15)
ϵ	Maximum electron consumption rate	1/s	Equation (13)
η	RuBisCO inorganic carbon binding probability	−	Equation (16)
ν	environmental photon flux	$\mu E/m^2$·s	−
ω	Electron demand: emoles needed to fix a cmole	emole/Cmole	Equation (9)

2.3.2. Photoinhibition and Oxidative Stress

In the case that $\epsilon > \delta$, i.e., the rate of the normalized electron production is greater than the rate of inorganic carbon inflow, excess electron production can lead to inhibition of photosynthesis machinery and other apparatus via saturation of electron transport structure and consequent formation of harmful radical oxygen species as well as other undesirable effects [11,22]. These effects have been modeled with an inhibition function [23,24] which allows for removal of excess electrons without detriment, to a point, after which reduction in growth rate occurs [22,25]. These inhibition models, however, are generally functions only of photon flux rate and not, for example, dependent on IC and O_2 concentrations or transport rates, though such dependence is likely important, at very least through any mismatch of electron production rate with inflow rate of electron donors, and is certainly central to the model presented here. Thus, we define a photoinhibition function to take account of electron-IC mismatch of the form

$$I(\epsilon - \delta, O_2) = \begin{cases} 1 & \epsilon - \delta \leq 0 \\ (1 + \gamma_2^{-2}(\epsilon - \delta)^2 O_2^2)^{-1} & \epsilon - \delta > 0 \end{cases} \tag{15}$$

where γ_2 is an excess capacity coefficient. (Note, from (14) that only the $\epsilon - \delta > 0$ definition is relevant.) The quadratic dependencies on O_2 and $\epsilon - \delta$ are ad hoc forms meant to model the capacity for cells to avoid or repair damage of small mismatches; small values of $\epsilon - \delta$ and O_2 do not result in significant net damage whereas large values might. The parameter γ_2 is chosen to provide a reasonable high light/oxygen cutoff on growth. As a consequence of the degree of arbitrariness in form and parameterization of I, we avoid conclusions which would appear to rely on its particulars beyond a general tendency to inhibit growth in carbon limited conditions.

We note here that at moderately high concentrations $O_2 \gtrsim 5 \cdot 10^{-4}$ Omoles/L, oxygen may come out of solution, providing effectively a method for limiting effects of oxygen stress. Such critical oxygen concentrations are not reached in computations shown here, but can occur at environmentally reasonable light intensities in some situations.

2.3.3. Photorespiration

A key step in dark reaction carbon fixation is binding of CO_2 to the enzyme RuBisCO. However, as it happens, O_2 competes for the same binding site as CO_2, and when a molecule of O_2 does in fact bind then glycolate ($CH_2O_{1.5}$, degree of reduction +3) is produced in the stead of further reduced biomaterial ($CH_{1.7}O_{0.5}N_{0.2}$, degree of reduction +4.7 for phototrophs). We refer to this process as photorespiration (though as noted earlier, photorespiration can be used as an umbrella term for a number of re-oxidizing processes). We denote the probability of CO_2 binding to RuBisCO by η, with

$$\eta = \frac{a_c IC}{a_c IC + a_o O_2} = \frac{1}{1 + \gamma_1 \frac{O_2}{IC}} \tag{16}$$

where a_c, a_o are binding affinities and γ_1, the ratio of those affinities, is the inverse specificity factor (with respect to IC versus O_2). Recall that we confuse inorganic carbon concentration IC here with CO_2 concentration, supposing that inorganic carbon in forms other than CO_2 can readily be converted into CO_2 via carbonic anhydrase enzymes.

Effectively, η is a branching function of O_2 and IC that determines how much photosynthetic product goes to synthesis of new biomaterial and how much to synthesis of soluble, excretable, organic carbon. Phototrophs may have a degree of control over the value of η either directly through the structure of RuBisCO itself [26,27] or through indirect machinery such as carbon capture mechanisms, so we treat γ_1 as a tunable parameter and study effects of its variation.

The purpose of photorespiration (oxygenase activity of RuBisCO, to be precise), if there is one, is uncertain. It is sometimes argued to be wasteful, e.g., [20], and possibly a relic of early earth history when levels of CO_2 were much higher than today, and levels of O_2 lower, so that the the ratio O_2/CO_2 was presumably small. However, observations suggest it is not superfluous [28] and the orders of magnitude variability of γ across different species [11,26,27] suggests that there may be selective pressure at work. Photorespiration diverts carbon fixing power away from new biomass, but also note in fact the following: though glycolate has a lower degree of reduction (+3) than biomaterial (+4.7), its production requires $1/2\, O_2$ mole per Cmole of glycolate and hence, balancing electrons, also removes an additional two electrons per glycolate. Thus, effectively, each Cmole of glycolate produced removes 5 electron moles from the system, more than the 4.7 electron moles removed per Cmole of biomaterial produced. Thus photorespiration serves to reduce electron pressure, particularly when oxygen pressure is high. At the same time, oxygen pressure is reduced. Also, photorespiration produces a supply of dissolved, reduced organic carbon, allowing the possibility of supplying a heterotroph population. Hence, accidental or not, photorespiration may have significant effects on population dynamics.

2.3.4. Fixation Stability

As a technical point that will be repeatably useful below but also seems reasonable biologically, we impose the condition

$$\eta_{IC} g_{1,O_2} - \eta_{O_2} g_{1,IC} \geq 0 \tag{17}$$

(with subscripts IC and O_2 denoting partial derivatives with respect to those quantities)) which, with η as in (16), reduces to IC $g_{1,IC} \geq -O_2\, g_{1,O_2}$. (In fact we will only really require (17) to hold at steady state.) This condition can be appreciated through linearization of the carbon fixation process applied to inorganic carbon and oxygen, i.e., linearization of the subsystem

$$\frac{d}{dt}IC = -Y_{P_1,IC}^{-1} g_1(IC, O_2; \nu)P_1$$
$$\frac{d}{dt}O_2 = f(\eta)g_1(IC, O_2; \nu)P_1$$

around a state (P_1, IC, O_2), with associated Jacobian matrix

$$J = \begin{pmatrix} -Y_{P_1,IC}^{-1} g_{1,IC}P_1 & -Y_{P_1,IC}^{-1} g_{1,O_2}P_1 \\ (fg_1)_{IC}P_1 & (fg_1)_{O_2}P_1 \end{pmatrix}.$$

The eigenvalues of J have non-positive real part as long as derivatives with respect to IC are non-negative, derivatives with respect to O_2 are non-positive, and condition (17) holds. In the case that (17) is false, then J has an unstable eigendirection that corresponds to an instability in the fixation process: a simultaneous increase in IC and O_2 levels can lead to simultaneous decrease in net fixation rate and in photorespiration, thus further amplifying IC and O_2 levels, etc. Such dynamics are unsustainable. Equivalently, it can be seen that, if (17) is false, then an increment in available inorganic carbon actually reduces photosynthesis rate, see Appendix C.

Condition (17) is satisfied for reasonable choices of η and g_1, with one caveat, see below. We divide into two cases based on (14). In the light limited regime,

$$\eta_{IC} g_{1,O_2} - \eta_{O_2} g_{1,IC} = \eta_{IC}\epsilon_{O_2} - \eta_{O_2}\epsilon_{IC} = 0,$$

satisfying (17). In the carbon limited regime

$$\eta_{IC} g_{1,O_2} - \eta_{O_2} g_{1,IC} = \eta_{IC}(\delta I)_{O_2} - \eta_{O_2}(\delta I)_{IC} = \eta_{IC}\delta I_{O_2} - \eta_{O_2}(\delta I)_{IC} \tag{18}$$

The first term on the far right hand side is generically non-positive, while the second is generically non-negative. Note the key controlling function, I_{O_2}, indicates the rate at which increasing oxygen levels increases oxidative stress; only if this rate is too large can (18) be negative. Otherwise, fixation stability condition (17) also holds in the carbon limited regime.

2.4. Heterotrophic Biosynthesis

The third pathway in the model system is a simplified heterotrophic anabolysis described by

$$2CH_2O_{1.5} + 0.475O_2 \longrightarrow CH_{1.7}O_{0.5}N_{0.2} + CO_2, \tag{19}$$

with stoichiometry constrained to balance carbon and degree of reduction (using degree of reduction of $(CH_{1.7}O_{0.5}N_{0.2}) = +4.1$ for heterotrophs, see Appendix A). As with the photosynthesis pathway, oxygen and nitrogen are not balanced; to do so would require introduction to the model of new details of secondary interest. Note that the stoichiometry determines yield coefficients $y_{P_2,OC} = 1/2$, $y_{P_2,O_2} = 4/1.9$, and $y_{P_2,IC} = 1$ in Equations (3)–(5).

Reaction (19) indicates that organic carbon in the form of glycolate is further reduced to biomaterial. The increased degree of reduction is accomplished by sacrificing some of the glycolate for its electrons, a portion of which go to biomaterial and a portion of which are shunted off to carbon dioxide to maintain carbon balance.

The rate at which (19) proceeds is given by $g_2(OC, O_2)$ as

$$g_2(OC) = r_h \min\left(\frac{OC}{K_{OC} + OC}, \frac{O_2}{K_{O_2} + O_2}\right), \tag{20}$$

based on the assumption that the rate of biosynthesis is controlled by the minimum rate at which biosynthesis components can be transported to biosynthesis machinery. Note as well that as a result of reaction (19), a source term for P_2 appears in (2) and sink terms for OC and O_2, appear in (4) and (5).

2.5. Equations

Pathway stoichiometry can now be incorporated into Equations (1)–(5). Yields $Y_{\alpha,\beta}$ parameterize autotroph pathways (and damage) and yields $y_{\alpha,\beta}$ parameterize the heterotroph pathway (and damage). In units of Cmoles and oxygen moles, $Y_{P_1,Q_1} = y_{P_2,Q_2} = 1$ (see Section 2.3.2), $Y_{P_1,IC} = 1$ (see Section 2.3). Also, $y_{P_2,IC} = 1$, $y_{P_2,OC} = 1/2$ (see Section 2.4), and $Y_{IC,OC} = Y_{OC,O_2} = 2$ (see Section 2.3.3). Altogether

$$\frac{d}{dt}P_1 = (\eta g_1(IC, O_2; v) - D)P_1, \tag{21}$$

$$\frac{d}{dt}P_2 - (g_2(OC, O_2) - D)P_2, \tag{22}$$

$$\frac{d}{dt}IC = -g_1(IC, O_2; v)P_1 + g_2(OC, O_2)P_2 + D(IC_0 - IC), \tag{23}$$

$$\frac{d}{dt}OC = (1 - \eta)g_1(IC, O_2; v)P_1 - 2g_2(OC, O_2)P_2 - D\,OC, \tag{24}$$

$$\frac{d}{dt}O_2 = \left(\frac{3}{4} + \frac{17}{40}\eta\right)g_1(IC, O_2; v)P_1 - \frac{1.9}{4}g_2(OC, O_2)P_2 + D(O_{2,0} - O_2). \tag{25}$$

We track two key quantities, carbon and electrons, through the system. Set

$$C = P_1 + P_2 + IC + OC$$

to be total Cmole concentration in the chemostat to obtain

$$\frac{d}{dt}C = D(IC_0 - C), \tag{26}$$

with solution $C(t) = IC_0 + C(0)e^{-Dt}$. So, after a chemostat turnover time D^{-1} or so, $C(t)$ approaches the constant value $C = IC_0$, the inflow Cmole concentration, to exponentially small error in time. Effectively, thus, the chemostat conserves total Cmoles. Similarly, set the total degree of reduction (DoR) of the system to be

$$\begin{aligned} DoR = & (4.7\,\text{emole/Cmole})P_1 + (4.1\,\text{emole/Cmole})P_2 \\ & + (0.0\,\text{emole/Cmole})IC + (3.0\,\text{emole/Cmole})OC - (4.0\,\text{emole/mole})O_2, \end{aligned}$$

Note that

$$\frac{d}{dt}DoR = -D(4O_{2,0} + DoR), \tag{27}$$

with solution $\mathrm{DoR}(t) = -4O_{2,0} + \mathrm{DoR}(0)e^{-Dt}$, and thus, after a chemostat turnover time or so, $\mathrm{DoR}(t) = -4O_{2,0}$, the degree of reduction of the inflow, to exponentially small error in time. Hence, effectively, the chemostat conserves DoR.

In a chemostat, degree of reduction (at least as calculated here) is dictated by the inflow environment, since all reactions conserve it in detailed balance and since all material flows out of the chemostat at the same rate. This contrasts with a biofilm or sparged system where insoluble, soluble, and volatile material may leave the system at different rates. Note that a biofilm can thus have some local control over DoR.

3. Results

3.1. Single Species Chemostat Community

To begin, we consider first the case of a chemostat community of phototrophs only, i.e., $P_2(t) = P_2(0) = 0$. Note that the complementary case of a community of heterotrophs only, i.e., $P_1(t) = P_1(0) = 0$, is not sustainable: $P_1 = 0$ has the consequence that soluble organic carbon OC is not produced which results in $OC(t) \to 0$ which, in turn, results in $g_2(t) \to 0$ and hence, from (?), $P_2(t) \to 0$.

In the phototrophic (only) community case, Equations (21)–(25) reduce to

$$\frac{d}{dt}\mathrm{IC} = -g_1 P_1 + D(\mathrm{IC}_0 - \mathrm{IC}), \tag{28}$$

$$\frac{d}{dt}\mathrm{OC} = (1 - \eta)g_1 P_1 - D\mathrm{OC}, \tag{29}$$

$$\frac{d}{dt}O_2 = f(\eta)g_1 P_1 + D(O_{2,0} - O_2) \tag{30}$$

$$\frac{d}{dt}P_1 = (\eta g_1 - D)P_1, \tag{31}$$

with $f(\eta) = 3/4 + (17/40)\eta$ being the net yield of oxygen per carbon fixed, see (10). The coefficients $3/4$ and $17/40$ arise from degree of reduction details. Note that the equation order has been changed from earlier; the population equation is now listed after the chemical concentration equations for reasons of convenience in the following. The first term in (28) measures usage rate of inorganic carbon in photosynthesis, which produces new biomass (first term of (31)) and soluble organic carbon (first term of (29)) as well as oxygen, some of which is consumed, however, in the production of soluble organic carbon (first term of (30)). Terms involving D measure rates of wash in or out of the chemostat. Note that organic carbon (OC) decouples from the other quantities— the dynamics of IC, O_2, and P_1 are all independent of OC. Hence, system (28)–(31) is effectively three dimensional. We keep OC, though, because of its importance in the two species community dynamics to follow, and also because of its place in conservation of carbon and of degree of reduction.

3.1.1. Steady States

Our interest is in the role of photorespiration in long time community behavior. As is often the case in chemostat models, long time behavior reduces here to the study of steady state solutions. Equations (28)–(31) have two possible types of steady states: (1) the washout solution $P_1(t) = 0$ with $\mathrm{IC}(t) = \mathrm{IC}_0$, $\mathrm{OC}(t) = 0$, and $O_2(t) = O_{2,0}$, which exists for all parameter choices though is not always stable, and (2) the viable solution $P_1(t) = P_1^* > 0$ with $\mathrm{IC}(t) = \mathrm{IC}^*$, $\mathrm{OC}(t) = \mathrm{OC}^*$, and $O_2(t) = O_2^*$. For a viable solution, (31) requires that

$$\eta(\mathrm{IC}, O_2)g_1(\mathrm{IC}, O_2) = D \tag{32}$$

have a nonnegative solution (IC^*, O_2^*) indicating that biomass production rate balance with washout. Also, by combining Equations (28) and (30), a second equation,

$$f(\eta)(IC_0 - IC) = O_2 - O_{2,0} \tag{33}$$

is obtained relating IC^* and O_2^*. Equation (33), an equilibrium relationship constraining the ratio of surplus O_2 to IC deficit, is a consequence of carbon fixation stoichiometry combined with degree of reduction balance. If (32) and (33) can be solved with non-negative values IC^* and O_2^* then the remainder of a viable state steady state is given by

$$
\begin{aligned}
P_1^* &= \eta(IC_0 - IC^*), & (34)\\
OC^* &= (1 - \eta)(IC_0 - IC^*), & (35)
\end{aligned}
$$

using (28) and (29).

Thus, existence and uniqueness of viable solutions reduces to existence and uniqueness of solutions to (32) and (33). This appears to provide two conditions for viability; in fact, though, (32) and (33) can be solved under the single condition that $\eta(IC_0, O_2)g_1(IC_0, O_2) = D$ has a solution with $O_2 > O_{2,0}$, that is, under the condition that the organism-free ($P_1 = 0$, $IC = IC_0$) chemostat is capable of supporting growth under its given dilution rate D and ambient oxygen level $O_{2,0}$. For the particular choices of g_1, f, and η made here, either one or no viable solutions exist, depending on choice of environmental conditions IC_0, $O_{2,0}$, and D. See Appendix C for details.

3.1.2. Stability of Steady States

To characterize stability, we add a small pertubation to a steady state solution and then watch ensuing dynamics. We summarize results here; see Appendix D for details. Generally, any component of a perturbation to a steady state that introduces excess or deficient total carbon or degree of reduction is washed out of the system on the chemostat turnover time scale D^{-1}. Thus understanding perturbation dynamics of the four dimensional system (28)–(31) reduces to understanding dynamics on a two dimensional subsystem, in fact a system that can be interpreted as the phototroph flux mode space and is spanned by the vectors

$$\mathbf{EFM}_1 = \begin{pmatrix} -1 \\ 0 \\ 4.7/4 \\ 1 \end{pmatrix}, \quad \mathbf{EFM}_2 = \begin{pmatrix} -1 \\ 1 \\ 3/4 \\ 0 \end{pmatrix},$$

that encode the two phototroph elementary flux modes. Recall Figure 3: perturbation of the viable steady state by increasing or decreasing flux through the photosynthesis-driven biosynthesis mode corresponds to perturbation of the viable steady state solution in the direction $\mathbf{EFM}_1$ (one Cmole biomass and $4.7/4$ Omoles produced per Cmole inorganic carbon consumed) and, likewise, perturbation of the viable steady state by increasing or decreasing flux through the photorespiration mode corresponds to perturbation of the viable steady state solution in the direction $\mathbf{EFM}_2$ (one Cmole organic carbon and $3/4$ Omoles produced per Cmole inorganic carbon consumed). Stability in this flux mode space will be discussed for particular steady states below.

Washout State (One Species System). The washout state ($P_1 = 0$) is stable or unstable depending on sign of the quantity $\lambda = \eta(IC_0, O_{2,0})g_1(IC_0, O_{2,0}) - D$. If negative then the steady state is stable, i.e., phototrophs cannot invade, while is positive, then invasion can occur. Note that λ is the net intrinsic biomass production rate at inflow conditions. When a small quantity of phototrophs are added to the system, in the unstable case $\lambda > 0$ dynamics of the linearized system effectively reduce to exponential growth on the one dimensional space $\eta \mathbf{EFM}_1 + (1 - \eta) \mathbf{EFM}_2$, indicating that the linearized growth dynamics occurs, as to be expected, as a combination of the photosynthesis mode and the photorespiration mode weighted by the branching parameter $\eta(IC_0, O_{2,0})$.

Viable State (One Species System). For the viable state ($P_1 > 0$), dynamics are again characterized by the basis formed by the two mode vectors $\mathbf{EFM}_1$ and $\mathbf{EFM}_2$ and in this case are always stable (i.e., perturbations decay) under the assumptions that derivatives with respect to IC are non-negative, derivatives with respect to O_2 are non-positive, and condition (17) holds. That is, the viable state is stable under the conditions that we consider biologically reasonable.

3.1.3. Viability and Light-Limited Ranges

We suppose that the RuBisCO inverse specificity factor γ_1, see Section 2.3.3, is subject to some influence by the organism itself, at least adaptively if not through direct regulation, leading to some control over the branching function η. Recall

$$\eta = \frac{1}{1 + \gamma_1 \frac{O_2}{IC}} \tag{36}$$

and note that $\gamma_1 = 0$ would correspond to the extreme of $\eta = 1$ (all fixed carbon goes to biosynthesis) and that $\gamma_1 = \infty$ would correspond to $\eta = 0$ (all fixed carbon goes to photorespiration). Increasing inverse specificity γ_1 corresponds to increasing, relatively speaking, RuBisCO oxygen affinity and hence increasing photorespiration rates. So then which factors might determine, or at least influence, the value of γ_1?

We show in Appendix E that, for the single species solution as described above (including the assumption (17)), the choice $\gamma_1 = 0$ is favored in the following sense: for any fixed, positive value of γ_1 and the resulting steady state population $P_1^*(\gamma_1) > 0$, it is in fact the case that $(d/d\gamma_1)P_1^* < 0$. That is, the autotroph population decreases with increasing inverse specificity factor, see, e.g., Figure 4, left panel, for example. Hence, as a larger affinity factor corresponds to increased photorespiration, in the single species, static chemostat environment the autotrophs are always disadvantaged by photorespiration in terms of total biomass.

However, maximizing biomass is not necessarily the only consideration. Another important factor might be viability range—solar light intensity varies significantly over the course of a day (or a year) so that capacity to efficiently function over a wide range of photon flux intensities may also be valuable. High light can cause damage and hence require extra resources, and thus is desirable to avoid or mitigate. In this context, non-zero inverse specificity has competing impacts. First, larger inverse specificity increases, per unit inorganic carbon, the usage of photosynthetically generated electrons and oxygen, thus decreasing rate of damage. Second, larger inverse specificity diverts more fixed carbon from biomass, thus decreasing growth rate. Note though that decreased growth rate leads to reduced population biomass and hence increased available inorganic carbon—a smaller population can be a healthier one. Altogether, then, photorespiration can be expected to shift upwards in both the lower and upper photon intensity viability bounds. To understand how, see Figure 4 right panel, a central result of this study, which presents results of a number of solutions of the steady state Equations (32)–(35). Computations used parameters as described in Appendix B and in the caption.

Minimum Photon Flux. The lower-most curve in Figure 4, right panel, shows, as a function of γ_1, the minimum photon flux necessary for a viable population. This curve was computed analytically by using condition (32) to determine photon flux ν as a function of γ_1 at ambient inorganic carbon and oxygen levels IC=IC_0 and O_2=$O_{2,0}$, the limiting viability concentrations. (It was also checked against a numerical computation of minimum ν for viability as a function of γ_1.) Its form is easily understood in terms of the non-dimensional number $\gamma_1 O_{2,0}/IC_0$ which measures the ratio of likelihoods of O_2 versus IC RuBisCO binding in relation to ambient or near-ambient conditions, The ambient ratio $O_{2,0}/IC_0$ we use is 0.05 Omole/Cmole, i.e., 20 times more inorganic carbon than oxygen as measured in carbon and oxygen moles. Thus, for γ_1 less than approximately 20 Cmole/Omole, binding site competition is unimportant at ambient conditions and hence no penalty, at least with respect to minimum photon flux for viability, is paid. However, as γ_1 increases beyond 20, the minimum photon flux rapidly increases, see Figure 4, right panel.

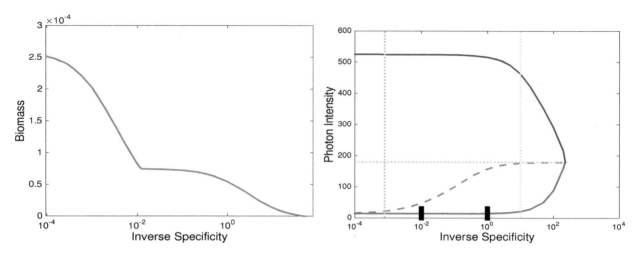

Figure 4. Plots of various steady state quantities arising from solutions of (32)–(35) as functions of inverse specificity γ_1 (Cmoles/Omoles) with $D = 10^{-4}$ s^{-1}, and other fixed parameters are as described in Appendix B. Increasing γ_1 corresponds to increasing importance of photorespiration. Left. Steady state biomass P_1^* (in Cmoles/L) versus inverse specificity γ_1, with $\nu = 50$ μE. Biomass decreases monotonically with increasing inverse specificity. The kink at $\gamma_1 \cong 10^{-2}$ occurs where the steady state transitions from carbon limitation to light limitation. Right. Solid curve (bottom): minimum photon intensity ν (μE) for population viability as a function of inverse specificity γ_1. Solid curve (top): maximum photon intensity for population viability as a function of inverse specificity γ_1. Dashed curve: boundary photon intensity separating light-limiting conditions (below the curve) from carbon limiting conditions (above the curve). Horizontal dotted line: boundary photon intensity asymptote (in large inverse specificity limit). Vertical dotted line (left): the inverse specificity value $\gamma_1 = IC_1/O_{2,1}$ beyond which the light-limiting range significantly expands. Vertical dotted line (right): the inverse specificity value $\gamma_1 = IC_0/O_{2,0}$ below which photorespiration does not significantly reduce the minimum range of light intensities that allow population viability. Note that, for larger γ_1, steady state biomass drops off, see left plot. Note that for the chosen set of parameters, washout occurs beyond inverse specificity of approximately 230 Cmoles/Omoles for all light intensities. Measured values of inverse specificity in a variety of organisms lie in the approximate range 10^{-2}–10^0 [11,26,27], delimited in the plot by the thick bars.

To summarize, the bottom solid curve in Figure 4, right panel, is important in that it shows minimum photon intensity for community viability as a function of γ_1. This curve is, roughly, described by two parameters: (1) the photon intensity at $\gamma_1 = 0$, which is determined by details of photosynthesis rate function g_1 as well as choice of chemostat turnover rate D, and more importantly (2) the value of $\gamma_1 = IC_0/O_{2,0}$ (right-most dotted vertical line) above which significant increase in photon intensity is required for viability.

A similar discussion applies for the upper-most curve in Figure 4, right panel, which shows as a function of γ_1 the maximum photon flux allowable for a viable population. Again, $IC \cong IC_0$, $O_2 \cong O_{2,0}$, at the viability boundary so that the viability photon flux upper bound is only weakly dependent on γ_1 for $\gamma_1 O_{2,0}/IC_0$ noticeably less than 1, i.e., γ_1 noticeably less than about 20 Cmole/Omole.

Light-Limited to Carbon-Limited Transition. The dashed curve in Figure 4, right panel, computed numerically, measures as a function of inverse specificity the boundary light intensity between light-limited (region below the curve) and carbon-limited (region above the curve) intensities. In the carbon-limited region, i.e., where photons are sufficiently abundant so that photosynthesis is limited by access to inorganic carbon rather than light, excess electrons are present leading to photoinhibition (recall (14)). The cross-over from light limitation to carbon limitation occurs when $\epsilon = \delta$, i.e., when electron production rate as measured in capability to process inorganic carbon is equal to cellular inflow rate for inorganic carbon. The right asymptote ($\gamma_1 \to \max(\gamma_1)$), shown as the horizontal line in Figure 4 is well approximated by setting $\gamma_1 = \infty$ and solving $\epsilon = \delta$ at IC = IC_0, $O_2 = O_{2,0}$, the viability values. More importantly,, though, we can understand the small γ_1 behavior of this curve as follows. If $\gamma_1 = 0$ then $\eta = 1$ so $g_1 = D$ and thus $D = g_1 = \epsilon = \delta$ which, upon solving for IC and O_2, results in values IC_1, $O_{2,1}$, the $\gamma_1 = 0$ boundary concentrations. When $\gamma_1 O_{2,1}/IC_1$ is significantly less than one, i.e., γ_1 noticeably less than $IC_1/O_{2,1}$ photorespiration remains relatively insignificant at cross-over and hence cross-over is only weakly dependent on γ_1. For the parameters used here, $IC_1/O_{2,1} \cong 8 \cdot 10^{-4}$, see left-most vertical dotted line in Figure 4, right panel. For larger values of γ_1, the range of light-limiting photon intensities expands significantly.

Summary. We summarize Figure 4, right panel, as follows.

- Setting $\gamma_1 = 0$, i.e., turning photorespiration off entirely, results in only a single light intensity with a viable, non carbon-limited steady state population. However, at ambient O_2 and IC concentration levels, competition for RuBisCO binding is insignificant for inverse specificities $\gamma_1 < IC_0/O_{2,0}$. Hence, from the point of view of population viability at least, there is no penalty for allowing RuBisCO oxygenase activity over this inverse specificity range.
- On the other hand, inverse specificities such that $IC_1/O_{2,1} < \gamma_1$ result in significantly enlarged light-limited intensity range, so that large enough inverse specificities may have some advantage.
- Assembled, the inverse specificity interval

$$\frac{IC_1}{O_{2,1}} < \gamma_1 < \frac{IC_0}{O_{2,0}},$$

for parameters used here (based on best approximations in comparison to known data) agrees well with measured values of inverse specificity [11,26,27].

It should be noted that while the upper bound $IC_0/O_{2,0}$ is a function of ambient IC and O_2 levels and is thus is somewhat context-independent at least in the absence of other organisms, the lower bound $IC_1/O_{2,1}$ does depend on specifics of the system like dilution rate D and hence may vary under different conditions. More particularly, IC_1 is found by equating $\delta = D$. For δ as defined here, the resulting concentration IC_1 is given by

$$IC_1 = \frac{D}{r_{IC} - D} K_{IC}. \tag{37}$$

Generally speaking, though, the solution of $\delta = D$ will result in a value of IC_1 as a function of some external rate of transport of IC in comparison to internal, cellular transport mechanisms. Given IC_1 then $O_{2,1}$ is determined stoichiometrically from (33). Hence the ratio $IC_1/O_{2,1}$ is essentially determined by properties of transport of IC to RuBisCO (relative to the rate of transport of IC into the system), with increased rates corresponding to smaller ratio and hence larger favorable inverse specificity range. Carbon concentration mechanisms, though not included here, might have a similar impact.

Altogether, then, the model suggests that there is possible advantage in the form of redox and oxygen stress control by allowing photorespiration with inverse specificity within the range (37), the lower bound of which is under some internal control. In particular, an expanded range of light-limiting photon intensities may result. This may be important as, typically, photon flux varies considerably over time. (It should be noted that our observation is based on steady state results in a time-independent model, though it seems possible that the idea extends to periodically varying systems.) Note that increased photorespiration results in reduced biomass, which may be considered a disadvantage. However, it is in part because of reduced biomass that the range of light-limiting photon intensities increases, as reduction in biomass is accompanied by increase in IC availability.

From the point of view of flux mode modeling, photorespiration provides a sort of rate synchronization mechanism; biosynthesis (left mode in Figure 3) is required to produce biomass, i.e., P_1, at rate dictated by chemostat dilution while photon input is independently, and likely conflictingly, determined by photon inflow rate, both of which are not controlled by the phototrophs themselves

(IC input rate can be controlled by the organisms including through varying total biomass). Biomass production rate must match chemostat dilution rate, however, so if the dilution and photon inflow rates differ then, in the absence of a photorespiration mode, excess electrons will be produced leading to damage. Presence of a photorespiration mode (center mode in Figure 3), however, allows some of those excess electrons to be shunted away in the form of reduced OC.

Note that biology-related parameters vary in value between different cyanobacterial species and even within the same species under different environmental conditions, e.g., [29]. We do not see this variability as a critical problem here, however, as our aim is to explore qualitative behavior of community interactions as inverse specificity varies from low (low photorespiration levels) to high (high photorespiration levels), regardless of parameter choices. The forms of the curves in Figure 4 are expected to hold under reasonable choices. In particular we are least confident about choices related excess electron damage, effecting mostly height of the top solid curve, Figure 4 right, and photon usage efficiency, effecting mostly height of the horizontal dotted curve Figure 4 right. The left vertical dotted curve in Figure 4 right, which indicates the approximate value of γ_1 above which photorespiration is significant, is dependent on properties of inflow of inorganic carbon about which we are also relatively uncertain, but because of the log scale used is unlikely, in our view, to move a lot under reasonable choice of parameters.

3.2. Two Species Community

Having explored the effects of photorespiration on steady state phototroph behavior in the one species model, we now add a second species, a heterotroph, in order to see if its addition, despite the resulting (indirect) competition for carbon, can in fact lead to an increase in phototroph steady state biomass. Heterotrophs offer two apparent direct benefits to phototrophs: (1) they use oxygen, thus reducing oxidative stress, and (2) they produce carbon dioxide, thus increasing the local inorganic carbon pool. The price paid is that the cyanobacteria must feed these heterotrophs as they cannot utilize inorganic carbon as a food source. Photorespiration provides a means to do so through production and secretion of soluble organic carbon, thus perhaps providing an additional advantage to its existence. Further, though secretion of organic carbon comes at the price of reduced production of new cyanobacterial biomass, doing so via photorespiration also provides additional control of redox balance through lowering net degree of reduction of the fixed carbon. In this section, then, we consider these combined effects, focusing on steady state cyanobacterial biomass as a metric.

The equations for the two species community are as in (21)–(25), rewritten as

$$\frac{d}{dt}\mathrm{IC} = -g_1(\mathrm{IC}, \mathrm{O}_2; \nu)\mathrm{P}_1 + g_2(\mathrm{OC}, \mathrm{O}_2)\mathrm{P}_2 + D(\mathrm{IC}_0 - \mathrm{IC}), \tag{38}$$

$$\frac{d}{dt}\mathrm{OC} = (1 - \eta)g_1(\mathrm{IC}, \mathrm{O}_2; \nu)\mathrm{P}_1 - 2g_2(\mathrm{OC}, \mathrm{O}_2)\mathrm{P}_2 - D\,\mathrm{OC}, \tag{39}$$

$$\frac{d}{dt}\mathrm{O}_2 = \left(\frac{3}{4} + \frac{17}{40}\eta\right)g_1(\mathrm{IC}, \mathrm{O}_2; \nu)\mathrm{P}_1 - \frac{1.9}{4}g_2(\mathrm{OC}, \mathrm{O}_2)\mathrm{P}_2 + D(\mathrm{O}_{2,0} - \mathrm{O}_2), \tag{40}$$

$$\frac{d}{dt}\mathrm{P}_1 = (\eta g_1(\mathrm{IC}, \mathrm{O}_2; \nu) - D)\mathrm{P}_1, \tag{41}$$

$$\frac{d}{dt}\mathrm{P}_2 = (g_2(\mathrm{OC}, \mathrm{O}_2) - D)\mathrm{P}_2. \tag{42}$$

These equations are the same as the single species ones (28)–(31) except with the addition of source/sink terms proportional to $g_2\mathrm{P}_2$ in each of (38)–(40) as well as the new Equation (42) describing heterotroph biomass. In effect we are adding the third elementary flux mode, recall Figure 3, into the system, with all of its component reactions occuring at rate g_2.

3.2.1. Steady States and Stability

Equations (38)–(42) have three types of steady state solutions: the washout solution with $P_1 = P_2 = 0$, the single species solution with $P_1 > 0$, $P_2 = 0$, and the coexistence solution $P_1, P_2 > 0$. Note that a fourth type of steady state with $P_1 = 0$, $P_2 > 0$ is not possible; if $P_1 = 0$ then the heterotrophs will be washed out of the system. We consider first the washout and single species states, each with $P_2 = 0$, and report results of stability analysis here, again referring to Appendix D for details. The coexistence steady state, with $P_2 > 0$, is explored numerically later. Note that if $P_2 = 0$ then (38)–(41) reduce, essentially, to (28)–(31), so that steady states for the washout and single species systems are the same as previously (with the addition that $P_2 = 0$).

Generally, as before, any component of a perturbation to a steady state that introduces excess or deficient total carbon or degree of reduction is washed out of the system on the chemostat turnover time scale D^{-1}. Thus, understanding perturbation dynamics of the five dimensional system (38)–(41) reduces to understanding dynamics on a three dimensional subsystem, now spanned by all three of the elementary flux modes, given in vector form by

$$\mathbf{EFM}_1 = \begin{pmatrix} -1 \\ 0 \\ 4.7/4 \\ 1 \\ 0 \end{pmatrix}, \quad \mathbf{EFM}_2 = \begin{pmatrix} -1 \\ 1 \\ 3/4 \\ 0 \\ 0 \end{pmatrix}, \quad \mathbf{EFM}_3 = \begin{pmatrix} 1 \\ -2 \\ -1.9/4 \\ 0 \\ 1 \end{pmatrix},$$

see Figure 3. Perturbation by increasing or decreasing flux through the photosynthesis-driven biosynthesis mode corresponds to perturbation in the direction $\mathbf{EFM}_1$ (one Cmole biomass and $4.7/4$ Omoles produced per Cmole inorganic carbon consumed) and perturbation by increasing or decreasing flux through the photorespiration mode corresponds to perturbation in the direction $\mathbf{EFM}_2$ (one Cmole organic carbon and $3/4$ Omoles produced per Cmole inorganic carbon consumed). The new vector $\mathbf{EFM}_3$ corresponds to perturbation that increases or decreases flux through the heterotroph biosynthesis mode (one Cmole biomass and 1 Cmole inorganic carbon produced per two Cmoles organic carbon and $1.9/4$ Omoles consumed).

Washout State (Two Species System). The two species washout state ($P_1 = P_2 = 0$) is, as in the one species washout case, unstable if $\lambda = \eta g_1 - D$ is positive and stable if λ is negative. As before, λ is the net intrinsic phototroph biomass production rate at inflow conditions. Also as before, when a small quantity of phototorphs are added to the system, in the unstable case $\lambda > 0$, dynamics of the linearized system effectively reduce to exponential growth on the one dimensional space $\eta \, \mathbf{EFM}_1 + (1 - \eta) \, \mathbf{EFM}_2$, indicating that the linearized growth dynamics occurs, as to be expected, as a combination of the photosynthesis mode and the photorespiration mode weighted by the branching parameter $\eta \, (\text{IC}_0, \text{O}_{2,0})$. Note that the heterotroph cannot invade as it requires an already established population of phototrophs (with corresponding finite supply of organic carbon) before it can become viable.

Single Species State: Invasion (Two Species System). We consider for several purposes the single species state ($P_1 > 0$, $P_2 = 0$). Note that this solution is identical to that in the single species case as discussed in Section 3.1.2 and Appendix C except with the additional component $P_2^* = 0$. We assume that this state is linearly stable to perturbations that do not introduce heterotrophs and ask what happens if a small amount of heterotrophs are added. In other words, how does the otherwise stable heterotroph-free system respond to a perturbation including heterotrophs? This is the invasion problem. In the case that invasion occurs, obviously there is some benefit from the phototrophic population to the invading, heterotrophs as they cannot survive in the chemostat by themselves. Linear analysis can provide some information on the specifics of this advantage.

A key observation here is that the five dimensional system (38)–(42) essentially reduces to the single species, four dimensional one (28)–(31) when $P_2 = 0$. In this four dimensional reduced system, dynamics of the single species state are stable, so that the full invasion dynamics are effectively restricted to a complementary one dimensional space. This space is necessarily a linear combination of the three mode vectors $\mathbf{EFM}_1$, $\mathbf{EFM}_2$, and $\mathbf{EFM}_3$. Notably, in the case of large $g_2(OC^*, O_2^*)$, instability dynamics are dominated by the heterotrophic growth mode $\mathbf{EFM}_3$. When growth is not as dominant, the role of phototroph flux modes in maintaining carbon and DoR balance is more evident. The governing quantity is $\lambda = g_2 - D$; if $\lambda > 0$ then heterotroph invasion occurs, and if $\lambda < 0$ then heterotrophs are unable to invade.

An interesting question here is whether perturbations that include introduction of heterotrophs, i.e., positive perturbation of P_2, result in both successful invasion of heterotrophs as well as, simultaneously, increase in phototroph biomass. We consider this question in the case of large $g_2(OC^*, O_2^*)$, see Appendix F for details. Note that, as dynamics are dominantly in the direction of $\mathbf{EFM}_3$, then the P_1 component of the perturbation dynamics is small. It is, however, positive as in this case the intuition that addition of heterotrophs, at least initially, increases inorganic carbon concentration and decreases oxygen concentration is correct. Hence, the immediate effects of heterotroph invasion on the phototrophs are mildly positive. Of course, the more important question is of long time effects, which will be considered next.

3.2.2. Two Species Consortium Steady State

We rely on numerical computations to investigate two species steady states. Parameters are as used previously for single species computations with the addition of parameters connected to the heterotrophic biomass mode, see again Appendix B. See Figure 5 for a typical numerical comparison of the single species steady state biomass (as in Figure 4, left panel) and two species steady state biomasses, as functions of inverse specificity. As in Figure 4 left panel, photon intensity is held at a fixed represntative level of 50 μE. For the parameters chosen, steady state conditions are carbon-limited for inverse specificity smaller than, approximately, 10^{-2}, and light-limited for larger inverse specificity.

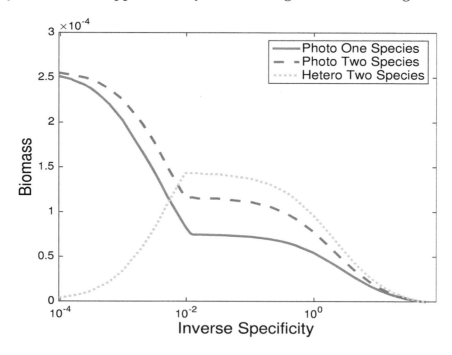

Figure 5. Steady state biomass (Cmoles/L) versus inverse specificity γ_1 (Cmoles/Omoles). Increasing γ_1 corresponds to increasing importance of photorespiration. $D = 10^{-4}\,\mathrm{s}^{-1}$, and other fixed parameters are as described in Appendix B and in Figure 4. Solid: steady state biomass P_1^*, single species community ($P_2 = 0$). Dashed: steady state biomass P_1^*, two species community. Dotted: steady state biomass P_2^*, two species community. The kinks at $\gamma_1 \cong 10^{-2}$ occur where the steady state transitions from carbon-limitation to light-limitation. Note that steady state phototroph biomass increases with addition of heterotrophs.

Note that phototroph biomass is larger in the two species community than in the one species community for all values of inverse specificity that allow a viable population. This can be understood as resulting from reconversion of some of the dissolved organic carbon back to inorganic form through heterotroph respiration, recall (19), where it is available for photosynthesis, as opposed to the single species system where all dissolved organic carbon is flushed from the chemostat. Increase in biomass is most noticeable in the interval corresponding to actual measured environmental values of γ_1 where, in the model results, light is limiting and heterotroph biomass is largest.

Intuition might suggest that introduction of heterotrophs into a phototoph population would result in increase in IC and decrease in O_2, both as a consequence of respiration. And indeed, such may initially be the case, see the invasion discussion above. However, for later times numerics suggest otherwise near steady state. Dissolved carbon and oxygen, for the same computation, are shown in Figure 6, left and middle respectively. Dissolved inorganic carbon levels are similar for both one and two species communities, with consequently matching protection from photorespiration against high light intensity in both communities. The similarity as well in inorganic carbon concentrations is a consequence of the steady state rate constraint $\eta g_1 = D$, see (41); in the small γ_1 carbon limited regime, $\eta \simeq 1$ so IC is determined by $\delta - g_1 \simeq D$ independent of presence or absence of heterotrophs, while in the light limited regime for relatively large γ_1 heterotroph population is low so has little effect and in the light regime with relatively low γ_1, g_1 is largely independent of IC and O_2 so that η must be approximately constant, again independently of presence or absence of heterotrophs, see Figure 6 right panel. Note that

$$\left| \frac{\eta_{IC}}{\eta_{O_2}} \right| = \left| \frac{O_2}{IC} \right|.$$

so that in this range, η sensitivity to change in IC is much larger than sensitivity to change in O_2. Hence steady state IC is largely unchanged between the one and two species communities. Dissolved organic carbon, however, is largely absent from the two species community, in contrast to the single species one, as organic carbon is limiting for heterotroph biomass production at all values of inverse specificity and thus is depleted in the two species community. This is perhaps the most dramatic change in chemical environment between the one species and two species environments.

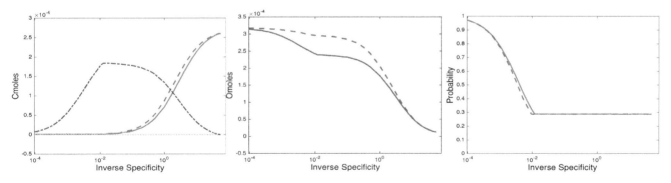

Figure 6. Steady state values versus inverse specificity γ_1 (Cmoles/Omoles) corresponding to the computations shown in Figure 5. Left: carbon concentrations (Cmoles/L) with (solid) single species community inorganic carbon IC, (dash) two species community inorganic carbon IC, (dash-dot) single species community organic carbon OC, (dot) two species community organic carbon OC. Middle: oxygen concentrations (Omoles/L) with (solid) single species community oxygen O_2 (dash) two species community oxygen O_2. Right: branching function probability η for (solid) the one species community and (dash) the two species community.

Interestingly, oxygen concentration levels are actually somewhat higher for the two species community, see Figure 6 middle panel, despite oxygen usage via heterotroph respiration. Higher oxygen concentrations can be understood to be a consequence of carbon fixation—increase in reduced carbon in the form of biomass and hence increase in net degree of reduction DoR must be balanced by something. In this model, the only possibility is oxygen. Thus, oxygen concentration increases if biomass does so. This is a generally applicable observation: photosynthetically fixed carbon will be accompanied by more material with low degree of reduction. In an ideal chemostat environment, this balance must be maintained. In other environments, biofilms for example, it may be possible for byproducts like excess oxygen to be transported out of the system while fixed carbon in the form of biomass remains behind.

Note that, in the two species community, as virtually all dissolved carbon goes to biomass for smaller values of γ_1, then in fact two species oxygen levels are approximately independent of inverse specificity γ_1 and this remains so up until the point that γ_1 becomes sufficiently large that significant amounts of inorganic carbon go unutilized. Only then does oxygen concentration significantly decrease as a function of γ_1, though it always remains larger than the corresponding concentration in the one species community.

4. Discussion

The one species model. We observe, in the model, that photorespiration shunts reduced carbon away from biomass production and into dissolved, secreted organic carbon, resulting, at least in a single species oxygenic phototroph population, in three principle effects:

1. decrease in population biomass,
2. increase in population light tolerance,
3. and decrease in oxygen concentration.

The first two are connected through inorganic carbon concentrations: decrease in population results in decrease in inorganic carbon demand resulting in increase in inorganic carbon concentration resulting in reduced inorganic carbon limitation at high light intensities. Decrease in oxygen occurs for two reasons: (1) reduced phototrophic biomass results in reduced oxygen production, and (2) photorespiration product is less reduced than biomaterial, so its production results in less oxygen as a consequence of degree of reduction balance.

The increase in light tolerance and decrease in oxygen concentration suggests an advantage to photorespiration. However, reduction in population size suggests the possibility of fitness deficit in comparison to a population that does not photorespire. We have not modeled such a competition here. However, it should be noted that we impose constant light intensity, and that it is not clear what effects variable light intensity, particularly transient peaks in intensity, might have on a competition of two species, one of which grows more efficiently in low light conditions and the other of which is better protected in high light conditions.

In our set-up, RuBisCO oxygenase activity (which we identify with photorespiration) can serve as a differential of sorts able to synchronize influx of photons with influx of inorganic carbon. Using estimates of parameters, we find an interval of values of the inverse specificity γ_1 which, on the one hand, result in population levels for which, over an increased range of photon intensities, light is limiting but also for which, on the other hand, biomass synthesis is not excessively quenched to the point of reducing the photon intensity range of viability. Though biomass is decreased, the increase in range of "healthy" light intensities might suggest more resilience to light intensity variations, i.e., increased ecological structural stability [30]. The upper bound on γ_1 is related to background inorganic carbon and oxygen concentrations (IC_0 and $O_{2,0}$ here) and thus may be relatively independent of model details. The lower bound on γ_1 is related to organismal transport rates for inorganic carbon and is perhaps more model dependent, though also allows the possibility of organismal control. In any case, the optimal interval we find for inverse specificities seems to be consistent with measured values over a range of organisms.

The two species model. A steady source of photorespiration-derived organic carbon begs the introduction of a heterotroph population to consume it, so we also modeled a phototroph-heterotroph consortium. Obviously, the heterotroph population benefits from the interaction as it cannot survive in the model chemostat without the organic carbon supplied by the phototrophs. The phototrophs, on the other hand, retain their added tolerance to light intensity but see three principle new effects:

1. biomass increase,
2. reduction in dissolved organic carbon,
3. and oxygen concentration increase.

Biomass increase occurs because of increased inorganic carbon availability as a consequence of heterotroph respiration. Note that organic carbon, here glycolate, could have inhibitory effects, so its consumption by heterotrophs might also potentially increase growth rates, though this effect has not been included in the model. Increase in oxygen concentration is somewhat surprising as heterotrophs consume oxygen during respiration, but occurs again as a consequence of reduction balance: overall increase in reduced biomass must be balanced within the model by increase in oxygen. More directly, the increase in phototroph biomass leads to increased oxygen production. This may be, at least to an extent, an artifact of model simplicity. A more complex model could retain reduction balance through oxidized material other than oxygen. Also, simplicity of the chemostat itself requires that all material be washed out at the same rate whether reduce or oxidized. A more complex system might do otherwise, for example removing dissolved oxygen (e.g., gas sparging that removes O_2) faster than particulate biomass (e.g., biomass fixed in a biofilm).

The phototroph-heterotroph consortium is a more efficient consumer of inorganic carbon than the photorespiring phototroph population alone, and presence of organic carbon suggests that heterotrophs could be expected to join a photorespiring phototroph population. Hence, it may be that the question of competing photorespiring vs. non-photorespiring phototrophs may be the wrong one. Rather, non-photorespiring phototrophs should be asked to compete against a combined photorespiring phototroph-heterotroph consortium.

Connecting flux mode models to population scale models. Mathematically and physically, rates are natural quantities at the flux mode level whereas concentrations, including biomass, are natural at the population and environmental level. We find here that rate functions (in the population model) serve to translate cell scale flux modes into the larger scale population level, where they then determine external concentrations in combination with large scale transport constraints. Flux modes themselves naturally appear, mathematically, in near-steady state dynamics and are relatable to eigenvectors which in turn are natural structures for dynamics. The process of converting flux modes to rate functions is in principle automatable and should be a part of the overall program of extracting information from 'omics data.

Conversely, the mathematical issues involved in the inverse process of determining how large scale effects influence flux mode regulation are interesting ones and only addressed indirectly here. Generally speaking, microbial communities can have metabolic capabilities available to organisms and to the overall community. This raises the question—how can these capabilities be best deployed to utilize available resources? Rate and stoichiometric constraints still apply, and steady states or, more generally, asymptotic states, can be computed though likely not in a unique way. However, there may be many branching type parameters over which the community has at least some control. Optimality becomes a question of distribution of resource flow (here carbon and electrons) between available pathways in the most efficient manner. Even in the system studied here, with a small number of well defined pathways and a relatively simple physical environment, the effects of that environment on pathway optimization are subtle and influential. The environment imposes rates at steady state and it also determines response to perturbation. These constraints as well as those arising from community ecology may easily be overlooked without considering the physical context of the biological system.

Acknowledgments: The authors would like to acknowledge support from NSF/DMS 1517100, 1022836, and 330367, the European Union Seventh Framework Programme (FP7-PEOPLE-2012-IOF-328315), PNNL, as well as assistance from Ashley Beck and Shane Nowack.

Author Contributions: F.E.M., I.K. and R.P.C. designed the model, F.E.M. and I.K. performed analyses and computations, and F.V. parameterized the model. I.K. wrote the paper.

Appendix A. Degree of Reduction

Degree of reduction of an atom or molecule is, roughly, the number of electrons that atom or molecule is apt to give away in a chemical reaction [31]. We use degree of reduction (DoR) here, essentially, as a convenient proxy for redox potential. Degree of reduction is computed using the values $DoR(C) = +4$, $DoR(H) = +1$, $DoR(O) = -2$. For nitrogen, we use $DoR(N) = 0$ for cyanobacteria (assuming nitrogen is extracted from N_2) and, effectively, $DoR(N) = -3$ for heterotrophs (assuming nitrogen is extracted from an organic source) [18]. This dichotomy for N is somewhat at odds with the definition given just above, but maintains consistency of degree of reduction balance by accounting for differences in biomaterial formation as explained below.

Degree of reduction for a molecule is estimated by summing degree of reduction of that molecule's individual atoms. Then the degrees of reduction for inorganic carbon (assumed of the form carbon dioxide CO_2), organic carbon (assumed of the form glycolate $CH_2O_{1.5}$) and biomass (assumed for both autotrophs and heterotrophs to be of the form $CH_{1.7}O_{0.5}N_{0.2}$) are estimated to be

$$
\begin{aligned}
DoR(CO_2) &= +0 \\
DoR(CH_2O_{1.5}) &= +3 \\
DoR(CH_{1.7}O_{0.5}N_{0.2}) &= +4.7 \quad \text{(autotroph)} \\
DoR(CH_{1.7}O_{0.5}N_{0.2}) &= +4.1 \quad \text{(heterotroph)}
\end{aligned}
$$

These are computed simply by adding values of the component atoms, though the nitrogen contribution introduces a small complication. Note that electrons have degree of reduction $+1$. Also note that, although the degree of reduction of glycolate is $+3$, in the context of the simplified model used here of photorespiration, $(1/2)O_2$ is removed from the system for each photorespiration reaction with the context that the degree of reduction of the entire system is increased by $+2$. Hence, effectively, formation of a unit of $CH_2O_{1.5}$ has the effect of changing the overall cell degree of reduction by $+5$. Biomass, represented by $CN_{1.7}O_{0.5}N_{0.2}$, comes with a DoR value of 4.7 computed on the basis of construction from molecular oxygen, hydrogen, carbon dioxide, and also molecular nitrogen (N_2), indicating that 4.7 moles of electrons are required to synthesize a mole of biomass, roughly. However, assuming heterotrophs are able to use an organic source of hydrogen, e.g., ammonia, rather than molecular nitrogen, then only approximately 4.1 electron moles are needed per biomass mole.

Appendix B. Parameter Estimation

Carbon moles per cell. We apply the following estimates for microbial cells:

$$
\begin{aligned}
\text{wet mass/volume} &\cong 1.1 \times 10^6 \text{g/m}^3, \\
\text{volume/cell} &\cong 5 \times 10^{-18} \text{m}^3, \\
\text{dry Cmass/wet mass} &\cong 1/10,
\end{aligned}
$$

where the last estimates carbon as comprising 10% of cells by mass. Using the fact that the mass of 1 carbon mole (Cmole) is 12 g, then the conversion parameter $c =$ Cmoles per cell can be approximated to be

$$
c = \left(\frac{\text{Cmole}}{\text{dry Cmass}} \right) \left(\frac{\text{dry Cmass}}{\text{wet mass}} \right) \left(\frac{\text{wet mass}}{\text{cell volume}} \right) \left(\frac{\text{cell volume}}{\text{cell}} \right) \cong 4.6 \times 10^{-14} \text{Cmole/cell.}
$$

Effective photon absorption rate. Approximating the volume of a cyanobacterial cell as a cylinder of radius 1 μm and length 4 μm, and assuming the cylinder to be randomly oriented with respect to the direction of light (or, alternatively, supposing light to be well scattered), then A, the average cross-sectional area exposed to light, is

$$
\begin{aligned}
A & \cong \text{(cylinder width)}\text{(average cylinder projected length)} \\
& = (1\ \mu m) \int_{-\pi/2}^{\pi/2} \ell(\theta) P(\theta) d\theta \\
& = (1\ \mu m) \int_{-\pi/2}^{\pi/2} (4\cos\theta\ \mu m)(\cos\theta)/2 d\theta \\
& = \pi\ \mu m^2
\end{aligned}
$$

where $\ell(\theta) = 4\cos\theta\ \mu m$ is the projected length of a cylinder of length $4\ \mu m$ and angle θ from transverse, and $P(\theta) = (1/2)\cos\theta$ is the probability of angle θ. This approximation underestimates slightly the contribution from the cylinder cap at the end of the cylinder pointing towards the light and overestimates slightly the contribution from the other cap. Note that assuming a cylindrical geometry (as opposed to a spherical one) may be an effective strategy to reduce light exposure in some situations.

Inorganic carbon transport parameters. We use values for *Synechocystis* sp. PCC6803, based on CO_2 and HCO_3^- uptake rate and half-saturations from [32] which reported the values of maximum inorganic carbon transport rate $V_{IC} \cong 391$ micromoles per milligram of chlorophyll per hour and approximately 1.03×10^{-9} milligrams chlorophyll per cell (*Synechocystis*) [32]. Converting, then, we obtain

$$
\begin{aligned}
V_{IC} & \cong 391 \frac{\mu mol\ CO_2}{mg\ Chl\ h} \\
& = \frac{391 \times 1.03 \times 10^{-9}}{3600} \frac{\mu mol\ IC}{cell\ s} \\
& = 1.12 \times 10^{-16} \frac{Cmol}{cell\ s}.
\end{aligned}
$$

Then

$$
\begin{aligned}
r_{IC} & = \frac{V_{trans}}{c} \\
& = \frac{1.12 \times 10^{-16}}{4 \times 10^{-14}} \frac{1}{s} \\
& = 2.80 \times 10^{-3} \frac{1}{s}.
\end{aligned}
$$

Also, from [32], $K_{IC} \cong 8.0 \times 10^{-5}$ in Cmoles. Note, perhaps as another indicator of the importance of community interactions and local environment, there is wide variation in mechanisms for inorganic carbon transport even among cyanobacteria [33], so that these parameters can be expected to vary between species.

Other parameters. Other parameter values used for numerics are tabulated below, together with literature references when appropriate. Yield parameters are fixed by stoichiometric and similar considerations. Inflow concentrations are estimated using Henry's law at standard atmospheric conditions. The true value of the photosynthetic efficiency parameter α is uncertain (though photosynthetic efficiencies have been estimated at the community level, e.g., [34], it is somewhat unclear how to translate to the cellular level) so we set $\alpha = 1$. Note that α can effectively be scaled into the photon flux, which is treated as an independent variable for computational purposes, so does not have independent effect on qualitative conclusions. Inflow concentrations IC_0, OC_0, and $O_{2,0}$ representative of environmental conditions are chosen. Background concentrations of these quantities can vary from one environment to another, but results are fairly insensitive to reasonable variations. Photon flux is given in terms of microeinsteins with 1 microeinstein = 1 μE = 10^{-6} moles of photons.

Symbol	Name	Unit	Value	Reference
γ_1	Inverse specificity	Cmol·Omol^{-1}	0.01–1	[11,26,27]
γ_2	Excess elec. rate capacity	s^{-1}	–	–
r_h	Maximal transport rate	s^{-1}	0.0225	Measured
r_{trans}	Maximal transport rate	s^{-1}	1.24×10^{-3}	[32]
K_{trans}	Half saturation	Cmol·L^{-1}	3×10^{-6}	[29]
K_{O_2}	Half saturation	Omol L^{-1}	8.1253×10^{-10}	[35,36]
K_{oc}	Half saturation	Cmol·L^{-1}	4.6022	[37]
ν	Photon flux	μE·m^{-2}·s^{-1}	0–2000	[38]
α	Efficiency	-	1	–
IC_0	Inflow IC concentration	Cmol·L^{-1}	2.6×10^{-4}	–
OC_0	Inflow OC concentration	Cmol·L^{-1}	0	–
$O_{2,0}$	Inflow O_2 concentration	Omol·L^{-1}	1.3×10^{-5}	–
D	Chemostat turnover rate	s^{-1}	various	–
Y_{ic}	Yield	Cmol·cell^{-1}	z	Yield
Y_{oc}	Yield	Cmol·cell^{-1}	z	Yield
$Y_{o_{21}}$	Yield	Omol·ph^{-1}	1/8	Yield
$Y_{o_{22}}$	Yield	Omol·cell^{-1}	$z/2$	Yield
$Y_{o_{23}}$	Yield	Omol·cell^{-1}	$1.9\,z/4$	Yield
$Y_{o_{24}}$	Yield	Omol·electron^{-1}	$z/4$	Yield
Y_{light}	Yield	Electron·ph^{-1}	1/2	Yield

Appendix C. Existence and Uniqueness of Single Species Viable State Solutions

First we show that Equations (28)–(31) either have a unique steady state solution $(IC^*, OC^*, O_2^*, P_1^*)$ with $IC^*, OC^*, O_2^* \geq 0$ and $P_1^* > 0$ (a viable solution) or no steady state solution with $P_1^* > 0$ at all, depending on choice of parameters IC_0, $O_{2,0}$ and D. The argument depends on the forms of rate function $g_1(IC, O_2)$ and branching function $\eta(IC, O_2)$. Specific forms for η and g_1 are supplied in (14) and (16), but for generality we will only require here that

1. η and g_1 are smooth.

2. Monotonicity in O_2: for fixed value of IC, $g_1(IC, O_2)$ is monotonically non-increasing in O_2 with values decreasing from $g_1(IC, 0)$ to 0 as O_2 varies from 0 to ∞, and $\eta(IC, O_2)$ is monotonically decreasing in O_2 with values decreasing from 1 to 0 as O_2 varies from 0 to ∞. Roughly speaking, increasing oxygen concentration if anything inhibits photosynthesis and always shifts photosynthetic product from biosynthesis to photorespiration.

3. Monotonicity in IC: for fixed value of O_2, $g_1(IC, O_2)$ is monotonically non-decreasing in IC with values increasing from 0 to $g_1(IC_0, O_2)$ as IC varies from 0 to IC_0, and $\eta(IC, O_2)$ is monotonically increasing in IC with values increasing from 0 to $\eta(IC_0, O_2)$ as IC varies from 0 to IC_0. (In fact, η should tend to 1 as $IC \to \infty$). Roughly speaking, increasing inorganic carbon concentration if anything promotes photosynthesis and always shifts photosynthetic product from photorespiration to biosynthesis.

4. Fixation stability: we assume that condition (17), namely $\eta_{IC}\,g_{1,O_2} - \eta_{O_2}\,g_{1,IC} \geq 0$, holds.

5. Note as well that the function $f(\eta)$ is necessarily a linear function with parameterization determined by stoichiometry and degree of reduction values. In fact, for the particular choices we use, $f(\eta) = (3/4) + (17/40)\eta$, however we here need only suppose that $f(\eta) = a + b\eta$ for some $a, b > 0$.

We consider Equations (28)–(31) in steady state, i.e.,

$$0 = -g_1 P_1 + D(IC_0 - IC), \tag{A1}$$

$$0 = (1 - \eta)g_1 P_1 - DOC, \tag{A2}$$

$$0 = f(\eta)g_1 P_1 + D(O_{2,0} - O_2), \tag{A3}$$

$$0 = (\eta g_1 - D)P_1, \tag{A4}$$

A viable steady state requires that equation

$$\eta(IC, O_2)g_1(IC, O_2) = D \tag{A5}$$

have a nonnegative solution (IC^*, O_2^*). By combining Equations (28) and (30), a second equation,

$$f(\eta)(IC_0 - IC) = O_2 - O_{2,0}, \tag{A6}$$

is obtained relating IC and O_2. As a consequence of Cmole conservation, see (26), it is evident that the steady state value IC^* is bounded from above by IC_0, i.e., $0 \le IC^* \le IC_0$. Note that (A6) has a unique positive solution $O_2(IC)$ for each value IC in the interval $[0, IC_0]$, with in fact $O_2(IC) \ge O_{2,0}$. Hence, any viable solution (IC^*, O_2^*) to Equations (A5) and (A6) must lie in the infinite half-strip solvability region $0 \le IC^* < IC_0$, $O_2^* \ge O_{2,0}$. (If $IC^* = IC_0$, then, from (A1), necessarily $P_1^* = 0$). In the case that (A5) and (A6) have a solution (IC^*, O_2^*), then P_1^* and OC^* can be recovered as

$$P_1^* = \eta(IC_0 - IC^*), \tag{A7}$$

$$OC^* = (1 - \eta)(IC_0 - IC^*),$$

with η evaluated at (IC^*, O_2^*). Thus, the problem essentially reduces to solving (A5) and (A6) for IC and O_2.

Note that, as a consequence of monotonicity and smoothness, the maximum value of g_1 for $O_2 \ge 0$ and $0 \le IC \le IC_0$ is $g_1(IC_0, 0)$. Recalling $0 \le \eta \le 1$, if $D > g_1(IC_0, 0)$ then (A5) has no solution. If $D \le g_1(IC_0, 0)$ then, under the assumptions made on g_1 and η, there is a value $0 < \widehat{IC} \le IC_0$ with $g_1(\widehat{IC}, 0) = D$, $\eta(\widehat{IC}, 0) = 1$, and (A5) has a one parameter set of solutions $(IC, O_2) = (IC, h(IC))$ over $\widehat{IC} \le IC \le IC_0$ where h is non-decreasing with $h(\widehat{IC}) = 0$. Since $\nabla(\eta g_1)$, by the requirements above, lies in the fourth quadrant (IC component is positive, O_2 component is negative) then the tangent to the curve $(IC, h(IC))$ in the increasing IC direction lies in the first quadrant. Also, since $\eta(IC_0, O_2)g_1(IC_0, O_2) = D$ has a finite solution, then the curve $(IC, h(IC))$ appears as one of the forms in Figure A1. If this curve has no segment in the half-strip solvability region (lower curve), then there is no viable solution. Conversely, if there is a segment in the solvability region (upper curve), then we will show that, under the above requirements, there is a unique viable solution.

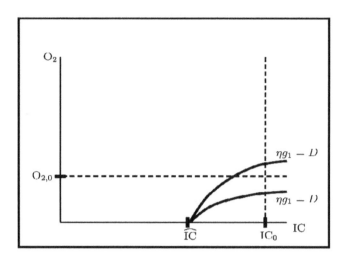

Figure A1. Two different possible curves $(IC, h(IC))$, where $\eta(IC, hIC)g_1(IC, hIC) = D$. The upper curve, which intersects the half-strip $0 \leq IC < IC_0, O_2 \geq O_{2,0}$, allows a viable solution; the lower curve does not.

Consider lines of the form

$$C(IC_0 - IC) = O_2 - O_{2,0}, \tag{A8}$$

cf. (A6), where C is a constant within the range $f_{min} \leq C \leq f_{max}$, with $f_{min} = f(\min(a + b\eta)) = f(a)$, $f_{max} = f(\max(a + b\eta)) = f(a + b)$. These are lines with slopes $-C$ and O_2-intercepts $(0, O_{2,0} + CIC_0)$ that all intersect at the single point $(IC_0, O_{2,0})$, see Figure A2. Note that lines with larger C have larger O_2-intercept than lines with smaller C, i.e., lines move upward with increasing C. In the case that the curve $(IC, h(IC))$ intersects the viable region, then it must also intersect each of the lines (A8) exactly once. Since the lines correspond to $f|_{\eta=0}$ running to $f|_{\eta=1}$, then there must be at least one intersection point where both (A5) and (A6) are simultaneously satisfied. Each such point provides a solution (IC^*, O_2^*).

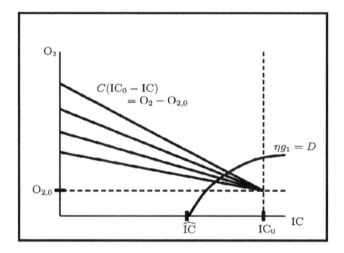

Figure A2. Illustration of the intersection of allowable lines of the form $C(IC_0 - IC) = O_2 - O_{2,0}$ with the curve $(IC, h(IC))$ on which $\eta g_1 = D$ is satisfied.

It is not clear that such a solution (IC^*, OC^*) would be unique in general. However, stability requirement (4) above is sufficient to guarantee uniqueness, argued as follows. The lines (A8) correspond to increasing η moving from bottom to top. Under the given requirements above, requirement (4) in particular, we claim that the value of η is non-increasing along the curve $(IC, h(IC))$

in the increasing IC direction. That is, when moving along the curve $(\text{IC}, h(\text{IC}))$ in the increasing IC direction, the left-hand side of (A6) decreases and the right-hand side increases. Hence there can be no more than one intersection point where both (A5) and (A6) are simultaneously satisfied.

Proof of claim. First, note that $\nabla \eta$, ∇g_1, and $\nabla(\eta g_1)$ all lie in the fourth quadrant of the (IC, O_2) plane. Further, the normal to the $\eta g_1 = D$, i.e., to the curve $(\text{IC}, h(\text{IC}))$, given by $\nabla(\eta g_1) = \eta \nabla g_1 + g_1 \nabla \eta$ is a positive linear combination of ∇g_1 and $\nabla \eta$, so in fact lies between ∇g_1 and $\nabla \eta$, see Figure A3. Stability requirement $\eta_{\text{IC}} g_{1,O_2} - \eta_{O_2} g_{1,\text{IC}} \geq 0$ guarantees that the geometry of the three gradient vectors is as in Figure A3 (as opposed to the one where $\nabla \eta$ and ∇g_1 are exchanged), except in the equality case $\eta_{\text{IC}} g_{1,O_2} - \eta_{O_2} g_{1,\text{IC}} = 0$ in which case all three vectors are parallel. This shows that the directional derivative of η is non-positive along the curve $(\text{IC}, h(\text{IC}))$ in the increasing IC direction, as was claimed. Note, further, that η is constant if $\eta_{\text{IC}} g_{1,O_2} - \eta_{O_2} g_{1,\text{IC}} = 0$ and strictly decreasing if $\eta_{\text{IC}} g_{1,O_2} - \eta_{O_2} g_{1,\text{IC}} > 0$. $\square$

As a side remark, reversing the geometry in Figure A3 (where $\nabla \eta$ and ∇g_1 are exchanged) would result in a situation such that photosynthesis rate g_1 actually decreases with increasing IC. The unlikeliness of such behavior provides another intuition for the necessity of the fixation stability condition (17).

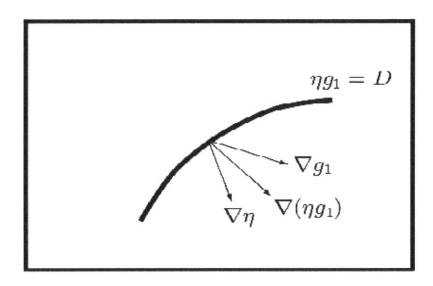

Figure A3. Under requirement (4), the vectors $\nabla(\eta g_1)$, ∇g_1, and $\nabla \eta$ are oriented relative to each other as shown.

Appendix D. Linearization and Stability

We consider here stability of steady states of the chemostat system, beginning with the single species community model (28)–(31), which has two possible steady states, namely washout $(P_1(t) = 0)$ and viable $(P_1(t) = P_1^* > 0)$. Writing $\text{IC}(t) = \text{IC}^* + \widetilde{\text{IC}}(t)$, $\text{OC}(t) = \text{OC}^* + \widetilde{\text{OC}}(t)$, $O_2(t) = O_2^* + \widetilde{O}_2(t)$, $P_1(t) = P_1^* + \widetilde{P}_1(t)$, where tilded quantities are small perturbations to steady state values, then system (28)–(31) linearizes to

$$\frac{d}{dt}\begin{pmatrix} \widetilde{\text{IC}}(t) \\ \widetilde{\text{OC}}(t) \\ \widetilde{O}_2(t) \\ \widetilde{P}_1(t) \end{pmatrix} = J^{(1)}(\text{IC}^*, \text{OC}^*, O_2^*, P_1^*)\begin{pmatrix} \widetilde{\text{IC}}(t) \\ \widetilde{\text{OC}}(t) \\ \widetilde{O}_2(t) \\ \widetilde{P}_1(t) \end{pmatrix} \tag{A9}$$

with

$$J^{(1)} = \begin{pmatrix} -g_{1,\text{IC}}P_1 - D & 0 & -g_{1,O_2}P_1 & -g_1 \\ ((1-\eta)g_1)_{,\text{IC}}P_1 & -D & ((1-\eta)g_1)_{,O_2}P_1 & (1-\eta)g_1 \\ (f(\eta)g_1)_{,\text{IC}}P_1 & 0 & (f(\eta)g_1)_{,O_2}P_1 - D & f(\eta)g_1 \\ (\eta g_1)_{,\text{IC}}P_1 & 0 & (\eta g_1)_{,O_2}P_1 & \eta g_1 - D \end{pmatrix}, \qquad (A10)$$

all quantities evaluated at the steady state solution.

From (26) and (27)

$$\frac{d}{dt}\begin{pmatrix} \tilde{C} \\ \tilde{\text{DoR}} \end{pmatrix} = \frac{d}{dt}\begin{pmatrix} \tilde{\text{IC}} + \tilde{\text{OC}} + \tilde{P}_1 \\ 3\tilde{\text{OC}} - 4\tilde{O}_2 + 4.7\tilde{P}_1 \end{pmatrix} = \begin{pmatrix} 1 & 1 & 0 & 1 \\ 0 & 3 & -4 & 4.7 \end{pmatrix} \frac{d}{dt}\begin{pmatrix} \tilde{\text{IC}} \\ \tilde{\text{OC}} \\ \tilde{O}_2 \\ \tilde{P}_1 \end{pmatrix} = -D\begin{pmatrix} \tilde{C} \\ \tilde{\text{DoR}} \end{pmatrix}$$

Hence, for any perturbation, its component normal to the null space of

$$A = \begin{pmatrix} 1 & 1 & 0 & 1 \\ 0 & 3 & -4 & 4.7 \end{pmatrix} \qquad (A11)$$

is damped (eventually) as e^{-Dt}. That is, excess or deficience in the initial perturbation of total carbon and degree of reduction is removed from the system through outflow on the chemostat turnover time scale. In fact, the row vectors of A are eigenvectors of the transpose of $J^{(1)}$ with eigenvalue $-D$ and so $-D$ is a multiplicity 2 (at least) eigenvalue of $J^{(1)}$. Note, thus, that we can therefore characterize the dynamics described by the four dimensional system (A9) if we can characterize the dynamics on a two dimensional subspace consisting of the null space of A, i.e., the subspace defined by $\tilde{C} = 0$, $\tilde{\text{DoR}} = 0$ (no net perturbation of total carbon or degree of reduction).

In fact, the null space of A can be interpreted as the phototroph flux mode space and is spanned by the vectors

$$\mathbf{EFM}_1 = \begin{pmatrix} -1 \\ 0 \\ 4.7/4 \\ 1 \end{pmatrix}, \quad \mathbf{EFM}_2 = \begin{pmatrix} -1 \\ 1 \\ 3/4 \\ 0 \end{pmatrix},$$

that encode the two phototroph elementary flux modes, recall Figure 3, with vector entries describing changes to concentrations of the corresponding external quantities. Perturbation of the viable steady state by increasing or decreasing flux through the photosynthesis-driven biosynthesis mode corresponds to perturbation of the viable steady state solution in the direction $\mathbf{EFM}_1$ (one Cmole biomass and 4.7/4 Omoles produced per Cmole inorganic carbon consumed) and, likewise, perturbation of the viable steady state by increasing or decreasing flux through the photorespiration mode corresponds to perturbation of the viable steady state solution in the direction $\mathbf{EFM}_2$ (one Cmole organic carbon and 3/4 Omoles produced per Cmole inorganic carbon consumed).

The two eigenvalues of $J^{(1)}|_{P_1>0}$ that correspond to eigenvectors in the null space of A are given by

$$\lambda_{2,3} = \frac{1}{2}(((fg_1)_{O_2} - g_{1,\text{IC}})P_1 - 2D + \eta g_1) \qquad (A12)$$

$$\pm \frac{1}{2}\left[(((fg_1)_{O_2} - g_{1,\text{IC}})P_1 + \eta g_1)^2 - (4\eta_{\text{IC}} - 3\eta_{O_2})g_1^2 P_1 - \frac{17}{10}(\eta_{\text{IC}}g_{1,O_2} - \eta_{\text{IC}}g_{1,O_2})g_1 P_1 \right]^{1/2}$$

and will be discussed for particular steady states below.

Washout State (One Species System). For the washout state ($P_1 = 0$), (A10) becomes

$$J^{(1)}\big|_{P_1=0} = \begin{pmatrix} -D & 0 & 0 & -g_1 \\ 0 & -D & 0 & (1-\eta)g_1 \\ 0 & 0 & -D & f(\eta)g_1 \\ 0 & 0 & 0 & \eta g_1 - D \end{pmatrix} \qquad (A13)$$

with g_1 and η evaluated at $IC = IC_0$ and $O_2 = O_{2,0}$. Note that (A13) has eigenvalues $\lambda_1 = -D < 0$ with multiplicity 3 and $\lambda_2 = \eta g_1 - D$, seen directly or by setting $P_1 = 0$ in (A13). Hence the washout state is stable if $\lambda_2 < 0$ and unstable if $\lambda_2 > 0$. Note that $\lambda_2 = \eta(IC_0, O_{2,0})g_1(IC_0, O_{2,0}) - D$ is the net intrinsic biomass production rate at inflow conditions.

The eigenspaces for (A13) are

$$E_1^0 = \text{span}\left\{ \begin{pmatrix} 1 \\ 0 \\ 0 \\ 0 \end{pmatrix}, \begin{pmatrix} 0 \\ 1 \\ 0 \\ 0 \end{pmatrix}, \begin{pmatrix} 0 \\ 0 \\ 1 \\ 0 \end{pmatrix} \right\}, \quad E_2^0 = \text{span}\left\{ \begin{pmatrix} -1 \\ 1-\eta \\ f(\eta) \\ \eta \end{pmatrix} \right\},$$

with $\eta = \eta(IC_0, O_{2,0})$, where superscript 0 indicates the washout state, and subscript j indicates eigenvalue λ_j. Perturbations of dissolved chemical concentrationss only, i.e., perturbations contained in the eigenspace E_1^0, decay at rate D since they are simply washed out of the chemostat. We can call E_1^0 the washout space. When a small quantity of cyanobacteria are added to the system, in the unstable case $\lambda_2 > 0$, dynamics of the linearized system thus effectively reduce to exponential growth on the one dimensional space E_2^0, with P_1, OC, and O_2 growing and IC decaying, in relative ratios as indicated by the entries of the eigenvector $\mathbf{v}_2$ for λ_2, where $\mathbf{v}_2$ is the basis vector shown above for R_2^0. Thus, the resulting invasion of cyanobacteria is accompanied by decrease in inorganic carbon concentration and increase in organic carbon and oxygen concentrations. Note that eigenvector $\mathbf{v}_2 = \eta\,\mathbf{EFM}_1 + (1-\eta)\,\mathbf{EFM}_2$, indicating that the linearized growth dynamics occurs, as to be expected, as a combination of the photosynthesis mode and the photorespiration mode weighted by the branching parameter $\eta(IC_0, O_{2,0})$.

Viable State (One Species System). For the viable state ($P_1 > 0$),

$$J^{(1)}\big|_{P_1>0} = \begin{pmatrix} -g_{1,IC}P_1 - D & 0 & -g_{1,O_2}P_1 & -g_1 \\ ((1-\eta)g_1)_{,IC}P_1 & -D & ((1-\eta)g_1)_{,O_2}P_1 & (1-\eta)g_1 \\ (f(\eta)g_1)_{,IC}P_1 & 0 & (f(\eta)g_1)_{,O_2}P_1 - D & f(\eta)g_1 \\ (\eta g_1)_{,IC}P_1 & 0 & (\eta g_1)_{,O_2}P_1 & \eta g_1 - D \end{pmatrix} \qquad (A14)$$

is evaluated at the viable state values of IC, O_2, and P_1. $J^{(1)}\big|_{P_1>0}$ has $\lambda_1 = -D$ as a multiplicity 2 eigenvalue with

$$E_1^1 = \text{span}\left\{ \begin{pmatrix} 0 \\ 1 \\ 0 \\ 0 \end{pmatrix}, \begin{pmatrix} -g_1\eta_{,O_2} \\ 0 \\ g_1\eta_{,IC} \\ (g_{1,IC}\eta_{,O_2} - g_{1,O_2}\eta_{,IC})P_1 \end{pmatrix} \right\}.$$

where superscript 1 refers to the viable state and subscript 1 to eigenvalue λ_1. As noted previously, dynamics of (A9) include the null space of A, recall (A11), as an invariant region, with components of the solution outside of this region damped at rate e^{-Dt}. Note that $E_1^1 \cap \text{null}(A) = \{\mathbf{0}\}$; E_1^1 can be considered to be the washout space. Decomposing $J^{(1)}\big|_{P_1>0} = K^{(1)}\big|_{P_1>0} - DI$ where $K^{(1)}\big|_{P_1>0}$ can be regarded as the kinetics portion of $J^{(1)}\big|_{P_1>0}$, note that eigenspace E_1^1 is the null space of $K^{(1)}\big|_{P_1>0}$ and can hence be interpreted as the space of community-level kinetically neutral perturbations.

Dynamics in the null space of A are characterized by the basis formed by the two mode vectors $\mathbf{EFM}_1$ and $\mathbf{EFM}_2$ as well as the two eigenvalues (A13). Using the viable state condition $\eta g_1 = D$, (A13) reduces to

$$\lambda_{2,3} = \frac{1}{2}(((fg_1)_{O_2} - g_{1,IC})P_1 - D)$$

$$\pm \frac{1}{2}\left[(((fg_1)_{O_2} - g_{1,IC})P_1 + D)^2 - (4\eta_{IC} - 3\eta_{O_2})g_1^2 P_1 - \frac{17}{10}(\eta_{IC}g_{1,O_2} - \eta_{IC}g_{1,O_2})g_1 P_1\right]^{1/2}$$

Note that the real parts of both λ_1 and λ_2 are negative under the assumptions that derivatives with respect to IC are non-negative, derivatives with respect to O_2 are non-positive, and condition (17) holds, i.e., the viable state, when it exists, is stable under the conditions that we consider biologically reasonable.

Next we present stability analyses for steady states of the two species system (38)–(42), which has three types of steady state solutions: the washout solution with $P_1 = P_2 = 0$, the single species solution with $P_1 > 0$, $P_2 = 0$, and the coexistence solution $P_1, P_2 > 0$. We present stability analyses only for the washout and single species states. (The coexistence steady state was explored numerically instead). Note that if $P_2 = 0$ then (38)–(41) reduce, essentially, to (28)–(31), so that steady states for the washout and single species systems are the same as previously (with the addition that $P_2 = 0$), though their stability status in principle might be different. System (38)–(41) linearizes to

$$\frac{d}{dt}\begin{pmatrix} \tilde{IC}(t) \\ \tilde{OC}(t) \\ \tilde{O}_2(t) \\ \tilde{P}_1(t) \\ \tilde{P}_2(t) \end{pmatrix} = J^{(2)}(IC^*, OC^*, O_2^*, P_1^*, P_2^*)\begin{pmatrix} \tilde{IC}(t) \\ \tilde{OC}(t) \\ \tilde{O}_2(t) \\ \tilde{P}_1(t) \\ \tilde{P}_2(t) \end{pmatrix} \qquad (A15)$$

In the cases under consideration of steady state solutions with $P_2^* = 0$, the Jacobian matrix takes the form

$$J^{(2)} = \begin{pmatrix} -g_{1,IC}P_1 - D & 0 & -g_{1,O_2}P_1 & -g_1 & g_2 \\ ((1-\eta)g_1)_{,IC}P_1 & -D & ((1-\eta)g_1)_{,O_2}P_1 & (1-\eta)g_1 & -2g_2 \\ (f(\eta)g_1)_{,IC}P_1 & 0 & (f(\eta)g_1)_{,O_2}P_1 - D & f(\eta)g_1 & -\frac{1.9}{4}g_2 \\ (\eta g_1)_{,IC}P_1 & 0 & (\eta g_1)_{,O_2}P_1 & \eta g_1 - D & 0 \\ 0 & 0 & 0 & 0 & g_2 - D \end{pmatrix} = \begin{pmatrix} & & & & g_2 \\ & & J^{(1)} & & -2g_2 \\ & & & & -\frac{1.9}{4}g_2 \\ & & & & 0 \\ 0 & 0 & 0 & 0 & g_2 - D \end{pmatrix}$$

Much of our stability results for the one species case are still of use here. Note that $J^{(2)}$ shares the same eigenvalues (and multiplicities) as $J^{(1)}$ with the addition of an extra eigenvalue $g_2 - D$. Eigenvectors of $J^{(1)}$ are also eigenvectors of $J^{(2)}$, corresponding to the same eigenvalues, with a 0 in the fifth component corresponding to P_2 concentration perturbations. The only remaining item to be determined is the eigenvector corresponding to the new eigenvalue $g_2 - D$.

Proceeding as in the one species case, from (26) and (27)

$$\frac{d}{dt}\begin{pmatrix} \tilde{C} \\ \tilde{DoR} \end{pmatrix} = \frac{d}{dt}\begin{pmatrix} \widetilde{IC} + \widetilde{OC} + \tilde{P}_1 + \tilde{P}_2 \\ 3\widetilde{OC} - 4\tilde{O}_2 + 4.7\tilde{P}_1 + 4.1\tilde{P}_1 \end{pmatrix}$$

$$= \begin{pmatrix} 1 & 1 & 0 & 1 & 1 \\ 0 & 3 & -4 & 4.7 & 4.1 \end{pmatrix} \frac{d}{dt}\begin{pmatrix} \widetilde{IC} \\ \widetilde{OC} \\ \tilde{O}_2 \\ \tilde{P}_1 \\ \tilde{P}_2 \end{pmatrix} = -D \begin{pmatrix} \tilde{C} \\ \tilde{DoR} \end{pmatrix}$$

Hence, for any perturbation, its component normal to the null space of

$$B = \begin{pmatrix} 1 & 1 & 0 & 1 & 1 \\ 0 & 3 & -4 & 4.7 & 4.1 \end{pmatrix} \tag{A16}$$

is damped (eventually) as e^{-Dt}, that is, excess or deficience in the initial perturbation of total carbon and degree of reduction is removed from the system through outflow on the chemostat turnover time scale. Note again, thus, that we can therefore characterize the dynamics described by the five dimensional system (A15) if we can characterize the dynamics on a three dimensional subspace consisting of the null space of B, i.e., the subspace defined by $\tilde{C} = 0$, $\widetilde{DoR} = 0$ (no net perturbation of total carbon or degree of reduction).

Continuing to proceed as before, we note that the null space of B can be interpreted as the two species flux mode space and is spanned by the vectors

$$\mathbf{EFM}_1 = \begin{pmatrix} -1 \\ 0 \\ 4.7/4 \\ 1 \\ 0 \end{pmatrix}, \quad \mathbf{EFM}_2 = \begin{pmatrix} -1 \\ 1 \\ 3/4 \\ 0 \\ 0 \end{pmatrix}, \quad \mathbf{EFM}_3 = \begin{pmatrix} 1 \\ -2 \\ -1.9/4 \\ 0 \\ 1 \end{pmatrix},$$

that encode the effect of the three elementary flux modes shown in Figure 3 on external concentrations. As before, perturbation by increasing or decreasing flux through the photosynthesis-driven biosynthesis mode corresponds to perturbation in the direction $\mathbf{EFM}_1$ (one Cmole biomass and 4.7/4 Omoles produced per Cmole inorganic carbon consumed) and perturbation by increasing or decreasing flux through the photorespiration mode corresponds to perturbation in the direction $\mathbf{EFM}_2$ (one Cmole organic carbon and 3/4 Omoles produced per Cmole inorganic carbon consumed). The new vector $\mathbf{EFM}_3$ corresponds to perturbation that increases or decreases flux through the heterotroph biosynthesis mode (one Cmole biomass and 1 Cmole inorganic carbon produced per two Cmoles organic carbon and 1.9/4 Omoles consumed).

Washout State (Two Species System). For the two species washout state ($P_1 = P_2 = 0$) along with $IC = IC_0$, $OC = 0$, $O_2 = O_{2,0}$,

$$J^{(2)}|_{P_1,P_2=0} = \begin{pmatrix} -D & 0 & 0 & -g_1 & 0 \\ 0 & -D & 0 & (1-\eta)g_1 & 0 \\ 0 & 0 & -D & f(\eta)g_1 & 0 \\ 0 & 0 & 0 & \eta g_1 - D & 0 \\ 0 & 0 & 0 & 0 & -D \end{pmatrix} \tag{A17}$$

(note from (20) that $g_2|_{OC=0} = 0$) with g_1 and η evaluated at IC $=$ IC$_0$, OC $= 0$, O$_2$ $=$ O$_{2,0}$. Note that (A17) has, in common with (A13), eigenvalues $\lambda_1 = -D < 0$ (with multiplicity 4) and $\lambda_2 = \eta g_1 - D$. Hence, again, the washout state is stable if $\lambda_2 < 0$ and unstable if $\lambda_2 > 0$. The eigenspaces for (A17) are

$$
E_1^0 = \text{span} \left\{ \begin{pmatrix} 1 \\ 0 \\ 0 \\ 0 \\ 0 \end{pmatrix}, \begin{pmatrix} 0 \\ 1 \\ 0 \\ 0 \\ 0 \end{pmatrix}, \begin{pmatrix} 0 \\ 0 \\ 1 \\ 0 \\ 0 \end{pmatrix}, \begin{pmatrix} 0 \\ 0 \\ 0 \\ 0 \\ 1 \end{pmatrix} \right\}, \quad E_2^0 = \text{span} \left\{ \begin{pmatrix} -1 \\ 1 - \eta \\ f(\eta) \\ \eta \\ 0 \end{pmatrix} \right\}.
$$

Perturbations without introduction of phototrophs decay at rate $-D$. As before, when a small quantity of phototrophs are added to the system, in the unstable case, dynamics effectively reduce to exponential growth on the one dimensional space E_2^0, with P_1, OC, and O$_2$ growing and IC decaying, in relative ratios as indicated by the entries of the eigenvector for λ_2. Note in particular that the P_2-component of the λ_2-eigenvector is zero, indicating that the heterotroph is unable to invade. That is, the heterotroph requires an already established population of phototrophs (with corresponding finite supply of organic carbon) before it can become viable. Note as before that λ_2-eigenvector can be written $\eta \, \mathbf{EFM}_1 + (1 - \eta) \, \mathbf{EFM}_2$, indicating again that the linearized growth dynamics occurs as a combination of the photosynthesis mode and the photorespiration mode weighted by the branching parameter $\eta(\text{IC}_0, \text{O}_{2,0})$.

Single Species State: Invasion (Two Species System). The Jacobian matrix for the base steady state is

$$
J^{(2)} = \begin{pmatrix} & & & & g_2 \\ & J^{(1)}|_{P_1 > 0} & & & -2g_2 \\ & & & & -\frac{1.9}{4}g_2 \\ & & & & 0 \\ 0 \;\; 0 \;\; 0 \;\; 0 & & & & g_2 - D \end{pmatrix} \tag{A18}
$$

where $J^{(1)}|_{P_1 > 0}$ is as given in (A14) and, in addition, g_2 is evaluated at OC* and O$_2^*$. As previously, the eigenvalues of $J^{(1)}|_{P_1 > 0}$ are also eigenvalues of $J^{(2)}$ with identical eigenspaces, except with zeros in the new, fifth component of the two species system corresponding to perturbations in the P_2 component. Hence the dynamics in those eigenspaces are independent of perturbation to $\widetilde{P}_2$, and, by assumption, the dynamics on those eigenspaces are stable. The new eigenvalue is $\lambda_4 = g_2 - D$ with eigenspace $E_4^1 = \text{span}\{\mathbf{v}_4\}$, where $\mathbf{v}_4 \neq \mathbf{0}$ satisfied $J^{(2)}\mathbf{v}_4 = \lambda_4\mathbf{v}_4$. Note that $\mathbf{v}_4$ is necessarily a linear combination of the three mode vectors $\mathbf{EFM}_1$, $\mathbf{EFM}_2$, and $\mathbf{EFM}_3$. It is easily seen in the case of large $g_2(\text{OC}^*, \text{O}_2^*)$ that $\mathbf{v}_4 \cong \mathbf{EFM}_3$, that is, the instability dynamics are dominated by the heterotrophic growth mode. When growth is not as dominant, the relative role of phototroph flux modes in maintaining carbon and DoR balance is more significant.

Appendix E. Optimization in the Single Species Chemostat With Respect to Affinity

We consider a unique, viable solution $(\text{IC}^*, \text{OC}^*, \text{O}_2^*, \text{P}_1^*)$ to Equations (A1)–(A4), under the assumptions of Appendix C, as a function of affinity parameter γ_1. In particular, we show that $(d/d\gamma_1)\,\text{P}_1^* < 0$ for $\gamma_1 > 0$, i.e., steady state biomass increases with decreasing γ_1 To do so, we compute the variation with respect to γ_1 of the solution to Equations (A5) and (A6). In particular, perturbing $\gamma_1 \to \gamma_1 + \Delta\gamma_1$, then perturbed quantities $\text{IC}^* + \widetilde{\text{IC}}\,\Delta\gamma_1$ and $\text{O}_2^* + \widetilde{\text{O}}_2\,\Delta\gamma_1$ satisfy, to linear order,

$$
\begin{pmatrix} A & B \\ C & D \end{pmatrix} \begin{pmatrix} \widetilde{\text{IC}} \\ \widetilde{\text{O}}_2 \end{pmatrix} = \begin{pmatrix} E \\ F \end{pmatrix}, \tag{A19}
$$

where

$$\begin{aligned}
A &= (\eta g_1)^*_{,\text{IC}}, \\
B &= (\eta g_1)^*_{,\text{O}_2}, \\
C &= f'(\eta^*)\eta^*_{,\text{IC}}(\text{IC}_0 - \text{IC}^*) - f(\eta^*), \\
D &= f'(\eta^*)\eta^*_{,\text{O}_2}(\text{IC}_0 - \text{IC}^*) - 1, \\
E &= -\eta^*_{,\gamma_1} g_1^* \\
F &= -f'(\eta^*)\eta^*_{,\gamma_1}(\text{IC}_0 - \text{IC}^*).
\end{aligned}$$

Here, superscript $*$ corresponds to evaluation at $\text{IC}=\text{IC}^*, \text{O}_2=\text{O}_2^*$. System (A19) has solution

$$\begin{pmatrix} \widetilde{\text{IC}} \\ \widetilde{\text{O}}_2 \end{pmatrix} = \frac{1}{AD - BC} \begin{pmatrix} DE - BF \\ AF - CE \end{pmatrix}. \tag{A20}$$

Computing,

$$AD - BC = (\eta^*_{,\text{O}_2} g_{1,\text{IC}} - \eta^*_{,\text{IC}} g_{1,\text{O}_2}) f'(\eta^*)\eta^*(\text{IC}_0 - \text{IC}) + g_1^*(f(\eta^*)\eta^*_{,\text{O}_2} - \eta^*_{,\text{IC}}) + \eta^*(f(\eta^*)g^*_{1,\text{O}_2} - g^*_{1,\text{IC}})$$

which, assuming stability condition (17), is strictly negative. Hence, (A20) is the unique solution to (A19).

To compute the variation of $\widetilde{P}_1$, we take the the variation of Equation (A7) and, using solution (A20), obtain after some computation

$$\begin{aligned}
\frac{d}{d\gamma_1}P_1^* &= (\eta^*_{,\gamma_1} + \eta^*_{,\text{IC}}\text{IC}^*_{,\gamma_1} + \eta^*_{,\text{O}_2}\text{O}^*_{2,\gamma_1})(\text{IC}_0 - \text{IC}^*) - \eta^*\text{IC}^*_{,\gamma_1} \\
&= -\eta^*\eta^*_{,\gamma_1}\frac{\text{IC}_0 - \text{IC}^*}{AD - BC}\left(g^*_{1,\text{IC}} - \frac{3}{4}g^*_{1,\text{O}_2} + \frac{g_1^*}{\text{IC}^* - \text{IC}_0}\right) \\
&< 0,
\end{aligned}$$

as was to be shown.

Appendix F. Invasion Eigenvector

The eigenvector $\mathbf{v}_4$ for the invasion dynamics matrix (A18) can be computed from row reducing the equation $J^{(2)} - \lambda_4 I = 0$, leading to the diagonal system (solvable by back-substitution) for $\mathbf{v}_4 = (ic, oc, o_2, p_1, p_2)$

$$\begin{aligned}
0 &= (g_{1,\text{IC}}P_1^* + g_2)ic + g_{1,\text{O}_2}P_1^* o_2 + g_1 p_1 + g_2 p_2 \\
0 &= (((1 + \eta)g_1)_{\text{IC}}P_1^* + 2g_2)ic + g_2 oc + ((1 + \eta)g_1)_{\text{O}_2}o_2 + (1 + \eta)g_1 p_1 \\
0 &= ((Fg_1)_{\text{IC}}P_1 - (19/40)g_2)ic + ((Fg_1)_{\text{O}_2}P_1^* - g_2)o_2 + Fg_1 p_1 \\
0 &= \left((\eta g_1)_{\text{IC}}P_1^* + \frac{g_2 - \eta g_1}{Fg_1}((Fg_1)_{\text{IC}}P_1^* - (19/40)g_2)\right)ic \\
&\quad + \left((\eta g_1)_{\text{O}_2}P_1^* + \frac{g_2 - \eta g_1}{Fg_1}((Fg_1)_{\text{O}_2}P_1^* - (19/40)g_2)\right)o_2
\end{aligned}$$

with $F = (11/40) + (17/40)\eta$ and all quantities evaluated at the steady state values $\text{IC}^*, \text{OC}^*, \text{O}_2^*, P_1^*$, as well as $P_2^* = 0$. Recall that all non-differentiated quantities are non-negative, that all derivatives with respect to IC are non-negative (with $g_{1,\text{IC}}$ strictly positive) and all derivatives with respect to O_2 are non-positive, and that, evaluated at the starred quantities, $\eta g_1 = D$.

If the eigenvalue $g_2 - D > 0$, with g_2 evaluated at steady state values, then the phototroph-only steady state is unstable to perturbations along eigenvector $\mathbf{v}$. The case $g_2^* \gg D = \eta^* g_1^*$ (with $\eta^* = \eta(\mathrm{IC}^*, \mathrm{O}_2^*)$, $g_1^* = g_1(\mathrm{IC}^*, \mathrm{O}_2^*)$, $g_2^* = g_2(\mathrm{OC}^*, \mathrm{O}_2^*)$), i.e., heterotroph growth time scale short in comparison to washout time, is informative. Expanding all quantities in powers of $\eta^* g_1^* / g_2^*$, e.g.,

$$\mathrm{IC}(t) = \mathrm{IC}^{(0)}(t) + \frac{\eta^* g_1^*}{g_2^*} \mathrm{IC}^{(1)}(t) + \left(\frac{\eta^* g_1^*}{g_2^*} \right)^2 \mathrm{IC}^{(2)}(t) + \ldots,$$

and similarly for OC, O_2, P_1, and P_2, we can apply standard asymptotic methods to approximate solutions order by order. To leading order, we find

$$
\begin{aligned}
\mathrm{OC}^{(0)} &= -2\mathrm{IC}^{(0)} \\
\mathrm{O}_2^{(0)} &= -\frac{19}{40}\mathrm{IC}^{(0)} \\
\mathrm{P}_1^{(0)} &= 0 \\
\mathrm{P}_2^{(0)} &= \mathrm{IC}^{(0)}
\end{aligned}
$$

Note that during the transient period of the initial invasion, intuition for heterotroph benefit to phototrophs holds: introduction of heterotrophs, i.e., $\mathrm{P}_2^{(0)} > 0$, results in increase in inorganic carbon concentration, i.e., $\mathrm{IC}^{(0)} > 0$, and decrease in organic carbon and oxygen concentrations, i.e., $\mathrm{OC}^{(0)}$, $\mathrm{O}_2^{(0)} < 0$. Note that these perturbations are consistent with the stoichiometry of $\mathbf{EFM}_3$, and that there is no effect of phototroph population at this order, a consequence of the $g_2^* \gg \eta^* g_1^*$ asymptotics, but at the next order,

$$\mathrm{P}_1^{(1)} = \left((\eta^* g_1^*)_{\mathrm{IC}} - \frac{19}{40}(\eta^* g_1^*)_{\mathrm{O}_2} \right) \frac{\mathrm{P}_1^*}{\eta^* g_1^*} \mathrm{P}_2^{(0)}.$$

so that $\mathrm{P}_1^{(1)}$ is positive if $\mathrm{P}_2^{(0)} > 0$, i.e., phototroph population biomass increases with introduction of heterotrophs in the transitory invasion period.

References

1. Villa, F.; Pitts, B.; Lauchnor, E.G.; Cappitelli, F.; Stewart, P.S. Development of a laboratory model of a phototroph-heterotroph mixed-species biofilm at the stone/air interface. *Front. Microbiol.* **2015**, *6*, 1251.
2. Croft, M.T.; Lawrence, A.D.; Raux-Deery, E.; Warren, M.J.; Smith, A.G.. Algae acquire vitamin B_{12} through a symbiotic relationship with bacteria. *Nature* **2005**, *438*, 90–93.
3. Palsson, B.Ø., *Systems Biology: Simulation of Dynamics Network States*; Cambridge University Press: Cambridge, UK, 2011.
4. Klamt, S.; Stelling, J. Two approaches for metabolic pathway analysis? *Trends Biotechnol.* **2003**, *21*, 64–69.
5. Schilling, C.H.; Edwards, J.S.; Letscher, D.; Palsson, B.O. Combining pathway analysis with flux balance analysis for the comprehensive study of metabolic systems. *Biotechnol. Bioeng.* **2001**, *71*, 286–306.
6. Schuster, S.; Hilgetag, C. On elementary flux modes in biochemical reaction systems at steady state. *J. Biol. Syst.* **1994**, *2*, 165–182.
7. Carlson, R.P. Decomposition of complex microbial behaviors into resource-based stresses. *Bioinformatics* **2009**, *25*, 90–97.
8. Taffs, R.; Aston, J.E.; Brileya, K.; Jay, Z.; Klatt, C.G.; McGlynn, S.; Mallette, N.; Montross, S.; Gerlach, R.; Inskeep, W.P.; et al. In silico approaches to study mass and energy flows in microbial consortia: A syntrophic case study. *BMC Syst. Biol.* **2009** *3*, 114.
9. Phalak, P.; Chen, J.; Carlson, R.P.; Henson, M.A. Metabolic modeling of a chronic wound biofilm consortium predicts spatial partitioning of bacterial species. *BMC Syst. Biol.* **2016**, *10*, 90.
10. Smith, H. L.; Waltman, P. *The Theory of the Chemostat: Dynamics of Microbial Competition*; Cambridge University Press: Cambridge, UK, 1995.
11. Falkowski, P.G.; Raven, J.A. *Aquatic Photosynthesis*; Princeton University Press: Princeton, NJ, USA, 2007.

12. Beck, A.E.; Bernstein, H.C.; Carlson, R.P. Stoichiometric network analysis of cyanobacterial acclimation to photosynthesis-associated stresses identifies heterotrophic niches. *Processes* **2017**, submitted.

13. Knoop, H.; Grundel, M.; Zilliges, Y.; Lehmann, R.; Hoffmann, S.; Lockau, W.; Steuer, R. Flux balance analysis of cyanobacterial metabolism: The metabolic network of *Synechocystis* sp. PCC 6803. *PloS Comput. Biol.* **2013**, *9*, e1003081.

14. Nogales, J.; Gudmundsson, S.; Knight, E.M.; Palsson, B.O.; Thiele, I. Detailing the optimality of photosynthesis in cyanobacteria through systems biology analysis. *Proc. Natl. Acad. Sci. USA* **2012**, *109*, 2678–2683.

15. Vu, T.T.; Stolyar, S.M.; Pinchuk, G.E.; Hill, E.A.; Kucek, L.A.; Brown, R.N.; Lipton, M.S.; Osterman, A.; Fredrickson, J.K.; Konopka, A.E.; et al. Genome-scale modeling of light-driven reductant partitioning and carbon fluxes in diazotrophic unicellular cyanobacterium *Cyanothece* sp. ATCC 51142. *PloS Comput. Biol.* **2012**, *8*, e1002460.

16. Henry, C.S.; Bernstein, H.C.; Weisenhorn, P.; Taylor, R.C.; Lee, J.Y.; Zucker, J.; Song, H.S. Microbial community metabolic modeling: A community data-driven network reconstruction. *J. Cell. Physiol.* **2016**, *231*, 2339–2345.

17. Perez-Garcia, O.; Lear, G.; Singhal, N. Metabolic network modeling of microbial interactions in natural and engineered environmental systems. *Front. Microbiol.* **2016**, *7*, 673.

18. Roels, J.A. Application of macroscopic principles to microbial metabolism. *Biotechnol. Bioeng.* **1980**, *22*, 2457–2514.

19. White, D. *The Physiology and Biochemistry of Prokaryotes*, 3rd ed.; Oxford University Press: Oxford, UK, 2007.

20. Mangam, N.M.; Brenner, M.P. Systems analysis of the CO_2 concentrating mechanism in cyanobacteria. *eLIFE* **2014**, *3*, e02043.

21. Rae, B.D.; Long, B.M.; Whitehead, L.F.; Forster, B.; Badger, M.R.; Price, G.D. Cyanobacterial carboxysomes: Microcompartments that facilitate CO_2 fixation. *J. Mol. Microbiol. Biotechnol.* **2013**, *23*, 300–307.

22. Asada, K. The water-water cycle in chloroplasts: scavenging of active oxygens and dissipation of excess photons. *Annu. Rev. Plant Physiol. Plan Mol. Biol.* **1999**, *50*, 601–639.

23. Eilers, P.H.C.; Peeters, J.C.H. A model for the relationship between Light- Intensity and the rate of photosynthesis in Phytoplankton. *Ecol. Model.* **1988**, *42*, 199–215.

24. Wolf, G.; Picioreanu, C.; van Loosdrecht, M.C.M. Kinetic modeling of phototrophic biofilms—The PHOBIA model. *Biotechnol. Bioeng.* **2007**, *97*, 1064–1079.

25. Bailey, S.; Grossman, A. Photoprotection in cyanobacteria: Regulation of light harvesting. *Photochem. Photobiol.* **2008**, *84*, 1410–1420.

26. Gubernator, B.; Bartoszewski, R.; Kroliczewski, J.; Wildner, G.; Szczepaniak, A. Ribulose-1,5-bisphosphate carboxylase/oxygenase from thermophilic cyanobacterium *Thermosynechococcus elongatus*. *Photosynth. Res.* **2008**, *95*, 101–109.

27. Jordan, D.B.; Ogren, W.L. Species variation in the specificity of ribulose biphosphate carboxylase/oxygenase. *Nature* **1981**, *291*, 513–515.

28. Eisenhut, M.; Ruth, W.; Haimovich, M.; Bauwe, H.; Kaplan, A.; Hagemann, M. The photorespiratory glycolate metabolism is essential for cyanobacteria and might have been conveyed endosymbiontically to plants. *Proc. Natl. Acad. Sci. USA* **2008**, *105*, 17199–17204.

29. Miller, A. G.; Turpin, D. H.; Canvin, D. T. Growth and photosynthesis of the cyanobacterium *Synechococcus leopoliensis* in HCO_3^--limited chemostats. *Plant Physiol.* **1984**, *75*, 1064–1070.

30. Rohr, R.P.; Saavedra, S.; Bascompte, J. On the structural stability of mutualistic systems. *Nature* **2014**, *345*, 1253497.

31. Mayberry, W.R.; Prochacka, G.J.; Payne, W.J., Factors derived from studies of aerobic growth in minimal media. *J. Bacteriol.* **1968**, *96*, 1424–1426.

32. Benschop, J.J.; Badger, M.R.; Price, G.D.. Characterisation of CO_2 and HCO_3^- uptake in the cyanobacterium *Synechocystis* sp. PCC6803. *Photosynth. Res.* **2003**, *77*, 117–126.

33. Badger, M.R.; Hanson, D.; Price, G.D. Evolution and diversity of CO_2 concentrating mechanisms in cyanobacteria. *Funct. Plant Biol.* **2002**, *29*, 161–173.

34. Al-Najjar, M.A.A.; de Beer, D.; Jørgensen, B.B.; Kühl, M.; Polerecky, L. Conversion and conservation of light energy in a photosynthetic microbial mat ecosystem. *ISME J.* **2010**, *4*, 440–449.

35. Chen, J.; Tannahill, A.L.; Shuler, M.L. Design of a system for the control of low dissolved oxygen concentrations: critical oxygen concentrations for *Azotobacter vinelandii* and *Escherichia coli*. *Biotechnol. Bioeng.* **1985**, *27*, 151–155.
36. Shaler T.A.; Klecka G.M. Effects of dissolved oxygen concentration on biodegradation of 2,4-dichlorophenoxyacetic acid. *Appl. Environ. Microbiol.* **1986**, *51*, 950–955.
37. Füchslin, H.P.; Schneider, C,; Egli, T. In glucose-limited continuous culture the minimum substrate concentration for growth, Smin, is crucial in the competition between the enterobacterium *Escherichia coli* and *Chelatobacter heintzii*, an environmentally abundant bacterium. *ISME J.* **2012**, *6*, 777–789.
38. Nowack, S.; Olsen, M.T.; Schaible, G.A.; Becraft, E.D.; Shen, G.; Klapper, I.; Bryant, D.A.; Ward, D.M. The molecular dimension of microbial species: 2. *Synechococcus* strains representative of putative ecotypes inhabiting different depths in the Mushroom Spring microbial mat exhibit different adaptive and acclimative responses to light. *Front. Microbiol.* **2015**, *6*, 626.

4

An Integrated Mathematical Model of Microbial Fuel Cell Processes: Bioelectrochemical and Microbiologic Aspects[†]

Andrea G. Capodaglio [1,*] **, Daniele Cecconet** [1] **and Daniele Molognoni** [1,2]

[1]   DICAr, University of Pavia, 27100 Pavia, Italy; daniele.cecconet@unipv.it (D.C.);
    dmolognoni@leitat.org (D.M.)

[2]   LEITAT Technological Centre, 08225 Terrassa, Barcelona, Spain

*   Correspondence: capo@unipv.it

†   This paper is an extended version of our paper published in the 29th European Conference on Modelling
    and Simulation as titled "Formulation and Preliminary Application of an Integrated Model of Microbial Fuel
    Cell Processes".

Abstract: Microbial Fuel Cells (MFCs) represent a still relatively new technology for liquid organic waste treatment and simultaneous recovery of energy and resources. Although the technology is quite appealing due its potential benefits, its practical application is still hampered by several drawbacks, such as systems instability (especially when attempting to scale-up reactors from laboratory prototypes), internally competing microbial reactions, and limited power generation. This paper is an attempt to address some of the issues related to MFC application in wastewater treatment with a simulation model. Reactor configuration, operational schemes, electrochemical and microbiological characterization, optimization methods and modelling strategies were reviewed and have been included in a mathematical simulation model written with a multidisciplinary, multi-perspective approach, considering the possibility of feeding real substrates to an MFC system while dealing with a complex microbiological population. The conclusions drawn herein can be of practical interest for all MFC researchers dealing with domestic or industrial wastewater treatment.

Keywords: microbial fuel cells; biolectrochemical systems; mathematical model; heterotrophic bacteria; methanogenic archaea; exoelectrogenic bacteria; complex substrate

1. Introduction

Depletion of fossil fuel reserves and global warming concerns make it necessary to develop alternative, climate-neutral technologies for energy production; not just employing traditional renewable sources (solar, wind, etc.), but also tapping into non-conventional ones, such as wastes of different origin, to achieve established targets [1,2]. Renewable bioenergy from wastes, presenting a neutral or even negative carbon footprint, is also viewed as one of the ways to alleviate the climate change crisis [3–5]. In this context, a specific research track, concerning Microbial Electrochemical Technologies (METs), has been pursued by scientists for the last couple of decades [6–8]. METs in their generality represent a technology concerned with the recovery of energy and resources from waste streams [9], and comprise various sub-systems, targeted to different objectives, including microbial fuel cells (MFCs). These are a class of bioelectrochemical systems that directly transform the chemical energy contained in bioconvertible organic matter substrates into electrical energy, exploiting the biocatalytic effect of specific electroactive bacteria (EAB), acting on one or both reactions of substrate oxidation and oxidant reduction, composing a classical redox reaction [7,8]. When wastewater containing organic matter is used as anode fuel, a MFC effectively removes the latter, while recovering energy, leading to the future possibility of designing energy-producing wastewater treatment plants (WWTPs), or Water and Resource Recovery Facilities (WRRFs), as the new terminology is starting to denote such installations. It has, in fact, been estimated that urban wastewater contains more than 9 times the amount of energy currently consumed to treat it with state-of-the-art processing technology [10], while the best current treatment technologies allow the recovery of 1/4th to 1/3rd, at most, of that energy. MFCs could be one of the new technologies to increase that fraction, with applications for almost all types of different liquid wastes [11–14].

Bioelectrochemical systems may also have other useful environmental applications: if such processes were applied in their "inverse" configuration, usually referred to as Microbial Electrolytic Cells (MECs), with which they share the general design and basic processes, they could achieve, for example, autotrophic denitrification of contaminated groundwater by externally supplying an adequate voltage to the system [15,16]. Additionally, in this case, these processes turn out to be particularly efficient from an energetic point of view, with lower specific energy consumption that other currently used denitrification systems [17,18].

Practical full-scale MFC application in WRRF design has been long delayed by the instability of full-scale engineered systems, low achieved power densities and output voltages practically achievable so far. Several practical issues remain to be solved before MFC systems could be deemed ready for full-scale applications; among them, reduction of the systems' internal resistance, which would allow higher substrate-electricity conversion rates, cathode technology improvements, efficient, scalable, design, and reduction of electrochemical losses. Undesirable anodic side-reactions, such as methanogenesis, aerobic or anoxic respiration by competitive microorganisms, represent some of the drawbacks of the process, and also need structural address, even though they can be partly limited by appropriate operational strategies [19–21]. Deeper process understanding and its mathematical reproducibility can also play an important role in the quest for improvement of this technology.

Since the mid-90s, researchers have attempted to simulate the bioelectrochemical activity of MFCs, as summarized in Table 1. This table does not consider applications of soft simulation methods such as genetic programming (GP), artificial intelligence (AI), fuzzy logic and neural networks, which are sometimes used as an alternative for deterministic mathematical modeling of complex physical non-linear systems, such as a MFCs [22] or conventional-technology WWTPs [23,24].

Zhang and Halme [25] proposed a simple model based on a single anodic population and focused on the generated power in relationship to substrate concentration and cathodic-chamber mediator. Later, models by Kato Marcus et al. [26] were developed, neglecting the contribution of the mediator, but considering a complex bacterial population composed by exoelectrogen and non-exoelectrogen species. In the same year, Picioreanu et al. [27] proposed a 3-dimensional model considering both adhese and suspended microorganisms. Zheng et al. [28] developed a dual-chamber MFC model that simulated transient conditions, including cathodic compartment reactions, while Pinto et al. [29] published a 2-population, anodic dynamic model representing the competition between exoelectrogens and methanogens. Later, Oliveira et al. [30] proposed a steady-state MFC model, focusing on the effect of some parameters such as: cell temperature, substrate concentration, biofilm thickness and current density.

In order to develop an enhanced model capable of describing complex bacterial communities such as those present in a MFC, as well as the complexity of feed substrates, the Pinto model [26] was integrated with the ASM2d model [31]. The resulting model, an improvement of a previous work [32], is herein discussed, together with its application to longer series of MFC operational data. Results are discussed, confirming the good performance of the new model.

Table 1. Summary of published MFC models.

Model	Compartment	Mediator	Species	Time Resolution	Space Resolution
Zhang e Halme,1995 [25]	Anodic	Yes	Single	Dynamic	1D
Kato Marcus et al., 2007 [26]	Anodic	No	Multiple	Dynamic	1D
Oliveira et al., 2013 [30]	Anodic/cathodic	No	Single	Steady st.	1D
Picioreanu et al., 2007 [27]	Anodic	Yes	Multiple	Dynamic	3D
Zheng et al., 2010 [28]	Anodic/cathodic	No	Single	Dynamic/Steady st.	1D
Pinto et al., 2010 [29]	Anodic	Yes	multiple	Dynamic	1D
Capodaglio et al., 2015 [32]	Anodic	Yes	Multiple	Dynamic	1D

2. MFC Integrated Model

Fuel cells are devices performing a combustion reaction without resorting to thermal processes, thus achieving direct conversion of chemical energy (of a generic "fuel", or "substrate") into electrical energy, through the mediation of exoelectrogenic bacteria that act as catalysers of the half-reaction of substrate oxidation [8,33]. The first evidence of this phenomenon was discovered in 1911 by Potter [33], but very few practical advances were achieved in the field until the first patent of mediator-less MFCs, dated 1999 [34].

The process' working principle relies on splitting the semireactions of oxidation and reduction that make up a typical redox reaction, allowing them to occur in two different compartments: in the anodic compartment, exoelectrogen bacteria catalyse substrate oxidation and transfer the electrons, released from cellular respiratory chain, to an electrode (i.e., anode). Electrons flow through an external electric circuit towards the cathodic compartment, where they reduce the terminal electron acceptor (TEA, usually oxygen) [35]. For each electron released at the anode, an H^+ ion must reach the cathode through the electrolytic in the cell, in order to internally close the circuit and reestablish neutrality. Electrons and protons thus react with oxygen at the cathode, generating H_2O [36].

The maximum current that can be produced by a MFC depends on the actual rate of substrate biodegradation, whereas maximum theoretical cell voltage (also called electromotive force, or *emf*) depends on Gibbs' free energy of the overall reaction, and can be calculated as the difference between the standard reduction potentials of the cathodic oxidant (oxygen) and the chosen anodic substrate, as described by Heijnen [37]. Since the cell's *emf* is a thermodynamic value that does not take into account any internal losses [36], measured current experimental values are always substantially lower than theoretical ones.

2.1. Model Assumptions

The model herein presented is based on the work by Pinto et al. [29], and the authors' previous work [32]; it is, as shown in Table 1, a dynamic, 1-dimensional (completely mixed), multi-species model. Recently, model results have also been tested against a full, separate MFC hydrodynamic study, showing good correlation with experimental observations [38].

The model considers the presence of two distinct microbial populations in the anodic chamber: exoelectrogen (a.k.a. anodophilic bacteria) and methanogenic microorganisms co-existing in competition for available substrate, as observed in previous studies [39]. It is known that methanogens compete with anodophiles for substrate, thus reducing power generation and overall coulombic efficiency (CE) of the cell. The presence of an endogenous mediator, either in reduced or oxidized form, is responsible for the extracellular electronic transfer by exoelectrogenic bacteria. It is assumed that these adhere to the anode as a biofilm, while methanogens can either be suspended or adhese. The model also assumed that dynamics at the cathode's end are non-limiting, and thus not considered for simulation purposes (Figure 1).

This model therefore describes:

-substrate (S_a) oxidation to CO_2 by exoelectrogen bacteria (X_a), with reduction of the mediator:

$$(M_{ox} \rightarrow M_{red}) \tag{1}$$

where M_{ox} and M_{red} represent the oxidized and reduced mediator, respectively.

-mediator reoxydation, with release of free electrons and protons:

$$M_{red} \rightarrow M_{ox} + e^- + H^+ \tag{2}$$

-methane and carbon dioxide production by methanogens:

$$S_a \rightarrow CH_4 + CO_2 \tag{3}$$

where S_a is the substrate, expressed by mass balance Equations (4)–(6):

$$\frac{dS_a}{dt} = -q_a x_a - q_m x_m + D(S_{a0} - S_a) \tag{4}$$

$$\frac{dx_a}{dt} = \mu_a x_a - K_{d,a} x_a - \alpha_a D x_a \tag{5}$$

$$\frac{dx_m}{dt} = \mu_m x_m - K_{d,m} x_m - \alpha_m D x_m \tag{6}$$

where $D = 1/(\mathrm{HRT}_{\mathrm{anode}})$ [$t - 1$]; q_a, q_m substrate conversion rates for exoelectrogens and methanogens; μ_a, μ_m Monod-type growth rates, and K_{da}, α_a bacterial endogenous decay and washout coefficients, respectively.

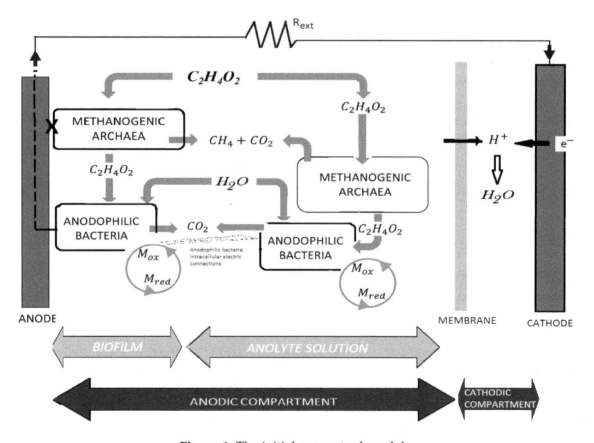

Figure 1. The initial conceptual model.

Monod kinetics are assumed for bacteria; specifically, exoelectrogens' growth is limited by both substrate (acetate) and oxidized mediator concentrations, while methanogens' only by acetate's. The Pinto model assumes that biomass growth occurs in two phases: growth, during which there is no microorganism dispersion/washout ($\alpha_a = 0$); and steady state, where a dynamic equilibrium between growth, endogenous decay and washout is established. An internal "switch" in the model converts between operating phases, depending on process conditions. Total mediator concentration (in reduced and oxidized forms) is assumed constant in the system.

Since one of the most important aspects characterizing MFC performance is the electric current produced, this is calculated from the cell's tension through Ohm's First Law:

$$E_{cell} = I_{MFC} R_{ext} \tag{7}$$

where E_{cell} is the cell's tension, R_{ext} the external resistance and I_{MFC} the current flowing between anode and cathode of the MFC.

The electromotive force (Equation (8)) is considered equal to the Open Circuit Voltage of the cell (E_{OCV}), neglecting activation losses:

$$I_{MFC} = \frac{(E_{OCV} - \eta_{conc})}{(R_{ext} + R_{int})} \left(\frac{M_{red}}{\varepsilon + M_{red}} \right) \tag{8}$$

where η_{conc} is the overpotential linked with concentrations, R_{int} is the internal resistance of the cell.

While competing methane production Q_a is calculated proportionally to acetate uptake through a specific yield coefficient, Y_{CH4}:

$$Q_a = Y_{CH4} q_m X_m V \tag{9}$$

2.2. Model Modification

While the Pinto model represents the ongoing competition between exoelectrogens and methanogens in the anodic chamber, at the same time it completely neglects other species (e.g., heterotrophs) that could be present in the cell, as well (Figure 2). Furthermore, the model considers acetate as the only substrate present, while, in reality, the composition of the incoming substrate will have a much more complex composition (Figure 3). In order to compensate for the above mentioned shortcomings, it was therefore decided to modify the model, by integrating in its structure specific elements of the well-known ASM2d model [32].

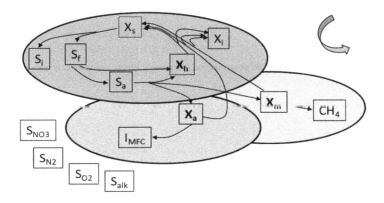

Figure 2. Schematics of the interaction between exoelectrogems (X_a), methanogens (X_m) and heterotrophs (X_h) populations in a MFC as represented in the modified integrated model. The reactions outside the shaded areas (unconnected) are not represented in the model.

The latter was designed to simulate the processes normally occurring in traditional activated sludge facilities, and considers basic substrate measured as COD (Chemical Oxygen Demand), although in diverse fractions, such as particulate (X) and soluble (S), as follows: S_f soluble substrate that can be fermented to S_a (acetate); inert soluble and particulate substrate, S_i and X_i; slowly degradable particulate, X_s; soluble nitrogenous, S_{NO3}, and ammonia, S_{NH4} matter. Ammonia and nitrogenous matter, as well as the influence of alkalinity and oxygen inhibition were, for the moment, neglected.

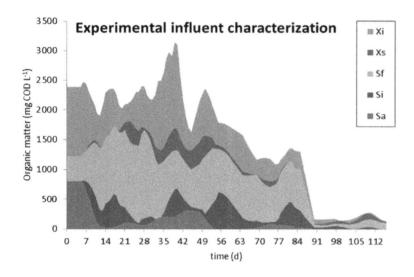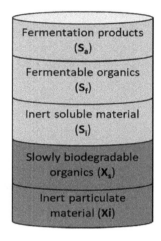

Figure 3. Results of experimental wastewater characterization over time.

Although, in theory, heterotrophic, autotrophic (nitrifiers) and phosphate-accumulating bacteria can all be present in wastewater treatment plants, the sole presence of heterotrophs was herein considered since, due to their characteristics, they are more likely to be actually present in a MFC's anodic chamber. The model as is, is therefore not applicable to MECs, but efforts are ongoing to study future changes appropriate to also representing this type of configuration and related processes. All degradation processes in ASM2d are represented by Monod-type kinetics:

$$\frac{dS_a}{dt} = -\frac{1}{Y_H}\rho_7 X_H + \rho_8 X_H \tag{10}$$

$$\frac{dX_S}{dt} = -\rho_2 X_H - \rho_3 X_H + (1 - f_{XI})\rho_9 \tag{11}$$

$$\frac{dS_{NH4}}{dt} = v_{NH4}(\rho_2 + \rho_3)X_H \tag{12}$$

$$\frac{dS_I}{dt} = f_{SI}(\rho_2 + \rho_3)X_H \tag{13}$$

$$\frac{dX_H}{dt} = \rho_6 X_H + \rho_7 X_H - \rho_9 \tag{14}$$

$$\frac{dS_f}{dt} = (1 - f_{SI})(\rho_2 + \rho_3)X_H - \frac{1}{Y_H}\rho_6 X_H - \rho_8 X_H \tag{15}$$

$$\frac{dX_I}{dt} = f_{XI}\rho_9 \tag{16}$$

$$\frac{dS_{NO3}}{dt} = -\frac{1 - Y_H}{2.86\, Y_H}(\rho_6 + \rho_7)X_H \tag{17}$$

$$\frac{dS_{N2}}{dt} = \frac{1 - Y_H}{2.86\, Y_H}(\rho_6 + \rho_7)X_H \tag{18}$$

where the ρ_i coefficients contained in Equations (10)–(18) are described in Table 2.

Table 2. Definition of the equations' coefficients.

Coefficient	Descriptive Equation
ρ_1	$K_h \dfrac{S_{O2}}{K_{O2}+S_{O2}} \dfrac{\frac{X_s}{X_H}}{K_{x,s}+\frac{X_s}{X_H}}$
ρ_2	$K_h \eta_{NO3,i} \dfrac{K_{O2}}{K_{O2}+S_{O2}} \dfrac{S_{NO3}}{K_{NO3}+S_{NO3}} \dfrac{\frac{X_s}{X_H}}{K_{X,S}+\frac{X_s}{X_H}}$
ρ_3	$K_h \eta_{fe} \dfrac{K_{O2}}{K_{O2}+S_{O2}} \dfrac{K_{NO3}}{K_{NO3}+S_{NO3}} \dfrac{\frac{X_s}{X_H}}{K_{X,s}+\frac{X_s}{X_H}}$

Table 2. *Cont.*

Coefficient	Descriptive Equation
ρ_4	$\mu_h \dfrac{S_{O2}}{K_{O2}+S_{O2}} \dfrac{S_f}{K_f+S_f} \dfrac{S_f}{S_f+S_a} \dfrac{S_{NH4}}{K_{NH4}+S_{NH4}} \dfrac{S_{alk}}{K_{alk}+S_{alk}}$
ρ_5	$\mu_h \dfrac{S_{O2}}{K_{O2}+S_{O2}} \dfrac{S_a}{K_a+S_a} \dfrac{S_a}{S_f+S_a} \dfrac{S_{NH4}}{K_{NH4}+S_{NH4}} \dfrac{S_{alk}}{K_{alk}+S_{alk}}$
ρ_6	$\mu_h \eta_{NO3} \dfrac{K_{O2}}{K_{O2}+S_{O2}} \dfrac{S_{NO3}}{K_{NO3}+S_{NO3}} \dfrac{S_f}{K_f+S_f} \dfrac{S_f}{S_f+S_a} \dfrac{S_{NH4}}{K_{NH4}+S_{NH4}} \dfrac{S_{alk}}{K_{alk}+S_{alk}}$
ρ_7	$\mu_h \eta_{NO3} \dfrac{K_{O2}}{K_{O2}+S_{O2}} \dfrac{S_{NO3}}{K_{NO3}+S_{NO3}} \dfrac{S_a}{K_a+S_a} \dfrac{S_a}{S_f+S_a} \dfrac{S_{NH4}}{K_{NH4}+S_{NH4}} \dfrac{S_{alk}}{K_{alk}+S_{alk}}$
ρ_8	$q_{fe} \dfrac{K_{O2}}{K_{O2}+S_{O2}} \dfrac{K_{NO3}}{K_{NO3}+S_{NO3}} \dfrac{S_f}{K_f+S_f} \dfrac{S_a}{S_f+S_a} \dfrac{S_{alk}}{K_{alk}+S_{alk}}$
ρ_9	$b_h X_H$

Integrating appropriately the above equations yields:

-a combined equation (from Equations (4)–(10)) describing S_a.

$$\frac{dS_a}{dt} = -q_a x_a - q_m x_m + D(S_{a0} - S_a) - \frac{1}{Y_H}\rho_7 X_H + \rho_8 X_H \tag{19}$$

-an equation in S_f, including the influent term for all COD components (S_a, S_i, S_f, X_i, X_s):

$$\frac{dS_f}{dt} = (1 - f_{SI})(\rho_2 + \rho_3)X_H - \frac{1}{Y_H}\rho_6 X_H - \rho_8 X_H + D\left(S_{f0} - S_f\right) \tag{20}$$

-an equation representing the lysis component for all microorganisms in the X_s and X_i mass balances, with addition of the washout coefficient for heterotrophs:

$$\frac{dX_H}{dt} = \rho_6 X_H + \rho_7 X_H - \rho_9 - \alpha_h D x_h \tag{21}$$

The effect of aerobic activity of heterotrophs has also been included in the model, by considering a small influent oxygen concentration (S_{O2} = 2 mg/L), and diffusive oxygen transfer from the cathode to the anodic chamber through the ion exchange membrane, by means of an oxygen mass balance equation:

$$\frac{dS_{O2}}{dt} = \left(1 - \frac{1}{Y_H}\right)(\rho_4 X_H + \rho_5 X_H) + D(S_{O20} - S_{O2}) \tag{22}$$

Also, in the integrated model, the dynamic formulas of internal resistance (R_{int}), and Open Circuit Voltage (E_{OCV}) are included, in order to better correlate their values with the actual concentration of exoelectrogens estimated at any time in the cell (in the original model, these were represented as constant values to be declared as initial conditions).

$$R_{int} = R_{min} + (R_{max} - R_{min})e^{-K_r X_a} \tag{23}$$

$$E_{OCV} = E_{min} + (E_{max} - E_{min})e^{-K_r^{-1} X_a^{-1}} \tag{24}$$

The resulting, integrated MFC model was then implemented in the MATLAB environment, and the representative differential equations solved by means of the MATLAB "ode23t" function.

It is clear that, by neglecting the presence of some entire components of the possible bacterial population of the cells, the increased bacterial community complexity of these system is partly lost, as well as some of the interrelated relationships among organisms that make it almost impossible to individually study the individual exoenergetic properties of each strain. In addition to exoelectrogen (a.k.a. anodophilic bacteria) and methanogenic microorganisms co-existing in competition for available substrate, an actual MFC would also contain nitrifiers, P-accumulating organisms, and others. Development of structured microbial communities within a cell's anode shows significant advantages compared to pure communities in the treatment of complex organic matter matrices, such as, for example, urban wastewater. In the model under present configuration, a direct competition among methanoges and anodophiles is represented (Figures 1 and 2). Even though both species contribute to the abatement of organic matter in the system, one does that by transforming it into methane and CO_2, the other into CO_2 and electrons, harvested at the cathode. As the design purpose of MFCs is actually to directly generate electric energy from the wastewater's organic matter, the former reaction is actually an undesirable by-product of poorly controllable circumstances, although it contributes to the organic matter removal efficiency of the system.

3. Results

The model was applied to the observations gathered from an intensely monitored, dual chamber, laboratory MFC with anodic volume of 0.42 L, continuously fed with swine wastewater at 1.5 L/day, operating in steady state at 21 °C for a prolonged period (over 110 days). The time series of the influent substrate (and its components) used for this purpose was previously shown in Figure 3. Following current modeling practice, a reduced subset of these data (30 days) was used to initially calibrate the integrated model. Initially, some available literature-reported parameter values were selected. If these were not available, "reasonable" best estimates (educated guesses from previous experience) were used. A Least Squares estimation method was subsequently applied to determine more fitting values based on those obtained from a different subset of experimental observations to verify the model. The results thus obtained from the present version of the model yield a much better representation of the original data, and are thus believed to be more representative of actual cell behavior than those obtained in the authors' previous work [32].

Once calibrated and verified, the integrated model was used to simulate the temporal trend of system's variables over time. Figure 4 shows the predicted behavior of exoelectrogen, methanogen and heterotropic populations in the MFC. After day 53 (when a steady decrease in COD load), exoelectrogens grow more rapidly than other groups, reaching a concentration of about 250 mg/L, against methanogens decreasing by half, and heterotroph concentrations remaining stable. A high concentration of exoelectrogenic biomass allows a higher production of electric current, from 4.4 mA during the previous period to 8.9 mA, when the organic load is lower (Figure 5). All the above results are in general agreement with the experimental observed trends of actual MFC electricity behavior, although some improvement in microbial population predictions of yield and development is still necessary, as there appears to be a lag of about 10 days between observed and predicted current maxima and minima, most likely linked to the dynamics of anodophiles and methanogens in the anodic cell volume and/or the yield coefficients of the former in the model.

Figure 6 shows predicted methane production over time. The simulated values agree well with actual measurements and with methanogenic population present in the system in time, showing also a good correlation ($r = 0.92$, according to Pearson) between simulated and measured values. This good correlation fit indicates that, in all likelihood, the differential between observed and predicted current productions is actually due in part to misestimated yield coefficients of the anodophile population, or to hydrodynamic factors in the anodic compartments.

Figure 7 shows the behavior of experimental and simulated soluble COD; results from the model showed a high level of correlation with measured ones ($r = 0.94$), while simulated data were still not fully convincing for total COD (data not shown), probably due to physical filtration effects of the granular graphite filling the anodic cell, trapping some of the small organic particulates. In the soluble COD case, however, maxima and minima of measured and simulated data are appropriately synchronous, showing that this organic matter removal is appropriately represented by the model.

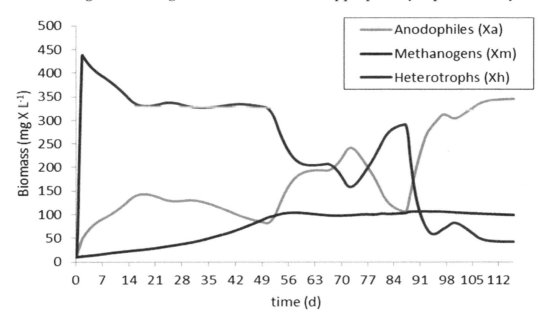

Figure 4. Predicted development of microbial populations in time.

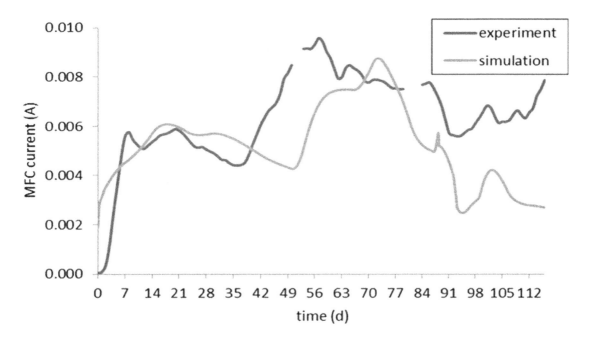

Figure 5. Simulated vs. observed current production by the cell.

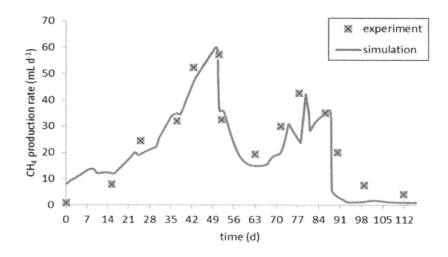

Figure 6. Methane production rate over time.

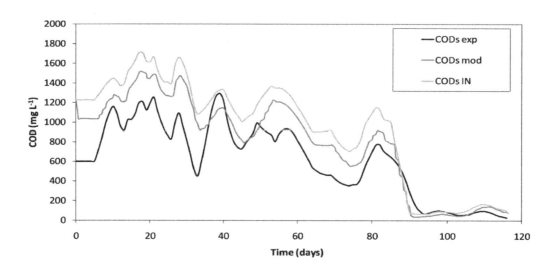

Figure 7. Simulation and observed soluble COD over time.

4. Discussion

Development of an integrated MFC description model based on deeper process understanding and allowing its mathematical reproducibility could play an important role in the quest for MFC technology improvement and industrialization, in the same way already observed for other types of similar processes [40]. The model herein presented is a small step forward in this effort, as it combines an existing model specifically developed for this type of system, but limited by the possibility of simulating only dual-component microbiodomes and mono-component (synthetic) substrate with a more general microbial population model. However, wastewater treatment by MFCs is characterized by several, simultaneous, multi-phase heterogeneous phenomena (e.g., biochemical reactions in the biofilm, electrochemical reactions at the electrodes, hydrodynamics of the bulk liquid, membrane polarization, etc.), that make development of a truly comprehensive mathematical model difficult (Figure 8). The proposed model considers an additional microbial component (heterotrophs) and the possibility of simulating complex substrates, including their intermediate transformations and interaction with microorganisms (Figure 2). It can therefore be considered a step forward, although not the final step, in the realization of a more complete MFC simulation model (Figure 8).

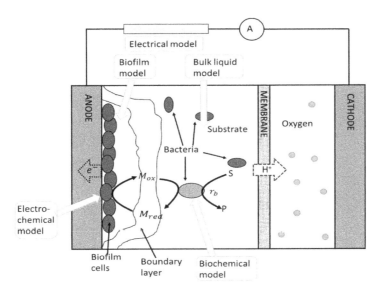

Figure 8. Integration of multiple models for comprehensive description of MFCs (redrawn from [41]).

When operating with complex substrate (i.e., other than acetate or glucose-media) applications, it could be of major interest to focus on microbial populations and organic matter dynamics within the anode chamber (or nitrogen dynamics in the cathode chamber) of the cell. In these conditions, MFCs show a much more complex microbiome than those operated with synthetic wastewater. These populations can show either syntrophic or competing behaviors. For example, heterotrophs could cut down complex, slowly biodegradable organic substrates, providing easily biodegradable molecules to EABs for electricity generation. In contrast, presence of methanogens could generate side-reaction compounds (i.e., methane) and reduce both coulombic and energy efficiencies of MFCs. Some of these phenomena are represented in the current model, but at the moment they can only be empirically controlled with external stimuli [21] in a reactive fashion. The model could be improved with appropriate algorithms representing the effects of implementable interventions. The present model also assumes completely mixed cells behavior. In view of system up-scaling, the influence of internal cell hydrodynamics (presence of shortcircuits, dead volumes, etc.) could become extremely relevant. For this reason, the model should be integrated with a hydrodynamic module, in order to better characterize and simulate overall cell performance; this could be obtained both by using computational fluid dynamics (CFD) [42] or tracer tests to assess the equivalent hydrodynamic configuration of a cell [43–45].

5. Conclusions

An integrated, dynamic, multi-species model for a completely mixed MFC is presented. The model was obtained by combining the Pinto model of an acetate-fed MFC, with the ASM2d activated sludge model, representing biological treatment systems fed by complex substrates. Hence, the presence of various microorganism species (exoelectrogens, methanogens and heterotrophs), feeding on a complex influent substrate, was able to be represented, and different methabolic processes simulated. The model was implemented on a MATLAB platform; its equations, solved by a numerical solutor, allowed the reproduction of the growth dynamics of microorganisms, organic matter degradation, current and methane production within a MFC. Monitoring observations from a MFC laboratory system operating for about four months were used to calibrate the model, and to compare results obtained from the simulations. Further validation from similar MFC systems, fed with different organic substrates is, however, necessary.

The model could be used to simulate scaled-up systems with the same physical configuration, keeping in mind, however, that the influence of the physical configuration and effects of these bioelectrochemical processes is far from being completely understood and replicable.

Acknowledgments: No funding was received for this study. At the time of the study, Molognoni was employed at the University of Pavia as a Postdoc researcher.

Author Contributions: The authors contributed paritetically to the writing of this manuscript.

References

1.   Capodaglio, A.; Callegari, A.; Lopez, M. European Framework for the Diffusion of Biogas Uses: Emerging Technologies, Acceptance, Incentive Strategies, and Institutional-Regulatory Support. *Sustainability* **2016**, *8*, 298. [CrossRef]

2.   Raboni, M.; Viotti, P.; Capodaglio, A.G. A comprehensive analysis of the current and future role of biofuels for transport in the European Union (EU). *Ambient. e Agu—Interdiscip. J. Appl. Sci. J. Appl. Sci.* **2015**, *10*, 9–21. [CrossRef]

3.   Capodaglio, A.G.; Callegari, A.; Dondi, D. Microwave-Induced Pyrolysis for Production of Sustainable Biodiesel from Waste Sludges. *Waste Biomass Valor.* **2016**, *7*, 703–709. [CrossRef]

4.   Capodaglio, A.G.; Ranieri, E.; Torretta, V. Process enhancement for maximization of methane production in codigestion biogas plants. *Manag. Environ. Qual. Int. J.* **2016**, *27*, 289–298. [CrossRef]

5.   Capodaglio, A.G.; Callegari, A. Feedstock and process influence on biodiesel produced from waste sewage sludge. *J. Environ. Manag.* **2017**, in press. [CrossRef] [PubMed]

6.   Du, Z.; Li, H.; Gu, T. A state of the art review on microbial fuel cells: A promising technology for wastewater treatment and bioenergy. *Biotechnol. Adv.* **2007**, *25*, 464–482. [CrossRef] [PubMed]

7.   Capodaglio, A.G.; Molognoni, D.; Dallago, E.; Liberale, A.; Cella, R.; Longoni, P.; Pantaleoni, L. Microbial Fuel Cells for direct electrical energy recovery from urban wastewaters. *Sci. World J.* **2013**, *2013*, 634738. [CrossRef] [PubMed]

8.   Rabaey, K.; Verstraete, W. Microbial fuel cells: Novel biotechnology for energy generation. *Trends Biotechnol.* **2005**, *23*, 291–298. [CrossRef] [PubMed]

9.   Logan, B.E.; Rabaey, K. Conversion of Wastes into Bioelectricity and Chemicals by Using Microbial Electrochemical Technologies. *Science* **2012**, *337*, 686–690. [CrossRef] [PubMed]

10. Shizas, I.; Bagley, D.M. Experimental Determination of Energy Content of Unknown Organics in Municipal Wastewater Streams. *J. Energy Eng.* **2004**, *130*, 45–53. [CrossRef]

11. Puig, S.; Serra, M.; Coma, M.; Balaguer, M.D.; Colprim, J. Simultaneous domestic wastewater treatment and renewable energy production using microbial fuel cells (MFCs). *Water Sci. Technol.* **2011**, *64*, 904–909. [CrossRef] [PubMed]

12. Puig, S.; Serra, M.; Coma, M.; Cabré, M.; Dolors Balaguer, M.; Colprim, J. Microbial fuel cell application in landfill leachate treatment. *J. Hazard. Mater.* **2011**, *185*, 763–767. [CrossRef] [PubMed]

13. Cecconet, D.; Molognoni, D.; Callegari, A.; Capodaglio, A.G. Agro-food industry wastewater treatment with microbial fuel cells: Energetic recovery issues. *Int. J. Hydrogen Energy* **2017**. [CrossRef]

14. Cecconet, D.; Molognoni, D.; Callegari, A.; Capodaglio, A.G. Biological combination processes for efficient removal of pharmaceutically active compounds from wastewater: A review and future perspectives. *J. Environ. Chem. Eng.* **2017**, *5*, 3590–3603. [CrossRef]

15. Capodaglio, A.G.; Hlavínek, P.; Raboni, M. Advances in wastewater nitrogen removal by biological processes: State of the art review. *Ambient. e Agua—An Interdiscip. J. Appl. Sci.* **2016**, *11*, 250. [CrossRef]

16. Molognoni, D.; Devecseri, M.; Cecconet, D.; Capodaglio, A.G. Cathodic groundwater denitrification with a bioelectrochemical system. *J. Water Process Eng.* **2017**, *19*, 67–73. [CrossRef]

17. Cecconet, D.; Devecseri, M.; Callegari, A.; Capodaglio, A.G. Effects of process operating conditions on the autotrophic denitrification of nitrate-contaminated groundwater using bioelectrochemical systems. *Sci. Total Environ.* **2018**, *613–614*, 663–671. [CrossRef] [PubMed]

18. Pous, N.; Puig, S.; Dolors Balaguer, M.; Colprim, J. Cathode potential and anode electron donor evaluation for a suitable treatment of nitrate-contaminated groundwater in bioelectrochemical systems. *Chem. Eng. J.* **2015**, *263*, 151–159. [CrossRef]

19. Kaur, A.; Boghani, H.C.; Michie, I.; Dinsdale, R.M.; Guwy, A.J.; Premier, G.C. Inhibition of methane production in microbial fuel cells: Operating strategies which select electrogens over methanogens. *Bioresour. Technol.* **2014**, *173*, 75–81. [CrossRef] [PubMed]

20. Capodaglio, A.G.; Molognoni, D.; Puig, S.; Balaguer, M.D.; Colprim, J. Role of Operating Conditions on Energetic Pathways in a Microbial Fuel Cell. *Energy Procedia* **2015**, *74*, 728–735. [CrossRef]

21. Molognoni, D.; Puig, S.; Balaguer, M.D.; Capodaglio, A.G.; Callegari, A.; Colprim, J. Multiparametric control for enhanced biofilm selection in microbial fuel cells. *J. Chem. Technol. Biotechnol.* **2015**, *91*, 1720–1727. [CrossRef]

22. Garg, A.; Vijayaraghavan, V.; Mahapatra, S.S.; Tai, K.; Wong, C.H. Performance evaluation of microbial fuel cell by artificial intelligence methods. *Expert Syst. Appl.* **2014**, *41*, 1389–1399. [CrossRef]

23. Capodaglio, A.G. Evaluation of modelling techniques for wastewater treatment plant automation. *Water Sci. Technol.* **1994**, *30*, 149–156.

24. Raduly, B.; Gernaey, K.V.; Mikkelsen, P.S.; Capodaglio, A.G.; Henze, M. Artificial neural networks for rapid WWTP performance evaluation: Methodology and case study. *Environ. Model. Softw.* **2007**, *22*, 1208–1216. [CrossRef]

25. Zhang, X.; Halme, A. Modelling of a microbial fuel cell process. *Biotechnol. Lett.* **1995**, *17*, 809–814. [CrossRef]

26. Kato Marcus, A.; Torres, C.I.; Rittmann, B.E. Conduction-based modeling of the biofilm anode of a microbial fuel cell. *Biotechnol. Bioeng.* **2007**, *98*, 1171–1182. [CrossRef] [PubMed]

27. Picioreanu, C.; Head, I.M.; Katuri, K.P.; van Loosdrecht, M.C.M.; Scott, K. A computational model for biofilm-based microbial fuel cells. *Water Res.* **2007**, *41*, 2921–2940. [CrossRef] [PubMed]

28. Zeng, Y.; Choo, Y.F.; Kim, B.-H.; Wu, P. Modelling and simulation of two-chamber microbial fuel cell. *J. Power Sources* **2010**, *195*, 79–89. [CrossRef]

29. Pinto, R.P.; Srinivasan, B.; Manuel, M.F.; Tartakovsky, B. A two-population bio-electrochemical model of a microbial fuel cell. *Bioresour. Technol.* **2010**, *101*, 5256–5265. [CrossRef] [PubMed]

30. Oliveira, V.B.; Simões, M.; Melo, L.F.; Pinto, A.M.F.R. A 1D mathematical model for a microbial fuel cell. *Energy* **2013**, *61*, 463–471. [CrossRef]

31. Henze, M.; Gujer, W.; Mino, T.; Matsuo, T.; Wentzel, C.M.; Marais, M.V.R. Activated sludge model no. 2d, ASM2d. *Water Sci. Technol.* **1999**, *39*, 165–182.

32. Capodaglio, A.G.; Molognoni, D.; Callegari, A. Formulation And Preliminary Application Of An Integrated Model Of Microbial Fuel Cell Processes. In Proceedings of the 29th European Conference on Modelling and Simulation (ECMS 2015), Varna, Bulgaria, 26–29 May 2015.

33. Potter, M.C. Electrical effects accompanying the decomposition of organic compounds. *Proc. R. Soc. B: Biol. Sci.* **1911**, *84*, 260–276. [CrossRef]

34. Kim, B.H.; Park, D.H.; Shin, P.K.; Chang, I.S.; Kim, H.J. Mediator-Less Biofuel Cell. U.S. Patent 5,976,719, 2 November 1999.

35. Rozendal, R.A.; Hamelers, H.V.M.; Rabaey, K.; Keller, J.; Buisman, C.J.N. Towards practical implementation of bioelectrochemical wastewater treatment. *Trends Biotechnol.* **2008**, *26*, 450–459. [CrossRef] [PubMed]

36. Logan, B.E.; Regan, J.M. Microbial fuel cells—Challenges and applications. *Environ. Sci. Technol.* **2006**, *40*, 5172–5180. [CrossRef] [PubMed]

37. Heijnen, J.J. Bioenergetics of Microbial Growth. In *Encyclopedia of Bioprocess Technology: Fermentation, Biocatalysis, and Bioseparation*; Flickinger, M.C., Drew, S.D., Eds.; John Wiley & Sons: New York, NY, USA, 1999; pp. 267–291.

38. Cecconet, D.; Molognoni, D.; Bolognesi, S.; Callegari, A.; Capodaglio, A.G. On the influence of reactor's hydraulics on the performance of Microbial Fuel Cells. *Bioresour. Technol.* **2017**, under review.

39. Molognoni, D.; Puig, S.; Balaguer, M.D.; Liberale, A.; Capodaglio, G.; Callegari, A.; Colprim, J. Reducing start-up time and minimizing energy losses of Microbial Fuel Cells using Maximum Power Point Tracking strategy. *J. Power Sources* **2014**, *269*, 403–411. [CrossRef]

40. Harremoës, P.; Capodaglio, A.G.; Hellström, B.G.; Henze, M.; Jensen, K.N.; Lynggaard-Jensen, A.; Otterpohl, R.; Søeberg, H. Wastewater Treatment Plants under Transient Loading—Performance, Modelling and Control. *Water Sci. Technol.* **1993**, *27*, 71–115.

41. Picioreanu, C.; Katuri, K.P.; Head, I.M.; Van Loosdrecht, M.C.M.; Scott, K. Mathematical model for microbial fuel cells with anodic biofilms and anaerobic digestion. *Water. Sci. Technol.* **2008**, *57*, 965–971. [CrossRef] [PubMed]

42. Vilà-Rovira, A.; Puig, S.; Balaguer, M.D.; Colprim, J. Anode hydrodynamics in Bioelectrochemical Systems. *RSC Adv.* **2015**, *5*, 78994–79000. [CrossRef]

43. Kim, K.-Y.; Yang, W.; Logan, B.E. Impact of electrode configurations on retention time and domestic wastewater treatment efficiency using microbial fuel cells. *Water Res.* **2015**, *80*, 41–46. [CrossRef] [PubMed]

44. Zhao, L.; Li, J.; Battaglia, F.; He, Z. Investigation of multiphysics in tubular microbial fuel cells by coupled computational fluid dynamics with multi-order Butler-Volmer reactions. *Chem. Eng. J.* **2016**, *296*, 377–385. [CrossRef]

45. Richardson, J.F.; Peacock, D.G. Flow Characteristics of Reactors. In *Coulson & Richardson's Chemical Engineering: Chemical & Biochemical Reactors & Process Control*, 3rd ed.; Pergamon Press: London, UK, 1994; Volume 3.

5

Stoichiometric Network Analysis of Cyanobacterial Acclimation to Photosynthesis-Associated Stresses Identifies Heterotrophic Niches

Ashley E. Beck [1], Hans C. Bernstein [2] and Ross P. Carlson [3],*

[1] Microbiology and Immunology, Center for Biofilm Engineering, Montana State University, Bozeman, MT 59717, USA; ashley.beck@montana.edu

[2] Biological Sciences Division, Pacific Northwest National Laboratory, Richland, WA 99352, USA; hans.bernstein@pnnl.gov

[3] Chemical and Biological Engineering, Center for Biofilm Engineering, Montana State University, Bozeman, MT 59717, USA

* Correspondence: rossc@montana.edu

Academic Editor: Michael Henson

Abstract: Metabolic acclimation to photosynthesis-associated stresses was examined in the thermophilic cyanobacterium *Thermosynechococcus elongatus* BP-1 using integrated computational and photobioreactor analyses. A genome-enabled metabolic model, complete with measured biomass composition, was analyzed using ecological resource allocation theory to predict and interpret metabolic acclimation to irradiance, O_2, and nutrient stresses. Reduced growth efficiency, shifts in photosystem utilization, changes in photorespiration strategies, and differing byproduct secretion patterns were predicted to occur along culturing stress gradients. These predictions were compared with photobioreactor physiological data and previously published transcriptomic data and found to be highly consistent with observations, providing a systems-based rationale for the culture phenotypes. The analysis also indicated that cyanobacterial stress acclimation strategies created niches for heterotrophic organisms and that heterotrophic activity could enhance cyanobacterial stress tolerance by removing inhibitory metabolic byproducts. This study provides mechanistic insight into stress acclimation strategies in photoautotrophs and establishes a framework for predicting, designing, and engineering both axenic and photoautotrophic-heterotrophic systems as a function of controllable parameters.

Keywords: cross-feeding; cyanobacteria; elementary flux mode analysis; irradiance; resource allocation; RuBisCO; stress acclimation

1. Introduction

Environmental stresses dictate competitive ecological strategies impacting nutrient and energy flows from the scale of individual cells to ecosystems [1,2]. Cyanobacteria are significant drivers of global nutrient and energy flows, accounting for ~10% of global primary productivity [3] and forming essential links in carbon and nitrogen biogeochemical cycles [4]. Cyanobacteria are also used in wastewater treatment and as bioprocess catalysts for bioproduction of specialty chemicals [5,6]. Cyanobacteria are deeply rooted in the tree of life and have adapted competitively to common stressors associated with photosynthesis and are model organisms for examining metabolic acclimation to these stresses.

Photoinhibition is a broad term encompassing different types of photosynthesis-associated stresses including photo-damage by excitation, damage by reactive oxygen species (ROS), and high localized O_2 concentrations [7]. Cyanobacteria can mitigate photo-damage by downregulating synthesis of photosystems, as well as adjusting relative photon absorption at photosystems I and II (PSI, PSII) to modulate ATP and NADPH regeneration (PSII extracts electrons from water which can be used in conjunction with PSI to regenerate NADPH and ATP, while PSI operating alone recycles electrons to regenerate ATP only; see Figure 1) [8]. High excitation can lead to oxidative damage at the photosystems and/or a highly reduced electron transport chain, which may also lead to cellular oxidative damage via ROS. Acclimation strategies include directing excess electrons toward alternative biochemical routes, such as reduction of O_2 (by either cellular respiration or the water-water cycle (photoreduction of O_2 to water)) or secretion of reduced carbon byproducts. High rates of oxygenic photosynthesis can also lead to locally high O_2 levels [9,10], and environments with high concentrations of O_2 relative to CO_2 can cause additional metabolic stress. Ribulose-1,5-bisphosphate carboxylase oxygenase (RuBisCO) is a dual-functioning enzyme which can react with either CO_2 or O_2. When RuBisCO reacts with O_2, 2-phosphoglycolate is produced, which is either secreted as the inhibitory compound glycolate or catabolized using one of three photorespiration pathways found in cyanobacteria [11]. Cyanobacteria have evolved mechanisms to reduce O_2 consumption at RuBisCO, including species-specific enzymes with varying affinities for CO_2 and O_2, as well as expression of carboxysomes to increase the relative CO_2 concentration in the vicinity of RuBisCO [12,13].

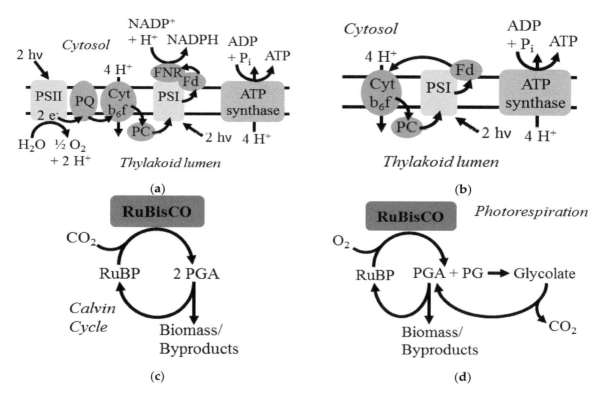

Figure 1. **Light and dark reactions of photosynthesis.** The role of photosystems I and II (PSI and PSII) in linear (**a**) and cyclic (**b**) photosynthesis and their relation to production of O_2 and regeneration of NADPH and ATP. Linear photosynthesis produces O_2 and regenerates both ATP and NADPH, whereas cyclic photosynthesis does not produce O_2 and regenerates ATP only. In the dark reactions, the bifunctional RuBisCO enzyme can incorporate inorganic carbon into biomass via the Calvin cycle (**c**) or can react with O_2 (**d**), resulting in a toxic byproduct and reducing incorporation of carbon into biomass. Abbreviations: hv, photons (photosynthetically active radiation); PQ, plastoquinone/plastoquinol; Cyt b_6f, cytochrome b_6f; PC, plastocyanin; FNR, ferredoxin-$NADP^+$ reductase; Fd, ferredoxin; RuBP, ribulose-1,5-bisphosphate; PGA, 3-phosphoglycerate; PG, 2-phosphoglycolate.

Stoichiometric modeling of metabolism enables prediction and interpretation of system-wide properties of complex metabolic networks, including community-level networks [14–20]. These systems biology approaches, such as flux balance analysis (FBA) and elementary flux mode analysis (EFMA), use genomic and physiological data to inform the construction of computational representations of metabolism. The application of a steady state assumption simplifies the mass-balanced metabolic reactions into a series of solvable linear equations, reducing the need for difficult-to-measure, condition-dependent enzyme kinetic parameters [16]. Whereas FBA uses objective functions such as biomass production to predict an optimal flux distribution under a specific set of conditions, EFMA calculates the complete set of minimal pathways (elementary flux modes, EFMs) through a metabolic network using steady state, reaction reversibility, and indecomposability constraints. Non-negative linear combinations of EFMs define the entire phenotypic solution space of a steady state metabolic network using a single simulation and can be used to examine all possible physiologies in an unbiased manner [19,21]. Similarities and differences in the output of EFMA versus other stoichiometric modeling techniques can be found in the review by Trinh et al. [19]. The enumerated EFMs can be evaluated by resource allocation theory, which quantitatively assesses the computational phenotypic space according to tradeoffs in consumption of different resources for the production of bioproducts [22–26]. Previous stoichiometric modeling studies of cyanobacterial metabolism have examined the occurrence of photorespiration as well as irradiance and carbon limitations [27–30].

The presented study analyzes metabolic acclimation to photosynthesis-associated stresses in the thermophilic, non-diazotrophic unicellular cyanobacterium *Thermosynechococcus elongatus* BP-1 (hereafter BP-1) and the formation of heterotrophic niches. BP-1 was isolated from the alkaline (pH 8.6) Beppu hot springs in Japan where temperatures range from 50–65 °C [31,32]. BP-1 is a major primary producer in its native hot springs where it often grows in bacterial mat communities with heterotrophs and is subject to high irradiance, high O_2, and low nutrient availability stresses. The objectives of this study were to (i) identify ecologically relevant acclimation strategies to high irradiance, O_2/CO_2 competition at RuBisCO, and nutrient limitation at varying degrees using a computational BP-1 stoichiometric model and EFMA combined with resource allocation theory, (ii) analyze BP-1 acclimation to high irradiance through controlled photobioreactors, (iii) compare general computational predictions to specific photobioreactor observations to interpret BP-1 acclimation strategies, and (iv) examine the impact of stress acclimation strategies on the ability of BP-1 to interact with heterotrophic partners. The presented study contributes to the understanding of cyanobacterial metabolism by examining specific photorespiration pathways, relative photon absorption of the photosystems, and byproduct secretion profiles under simultaneous stress conditions of high irradiance and O_2/CO_2 competition at RuBisCO, as well as by predicting cross-feeding photoautotrophic-heterotrophic interactions. The computational resource allocation-based modeling integrated with photobioreactor observations provides a rational basis for interpreting natural cyanobacterial behavior and a framework for controlling cyanobacteria for bioprocess applications.

2. Materials and Methods

2.1. Photobioreactor Culturing

T. elongatus BP-1 cultures were grown using modified BG-11 (mBG-11) medium [33,34], containing 17.6 mM $NaNO_3$, 0.304 mM $MgSO_4 \cdot 7H_2O$, 0.175 mM KH_2PO_4, 0.245 mM $CaCl_2 \cdot 2H_2O$, 0.0028 mM Na_2EDTA, and 0.0144 mM $FeCl_3$. A trace metal supplement was added (1 mL/L), comprised of 46.254 mM H_3BO_3, 9.146 mM $MnCl_2 \cdot 4H_2O$, 0.772 mM $ZnSO_4 \cdot 7H_2O$, 1.611 mM $Na_2MoO_4 \cdot 2H_2O$, 0.316 mM $CuSO_4 \cdot 5H_2O$, and 0.170 mM $Co(NO_3)_2 \cdot 6H_2O$. Inoculum cultures of BP-1 were initiated from frozen stocks into 150-mL sealed serum bottles filled with 50 mL mBG-11 amended with 15 mM sodium bicarbonate and adjusted to pH 7.5 under N_2 headspace containing 10% CO_2.

Photobioreactors were operated as turbidostats as described in Bernstein et al. [33], similar to Bernstein et al. and Melnicki et al. [35,36]. Reactors were inoculated with exponentially growing inoculum culture to $OD_{730nm} = 0.01$. All cultures were grown under continuous light of varying irradiances at 52 °C, pH 7.5, and were continuously sparged at 4 L min^{-1} with a 98% N_2 and 2%

CO_2 gas mixture. Incident and transmitted scalar irradiances were measured and used to adjust the turbidostat growth rate. The specific optical cross section (σ, m^2 (g CDW)$^{-1}$, CDW, cell dry weight) was determined according to a previously described method using a light diffuser and spectrophotometer [37]. The specific photon absorption rate was calculated by multiplying the specific optical cross section by the incident irradiance.

2.2. Biomass Composition Determination

Macromolecular composition was analyzed from turbidostat biomass samples (pelleted and frozen at Pacific Northwest National Laboratory and then shipped to Montana State University for subsequent analysis) according to the following procedures. DNA was quantified from alkali-lysed solutions with Hoechst 33258 fluorescent dye [38]. Glycogen was quantified by co-precipitation with sodium sulfate and detection with anthrone [39]. Lipids were quantified gravimetrically via chloroform-methanol extraction [40]. Total protein and amino acid distribution were quantified with HPLC fluorescence detection using o-phthalaldehyde (OPA) and 9-fluorenylmethylchloroformate (FMOC) derivatizations of acid-hydrolyzed protein [41]. Cysteine, methionine, and tryptophan were degraded, and asparagine and glutamine were converted to aspartate and glutamate, respectively, during hydrolysis [42]; therefore, abundances were predicted from protein-coding gene codon usage. RNA was quantified by lysis with potassium hydroxide, extraction into cold perchloric acid, and measurement of UV absorbance at 260 nm [43]. Appendix A contains detailed protocols for each method.

2.3. Model Construction

The metabolic network model for BP-1 was constructed in CellNetAnalyzer [44,45] from the annotated genome [46] with the aid of MetaCyc, KEGG, BRENDA, and NCBI databases [47–49]. Reversible exchange reactions were defined for protons and water. Irreversible exchange reactions defined bicarbonate, magnesium, nitrate, phosphate, photons, and sulfate as possible substrates and O_2, acetate, alanine, ethanol, formate, glycolate, lactate, pyruvate, and sucrose as possible byproducts. Biomass was also defined as a product.

Macromolecular synthesis reactions were defined for nucleic acids, glycogen (most common form of cyanobacterial carbohydrate storage [50]), lipid, and protein. Synthesis reactions utilized two phosphate bonds per nucleic acid monomer, one phosphate bond per glycogen monomer, and four phosphate bonds per protein monomer [51]. Nucleotide distributions were set based on percent GC content of the genome for DNA and nucleotide sequence of the rRNA genes for RNA. Fatty acid distribution was assigned based on literature values of fatty acid chain and lipid types measured for BP-1 [52–54]. The amino acid distribution was set using the experimentally measured values in the current study. Macromolecular composition (DNA, glycogen, lipid (including chlorophyll), protein, and RNA) was determined experimentally in the current study (see Section 2.2) and used to set the molar coefficients in the biomass synthesis reaction, normalized to 1 kg dry biomass (File S1 in the Supplementary Materials). Chlorophyll was also included in biomass synthesis using the mass fraction measured for *Synechococcus* sp. PCC 7002 [29], and the lipid mass fraction was adjusted to reflect the proportion of chlorophyll. The biomass composition was converted into an electron requirement using degree of reduction (moles of electrons per mole of carbon) calculations [55] with the assumption that each biosynthetic electron requires two photons (one absorbed at each PSII and PSI). Degree of reduction was calculated with respect to nitrate as a nitrogen source. To estimate photons necessary for ATP regeneration, the phosphate bond requirement for polymerization of monomers into macromolecules was converted into a photon requirement via a stoichiometry of four photons per phosphate bond (one photon absorbed at PSI per proton pumped, with four protons translocated per ATP molecule synthesized). Photon and proton stoichiometries remain active areas of research, and this estimate is recognized as an upper bound considering linear photosynthesis without a Q-cycle [56,57].

All reactions were balanced for elements, charge, and electrons. Thermodynamic considerations were built into the model via reaction reversibilities, based on data from BRENDA [49]; in the event that data for bacterial species were not available from BRENDA, thermodynamic calculations were performed with eQuilibrator (http://equilibrator.weizmann.ac.il/) to determine physiological reversibility, using a product concentration three orders of magnitude greater than the reactant concentration [58,59]. Nitrogen requirements were determined for each reaction by summing the number of nitrogen atoms specified by the enzyme amino acid sequences. Iron requirements were determined for central carbon metabolism and photosynthesis reactions based on metal requirements of similar cyanobacterial species in BRENDA [49]. For instances of missing or conflicting information in the database, literature values compiled for oxygenic photoautotrophs were used [60]. A one-to-one (minimal resource investment) correspondence of enzyme to reaction was used to calculate the total cost per EFM, as it has previously been shown to provide a good approximation of flux distributions in *Escherichia coli* [23,24]. EFMs were enumerated using EFMtool [61]. Resource allocation analysis (cost assessment) of the resulting EFMs was performed with MATLAB and Python. The metabolic model with supporting details and CellNetAnalyzer metabolite and reaction input, SBML model version, and documented analysis routines can be found in the Supplementary Materials (Files S1–S4).

3. Results

3.1. Computational BP-1 Metabolic Model and Photobioreactor Biomass Composition Measurement

The BP-1 computational metabolic model was constructed from the annotated genome [46]. Genetic potential was mapped to enzymes and metabolic reactions which encompassed photosynthesis, central metabolism, and biosynthetic reactions leading to biomass production according to a defined macromolecular composition reaction. Transport reactions were defined for nutrient uptake and product secretion. Subsequent EFMA resulted in a description of the phenotypic space spanning the range of possible nutrient uptake and product secretion rates, which could then be analyzed for ecologically relevant stress acclimation strategies. The model accounted for 334 metabolism-associated genes which were mapped to 279 metabolites and 284 reactions (File S1 in the Supplementary Materials). Photons were assumed to be within the spectrum of photosynthetically active radiation (PAR; 400–700 nm). A stoichiometrically balanced schematic demonstrating operation of the photosynthetic electron transport chain (linear and cyclic photosynthesis) and carbon flow in the model is shown in Figure 1. Nutrient substrates for the model were selected in alignment with the photobioreactor culturing medium. Bicarbonate was modeled as the sole carbon source based on culturing pH while interconversion with CO_2 was modeled via the carboxysomal carbonic anhydrase enzyme, and nitrate was modeled as the sole nitrogen source. Two of the three photorespiration pathways possible in cyanobacteria [11] were identified in BP-1, namely, the C2 cycle and the glycerate pathway. A variety of organic byproducts (Table 1) were considered based on previous genomic analysis of BP-1 [62] and culturing studies of related unicellular cyanobacteria [63,64]. Secretion of several different amino acids has been observed in BP-1 and related species [33,63,64]; alanine was included as a representative amino acid byproduct in the current model, closely linked to central metabolism.

Biomass composition impacts growth and byproducts [65], making appropriate composition parameters important for computational growth predictions. BP-1 macromolecular biomass composition was determined analytically from continuous culture samples and was used to parameterize the model growth reactions. The major measured macromolecule classes (DNA, glycogen, lipid (including chlorophyll), protein, and RNA) summed to 98.1% of cell dry weight (Table 2); the remaining 1.9% was assumed to be ash. Protein and lipid/chlorophyll comprised the largest mass fractions of biomass, accounting for 62.0% and 17.4%, respectively. Since protein comprises the largest fraction of biomass, amino acid monomer distribution was also determined analytically (Table A1 in Appendix C) and used to parameterize the model reaction for protein synthesis. A strong correlation was observed between the measured amino acid distribution and the distribution predicted from protein-coding gene sequences (Figure A1 in Appendix B).

Table 1. *T. elongatus* BP-1 metabolic model inputs and outputs, including potential reduced carbon byproducts, with corresponding degree of reduction.

	Compound	Formula	Charge	Degree of Reduction
Inputs	Carbon dioxide	CO_2	0	0
	Water	H_2O	0	0
	Photons	NA	NA	NA
	Nitrate	NO_3	−1	−8/−5/0
Outputs	Molecular oxygen	O_2	0	−4
	Biomass	$CH_{1.6}N_{0.2}O_{0.3}P_{0.01}S_{0.005}$	−0.7	4.3/5.0/6.1 [a]
	Acetate	$C_2H_3O_2$	−1	4
	Alanine	$C_3H_7NO_2$	0	4/5/6.7 [a]
	Ethanol	C_2H_6O	0	6
	Formate	CHO_2	−1	2
	Glycolate	$C_2H_3O_3$	−1	3
	Lactate	$C_3H_5O_3$	−1	4
	Pyruvate	$C_3H_3O_3$	−1	3.3
	Sucrose	$C_{12}H_{22}O_{11}$	0	4

[a] Degree of reduction calculated with respect to ammonia/molecular nitrogen/nitrate. NA, not applicable.

Table 2. Experimentally determined *T. elongatus* BP-1 biomass composition from turbidostat biomass samples grown under an irradiance of 2000 μmol photons m^{-2} s^{-1}.

Macromolecule	Mass Percent	Extraction Method/Analytical Method
DNA	0.4	Alkaline lysis/Hoechst 33258 fluorescence
Glycogen	2.0	Sodium sulfate co-precipitation/Anthrone detection
Lipid (including chlorophyll)	17.4	Chloroform-methanol/Gravimetric
Protein	62.0	Hydrochloric acid hydrolysis/OPA, FMOC derivatization
RNA	16.3	Alkaline lysis, perchloric acid/UV absorbance
Total	98.1	

3.2. Computational Analysis of Stress Acclimation

The computational BP-1 metabolic model was decomposed into 4,636,498 unique EFMs using EFMtool [61], with ~99.5% producing biomass. Each EFM, as well as any non-negative linear combination of multiple EFMs, represented a mathematically feasible phenotype and possible stress acclimation strategy. Competitive stress acclimation strategies were identified using ecological resource allocation theory. Resource allocation theory analyzes the amount of catabolic or anabolic resource required to synthesize a cellular product, often biomass. Minimizing the requirement of a limiting resource represents a competitive, cost-effective phenotype and is hypothesized to be a probable cellular strategy selected by evolution. When two or more resources are considered simultaneously, a multi-dimensional tradeoff surface is created that quantifies the utilization relationship between the limiting resources [22–25]. Biomass-producing EFMs were ranked quantitatively based on efficiency of resource use for biomass production under simulated environmental stresses including high irradiance, O_2/CO_2 competition at RuBisCO, and limited availability of dissolved inorganic carbon (DIC) as well as nitrogen or iron. The tradeoff between optimal use of two resources was quantified by simultaneously minimizing the cost of biomass production under two different stress factors. Similar methods have been applied to extend FBA to account for biosynthetic costs [26], but enumeration of complete EFMs combined with resource allocation theory allows exploration of the entire phenotypic space.

3.2.1. Irradiance and Photosynthetic Electron Flow

Photosynthetic electron flow was examined as a function of irradiance-induced stress to interpret relationships between photon absorption and photocatalytic water oxidation. Net O_2 production per carbon mole (Cmol) biomass produced was plotted as a function of photons absorbed per Cmol biomass produced, a metric of irradiance-induced stress (Figure 2). Each net O_2 molecule is the byproduct of four photosynthetically derived electrons extracted from water and requires eight total photons absorbed [66]; this relationship is reflected in the slope of the upper boundary of the phenotypic cone. Photons absorbed at PSI during cyclic photosynthesis are decoupled from O_2 production. Growth phenotypes were analyzed for the ability to direct electrons toward either biomass or reduced byproducts. The EFMs along the lowest boundary of the phenotypic cone in Figure 2 represented growth where all electrons were directed to biomass and no reduced byproducts were secreted, extending up to ~80 mol photons absorbed per Cmol biomass produced. Net O_2 production (~1.53 mol O_2 per Cmol biomass) at the lowest boundary corresponds to the biomass degree of reduction, ~6.1 mol electrons available to reduce O_2 per Cmol biomass (Table 1). EFMs with higher net O_2 production directed electrons to reduced carbon byproducts, such as formate or acetate.

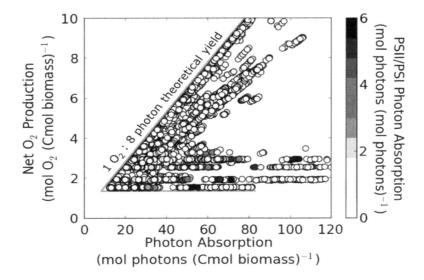

Figure 2. Computational analysis of irradiance and photosynthetic electron flow in cyanobacterium *T. elongatus* **BP-1.** Net O_2 production (net mol O_2 evolved (Cmol biomass produced)$^{-1}$) is plotted as a function of photon absorption (mol photons absorbed (Cmol biomass produced)$^{-1}$) for biomass-producing EFMs. Each point represents a unique EFM. The slope of the upper boundary of the phenotypic cone indicates maximum net moles of O_2 produced per mole of photons absorbed (eight photons required per molecule O_2 evolved; theoretical minimum quantum requirement). No byproducts were secreted on the lower boundary of the phenotypic cone minimizing net O_2 production per biomass produced; net O_2 production along this boundary was a direct result of electrons incorporated into biomass; secreted reduced carbon byproducts were predicted throughout the remaining phenotypic space. Color scale represents the photon absorption at PSII relative to PSI for each EFM (mol photons absorbed at PSII (mol photons absorbed at PSI)$^{-1}$). Relative contribution of PSII was predicted to increase as photon absorption increased. Less than 1% of the EFMs had a PSII/PSI ratio greater than 6 (with maximal value of 20) and were excluded from the plot to represent a more feasible phenotypic space [67–69]. Modeled biomass production did not include maintenance energy requirements. Points in the plot area shown are representative of 4,371,798 EFMs.

Biomass-producing EFMs were assessed for photon absorption at PSII relative to PSI to quantify the contribution of the two photosystems to photosynthetic electron flow (Figure 2, shaded color bar). A value less than one indicated elevated cyclic photosynthesis, a value greater than one indicated elevated operation of PSII independent of linear photosynthesis (i.e., reduction of O_2 through either cellular respiration or the water-water cycle), and a value equal to one indicated linear photosynthesis or, alternatively, equivalent cyclic photosynthesis and O_2 reduction. Figure 1 provides greater detail on cyclic and linear photosynthesis. An increase in photon absorption at PSII at a fixed net O_2 production indicated greater gross production of O_2, which was consumed by cellular respiration and/or the water-water cycle. In general, photon absorption at PSII relative to PSI was predicted to increase as irradiance-induced stress increased (Figure 2), indicating a greater contribution of PSII to photon absorption at high irradiance.

3.2.2. Irradiance, High O_2, and Nutrient Limitation

Computational BP-1 growth phenotypes were interrogated for stress acclimation strategies under a range of relative O_2 to DIC concentrations, represented by O_2/CO_2 competition at RuBisCO (Figure 3a). The tradeoff curve simultaneously minimizes the cost of biomass production under O_2/CO_2 competition (moles O_2 per mole CO_2 consumed at RuBisCO) and irradiance-induced stress; EFMs on the tradeoff curve (or non-negative linear combinations thereof) represent optimal predicted growth phenotypes under the combined stresses. Photorespiration, as opposed to secretion of glycolate, was predicted as an essential process on the tradeoff curve except at zero O_2/CO_2 competition, and utilization of photorespiration reactions increased with increasing stress. Utilization of the C2 photorespiration cycle was predicted to increase along the tradeoff curve, whereas use of the glycerate pathway remained minimal. Photon absorption at PSII relative to PSI was also predicted to increase along the tradeoff curve. Neither cellular respiration nor the water-water cycle was active along the tradeoff curve, indicating that all photosynthetically derived electrons were directed to either biomass or reduced carbon byproducts. The tradeoff curve was divided into four phenotypic zones based on the suite of byproducts predicted. Zone 1 phenotypes did not secrete reduced byproducts, but as O_2/CO_2 competition at RuBisCO increased, more energy from photons was required to mediate the stress as indicated by higher photon absorption per biomass. At high O_2/CO_2 competition (~0.8 mol O_2 (mol CO_2)$^{-1}$), byproduct secretion represented the most resource-efficient acclimation strategy under the combined stresses. Byproduct synthesis effectively consumed photosynthetically derived electrons at the expense of fixed DIC and, conditionally, reduced nitrogen, as seen in the transition in byproducts produced along the tradeoff curve. Formate was predicted to be the most resource-efficient byproduct (zone 2 phenotypes), followed by combinations of formate and amino acids, represented in the model as alanine (zone 3 phenotypes), and acetate and amino acids (zone 4 phenotypes). Secretion of glycolate was not the most competitive use of metabolic potential under the considered stresses. Net O_2 production of the tradeoff curve EFMs quantified the fraction of electrons directed to biomass and reduced byproducts as a function of stress acclimation (Figure 3b). A nonlinear increase in net O_2 production per Cmol biomass was predicted; the increase in net O_2 production correlated with the secretion of reduced byproducts (formate, acetate, and/or alanine).

In addition to DIC, nitrogen and iron are essential anabolic resources and place constraints on cellular functions such as growth or ATP regeneration [70]. Acclimation to nitrogen- or iron-limited growth, assessed by investment into enzymes, was analyzed in conjunction with O_2/CO_2 competition (Figure A2a,b in Appendix B). Increasing O_2/CO_2 competition at RuBisCO necessitated an increase in nitrogen and iron investments into metabolic enzymes due to the requirement to process 2-phosphoglycolate. Tradeoff curve analysis of simultaneous acclimation to O_2/CO_2 competition and nutrient limitation showed trends similar to those predicted under irradiance-induced stress in Figure 3a, and amino acid secretion was again predicted at the highest resource limitation stress. However, under nitrogen limitation, reduced byproduct secretion was required for the most competitive phenotypes over the entire range of resource-limited growth. BP-1 metabolism was predicted to be less robust to nitrogen-limited stress than irradiance-induced stress as indicated by relatively fewer suboptimal EFMs near the tradeoff curve (Figure A2a in Appendix B). Additional details on nitrogen and iron limitation are found in Appendix D. While the majority of EFMs produced biomass, energy-producing EFMs (not producing biomass) also showed similar optimal byproducts under irradiance-induced stress and O_2/CO_2 competition (data not shown).

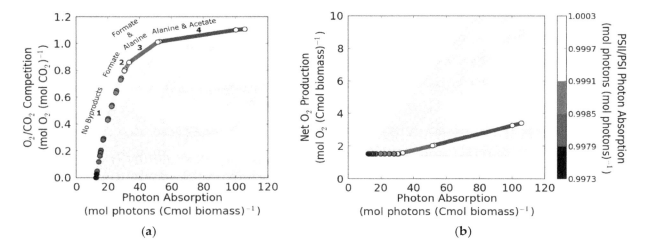

Figure 3. **Computational analysis of irradiance-induced stress and O_2/CO_2 competition at RuBisCO in cyanobacterium *T. elongatus* BP-1.** (a) O_2/CO_2 competition at RubisCO (mol O_2 (mol CO_2)$^{-1}$ consumed) as a function of photon absorption (mol photons absorbed (Cmol biomass produced)$^{-1}$) for biomass-producing EFMs are plotted. Each point represents a unique EFM. The tradeoff curve defining competitive strategies between O_2/CO_2 competition and irradiance-induced stress was divided into four distinct phenotypic regions based on byproduct secretion behavior, labeled accordingly (intensity of shading increases with increasing stress). The maximum amount of O_2 per CO_2 consumption at RuBisCO that can be sustained is two to one. Consumption of two O_2 molecules followed by photorespiration recycles 2-phosphoglycolate to regenerate the ribulose-1,5-bisphosphate precursor, but loses the single molecule of CO_2 that was consumed and thus cannot support biomass production. Points in the plot area shown are representative of 4,457,199 EFMs. (b) Net O_2 production (net mol O_2 evolved (Cmol biomass produced)$^{-1}$) as a function of photon absorption (mol photons absorbed (Cmol biomass produced)$^{-1}$) for biomass-producing EFMs are plotted. Colored points indicate net O_2 production of EFMs on the tradeoff curve in (**a**). Color scale represents the photon absorption at PSII relative to PSI (mol photons absorbed at PSII (mol photons absorbed at PSI)$^{-1}$). Modeled biomass production did not include maintenance energy requirements. Points in the plot area shown are representative of 4,355,094 EFMs.

3.3. Comparison of Computational Predictions with Photobioreactor Physiological Data

The optimal predicted growth phenotypes identified along the tradeoff curve (Figure 3a) were compared with data from turbidostat culturing experiments. Irradiance levels altered both specific growth rate and biomass yield during cultivation. Specific growth rates ranged from 0.06–0.29 h^{-1} at irradiances varying from 200–2000 μmol photons m^{-2} s^{-1} (Figure 4a). Specific growth rates increased linearly as a function of incident irradiance below 500 μmol photons m^{-2} s^{-1}. Above 500 μmol photons m^{-2} s^{-1}, irradiance became saturating, possibly inhibitory, and specific growth rate approached a maximum at 1800–2000 μmol photons m^{-2} s^{-1}. Conversely, biomass yield per photon absorbed had a maximum at low irradiance (200–300 μmol photons m^{-2} s^{-1}) and decreased nonlinearly as a function of incident irradiance (Figure 4b). Irradiance-induced stress at 2000 μmol photons m^{-2} s^{-1} reduced the biomass production efficiency by more than 50% compared to low irradiance conditions. The decrease in biomass per photon yield is consistent with predicted acclimation strategies, as is the nonlinear relationship between stress and biomass growth efficiency (Figure 3).

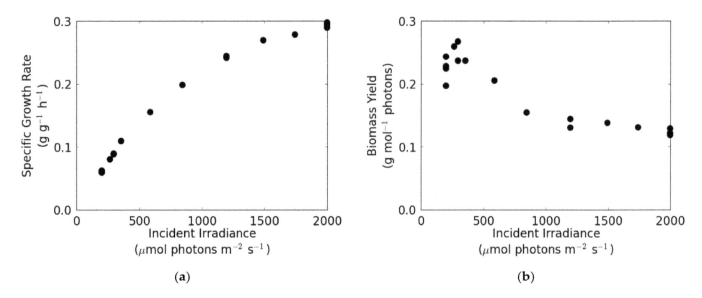

(a)   **(b)**

Figure 4. Photobioreactor impact of irradiance on specific growth rate and biomass production efficiency in *T. elongatus* BP-1 continuous culture. (a) Photobioreactor measurement of BP-1 specific growth rate (g CDW (g CDW)$^{-1}$ h^{-1}) as a function of incident irradiance (μmol photons m^{-2} s^{-1}). CDW, cell dry weight. (b) Photobioreactor measurement of BP-1 biomass yield (g CDW (mol photons absorbed)$^{-1}$) as a function of incident irradiance (μmol photons m^{-2} s^{-1}).

BP-1 maintenance energy requirements were estimated by analyzing specific photon absorption rate as a function of specific growth rate (Figure 5). The non-growth associated maintenance energy requirement (0.16 mol photons (g CDW)$^{-1}$ h^{-1}) was estimated by extrapolating the specific photon absorption rate data to a zero growth rate. Photon requirements for growth can be partitioned into the cellular energy required to (1) reduce nutrients such as DIC and nitrate into biomass monomers and (2) polymerize monomers into macromolecules. The photon requirement to reduce nutrient substrates, including bicarbonate and nitrate, to biomass monomers was calculated using the experimentally measured biomass composition. Macromolecular synthesis reactions in the model incorporated the energy cost of phosphate bonds required to polymerize monomers.

Photon requirement per Cmol biomass increased nonlinearly at higher growth rates (Figure 5), which corresponded to higher incident irradiance and represented successively increasing irradiance-induced stress and reduced biomass production efficiency.

The difference between the photon requirement for biomass and the experimentally measured photon requirement is hypothesized to be the photon requirement for growth-associated maintenance energy, including tasks such as general protein repair, enzyme turnover, and maintenance of gradients, or other drains such as non-photochemical quenching of absorbed photons [8,71–73].

Additionally, the repair and recycling of PSII due to increased photoinactivation at high irradiance requires a large investment of nitrogen and poses a significant limitation on growth [74,75]. The implications of nitrogen source degree of reduction were also factored into maintenance energy calculations. A comparison of the effects of different nitrogen sources on the photon requirement is shown in Figure A3 in Appendix B. Molecular nitrogen and ammonia required fewer photons per biomass since they are more reduced than nitrate. Nitrate may be a preferred nitrogen source for photoautotrophs under high irradiance conditions, likely because it represents a possible sink for electrons which can buffer over-reduced photosynthetic machinery.

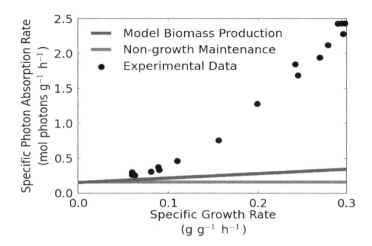

**Figure 5. Growth rate-dependent photon absorption rate and maintenance energy in cyanobacterium
T. elongatus BP-1.** Specific photon absorption rate (mol photons absorbed (g CDW)$^{-1}$ h^{-1}) is plotted as
a function of specific growth rate (g CDW (g CDW)$^{-1}$ h^{-1}) for experimental BP-1 turbidostat cultures
(black circles). Specific photon absorption rate is dependent on specific growth rate (μ) according to
the equation $0.16e^{9.47\mu}$ determined from an exponential regression of the photobioreactor data. The
non-growth associated maintenance energy requirement was extrapolated from a specific growth rate of
zero (light blue line). Measured photon absorption rates are contrasted with calculated requirements for
biomass synthesis, including polymerization (dark blue line). CDW, cell dry weight.

3.4. Comparison of Computational Predictions with Photobioreactor Transcriptomic Data

The optimal predicted growth phenotypes identified along the tradeoff curve (Figure 3a) were
compared with previously published BP-1 transcriptomic data [33]. The transcriptomic data were
analyzed for differentially expressed genes (two-fold or greater difference) between high and low
irradiance conditions (2000 versus 200 μmol photons m^{-2} s^{-1}). A change in expression of two-fold
or greater was observed for 1147 genes. Differentially expressed genes were examined according
to metabolic pathways and compared with the pathways utilized in the predicted optimal stress
acclimation phenotypes. Consistencies and inconsistencies between predicted and observed metabolic
functionalities were grouped into six categories (Table 3) and are discussed in detail below.

Table 3. Comparison between computational predictions of stress acclimations and previously
published photobioreactor gene expression data under high versus low irradiance conditions [33].
Metabolic functionalities were predicted from competitive pathways along the optimal tradeoff curve
for irradiance-induced stress and O$_2$/CO$_2$ competition (Figure 3a), and observations were made
from gene expression data comparing change in transcripts from high to low irradiance (2000 versus
200 μmol photons m^{-2} s^{-1}) [33].

	Prediction	Observation
1. Photosystem contribution	• Increased PSII photon absorption relative to PSI	• Upregulation of PSII-associated genes • No change in PSI-associated genes
2. Photorespiration pathways	• Use of photorespiration • Increase in photorespiration with higher O$_2$/CO$_2$ competition • Primarily C2 cycle, minimal glycerate pathway usage	• Transcription of photorespiration genes • Upregulation of C2 cycle genes • No change in glycerate pathway genes
3. Byproduct secretion	• Production of reduced byproducts• Formate and acetate production • Amino acid (alanine) secretion at highest stress	• Upregulation of formate, acetate, and sucrose synthesis genes • Upregulation of amino acid synthesis pathway and transporter genes

Table 3. *Cont.*

	Prediction	Observation
4. Glycolysis	• Increased use of lower portion of glycolysis	• Upregulation of genes in lower portion of glycolysis
5. TCA cycle	• No change in TCA cycle use	• Upregulation of TCA cycle genes leading to synthesis of α-ketoglutarate
6. Nitrate and sulfate assimilation	• Increased nitrate uptake in pathways that secrete amino acid byproducts • No change in sulfate uptake	• Upregulation of nitrate uptake and assimilation genes • Upregulation of sulfate uptake and assimilation genes

3.4.1. Photosynthesis, Photorespiration, and Byproducts

The predicted increase in photon absorption at PSII relative to PSI (Figure 3a) was reflected in the transcriptomic data [33] showing upregulation of genes coding for several PSII subunit and repair genes but no upregulation of PSI-associated genes (Table A2 in Appendix C). The increase in transcript level could be due to increased photon absorption, or it could reflect an increased turnover of PSII, which has been reported during culturing at high irradiance [74,75]. A relative increase in photon absorption at PSII would suggest an increased relative contribution of PSII to photosynthetic electron flow under irradiance-induced stress. Photorespiration was a predicted strategy under high irradiance and O_2/CO_2 competition, corresponding to upregulation of photorespiration pathway genes observed in the transcriptomic data [33]. Predicted pathways indicated preferential utilization of the C2 photorespiration cycle as opposed to the glycerate pathway, and transcriptomic data [33] indicated upregulation of C2 cycle genes with no change in expression of glycerate pathway genes (Figure 6). Byproduct secretion was predicted as a competitive strategy at high irradiance and O_2/CO_2 competition. Irradiance, nitrogen investment, and iron investment analyses in conjunction with O_2/CO_2 competition all predicted amino acid secretion as a resource-efficient strategy at the highest combined stress conditions (Figures 3a and A2). These predictions corresponded with observations of upregulated genes for synthesis pathways of organic compounds such as acetate and formate (Figure 6), as well as for more than 50 amino acid synthesis pathway and transporter genes (Table A3 in Appendix C). Altogether, these parallels with the transcriptomic data [33] suggest increased electron flow into the system, increased photorespiration, and reprocessing of salvaged carbon into other byproducts with greater degree of reduction (Table 1) at higher irradiance.

3.4.2. Central Metabolism and Nutrient Assimilation

Several glycolysis genes were observed to be upregulated under high irradiance conditions [33], primarily genes involved in the lower portion of glycolysis after glyceraldehyde-3-phosphate (glyceraldehyde-3-phosphate dehydrogenase, phosphoglycerate kinase, phosphoglycerate mutase, enolase, and pyruvate kinase) (Table A2 in Appendix C). Tradeoff curve analysis of the EFMs under high irradiance and O_2/CO_2 competition predicted that utilization of the reactions catalyzed by these enzymes increased with increasing stress except for glyceraldehyde-3-phosphate dehydrogenase. Both glyceraldehyde-3-phosphate and 3-phosphoglycerate intersect the Calvin cycle and glycolysis; thus, increased use of the lower portion of glycolysis suggested funneling of glyceraldehyde-3-phosphate from the Calvin cycle into glycolysis to produce pyruvate, which may be used to synthesize byproducts such as formate, acetate, and amino acids. Several TCA cycle genes were also observed to be upregulated under high irradiance conditions, predominantly genes catalyzing reactions up to the synthesis of α-ketoglutarate, from which several amino acids are synthesized (Table A2 in Appendix C). Tradeoff curve analysis of the EFMs under high irradiance and O_2/CO_2 competition predicted no change in utilization of any TCA cycle reactions. The BP-1 model utilized alanine as a representative amino acid which could be secreted as a byproduct; alanine is synthesized via pyruvate. However, if the computational model was modified to allow secretion of amino acids that are synthesized via TCA cycle intermediates, such as glutamate, it would lead to predictions of increases in some TCA cycle fluxes.

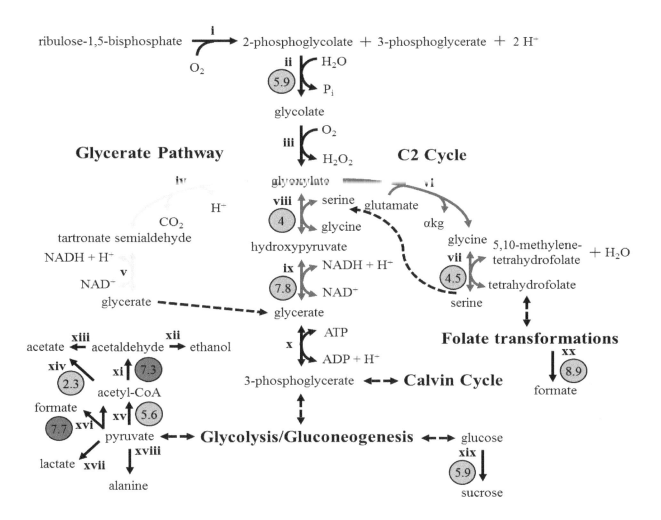

Figure 6.   Cyanobacterium *T. elongatus* BP-1 photorespiration and byproduct secretion pathways with transcriptomic data measured under high versus low irradiance. Genome-based photorespiration routes (C2 cycle, dark blue, and glycerate pathway, light blue) and byproduct secretion pathways included in the BP-1 model are illustrated. Green circles represent upregulated gene expression measured under high irradiance (2000 versus 200 μmol photons m^{-2} s^{-1}), and red circles represent downregulated gene expression, previously measured in [33]. Numbers indicate fold change for each gene. Enzymes coded in Roman numerals are: i, ribulose-1,5-bisphosphate carboxylase oxygenase; ii, phosphoglycolate phosphatase; iii, glycolate oxidase; iv, glyoxylate carboligase; v, tartronate semialdehyde reductase; vi, glycine transaminase; vii, serine hydroxymethyltransferase; viii, serine-glyoxylate transaminase; ix, glycerate dehydrogenase; x, glycerate 3-kinase; xi, acetaldehyde dehydrogenase; xii, alcohol dehydrogenase; xiii, succinate-semialdehyde dehydrogenase; xiv, acetyl-CoA synthetase; xv, pyruvate dehydrogenase; xvi, formate acetyltransferase; xvii, lactate dehydrogenase; xviii, alanine dehydrogenase; xix, sucrose phosphate synthase and sucrose phosphate phosphatase; xx, formyltetrahydrofolate deformylase. Other abbreviations: αkg, α-ketoglutarate.

Finally, the transcriptomic data [33] showed upregulation of genes involved in both nitrate and sulfate uptake and assimilation under high irradiance conditions (Table A2 in Appendix C). Tradeoff curve analysis of the EFMs under high irradiance and O_2/CO_2 competition predicted increased use of the nitrate uptake reaction for strategies that secreted amino acids; conversely, no change in use of the sulfate uptake reaction was predicted. Reduction of nitrate to ammonia for amino acid synthesis represents an effective strategy for using excess electrons from the photosynthetic electron transport chain, consuming 8 moles of electrons per mole of nitrate reduced.

Thus, at high irradiance and O_2 production, secretion of amino acids represents an economical stress acclimation strategy. Similarly, sulfate reduction also consumes 8 moles of electrons per mole of sulfate reduced to hydrogen sulfide, which is used in synthesis of cysteine and methionine. Permitting secretion of cysteine or methionine in the computational model would lead to predictions of increased sulfate uptake. Altogether, comparison of the predicted competitive strategies with transcriptomic data under high irradiance (Table 3) suggests overall consistency of the computational model with photobioreactor observations.

3.5. Stress Acclimation and Photoautotrophic-Heterotrophic Interactions

BP-1 acclimation to a variety of culturing stresses was predicted to result in secretion of reduced carbon byproducts including organic acids and amino acids (Figures 3a and A2a,b; illustrated in Figure 7a). These byproducts represent a nutritional niche for heterotrophs. Photoautotrophic-heterotrophic cross-feeding could represent a mutually beneficial mechanism for buffering a photoautotroph from environmental stresses.

Consumption of reduced carbon byproducts by the heterotroph would relieve potential inhibitory organic acid stress, as well as maximize the efficiency of total resource usage by the community (illustrated in Figure 7b). Cross-feeding of byproducts could also promote growth of the photoautotroph through consumption of O_2 by an aerobic heterotroph, thus decreasing local O_2 concentrations and lowering O_2/CO_2 competition.

The amount of heterotroph able to be supported by secreted byproducts was predicted as a function of stress using published heterotrophic biomass per byproduct yields [76–82] (Figure 7c, see File S5 in the Supplementary Materials for calculations).

The predicted amount of heterotrophic biomass that can be supported by BP-1 through cross-feeding of byproducts increased as stress increased due to higher byproduct yields at higher stress levels, as well as the varying heterotrophic biomass yields on different byproducts (Table A4 in Appendix C). The cross-feeding was also predicted to reduce local O_2 levels, which was calculated based on heterotrophic biomass O_2 requirements (Figure 7d, File S5 in the Supplementary Materials).

The predicted ratio of heterotroph to photoautotroph as a function of stress acclimation was compared to published photobioreactor co-culture data of BP-1 with the aerobic heterotroph *Meiothermus ruber* strain A [33]. Experiments reported heterotroph to photoautotroph ratios of ~1:10 [33]. This ratio, with some variation accounting for cell size differences between the two populations, falls within the range of heterotroph to photoautotroph ratios predicted at modest culturing stress (Figure 7c). These predictions considered autotrophic-heterotrophic interactions based on secreted carbon and not necessarily nitrogen source. Additional analysis of potential cross-feeding based on nitrogen or iron limitation can be found in Appendix B (Figure A2c–f).

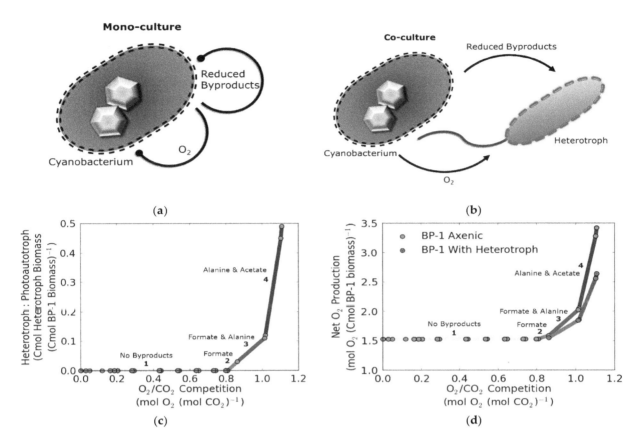

Figure 7. Byproduct secretion generates a heterotrophic niche and stimulates a mutually beneficial relationship. (**a**) Cyanobacterium BP-1 produces O_2 and reduced carbon compounds as metabolic byproducts during environmental stress, both of which are inhibitory to BP-1 growth. (**b**) The presence of reduced byproducts and O_2 forms a nutritional niche for heterotrophic organisms, which relieves inhibition for BP-1. (**c**) Heterotrophic biomass yield per BP-1 biomass (Cmol Cmol^{-1}) is presented as a function of O_2/CO_2 competition at RuBisCO (mol O_2 (mol CO_2)$^{-1}$ consumed) for the EFMs forming the optimal tradeoff with irradiance-induced stress. (**d**) Presence of a heterotroph lowers net O_2 production per Cmol BP-1 biomass as a function of O_2/CO_2 competition (mol O_2 (mol CO_2)$^{-1}$ consumed) for the EFMs forming the optimal tradeoff with irradiance-induced stress, which reduces O_2 inhibition. The distinct phenotypic regions defined by the tradeoff between O_2/CO_2 competition and irradiance-induced stress are labeled according to byproduct secretion patterns as in Figure 3a.

4. Discussion

Computational modeling was integrated with photobioreactor analyses to identify and interpret, from a systems perspective, the inferred mechanisms that underpin cyanobacterial acclimation to irradiance-associated stress. The combined results of this study show how different cyanobacterial systems, such as the photosynthetic apparatus and central carbon metabolism, can respond to environmentally induced stresses. Photobioreactor steady state growth of BP-1 showed decreased biomass production efficiency at high irradiance (Figure 4b), indicating that electrons were partitioned into non-biomass-producing alternative metabolic routes. Examination of transcriptomic data [33] comparing high to low irradiance conditions identified upregulation of genes involved in PSII operation, photorespiration, organic acid synthesis, and amino acid synthesis, among other pathways (Figure 6, Tables A2 and A3). Interrogation of BP-1 metabolic pathways with EFMA and resource allocation theory under conditions of high irradiance, high O_2, and limited nutrient availability provided a theoretical explanation for utilization of these pathways. Evolution has selected phenotypes which allocate limiting resources competitively. The origin of the stresses is the imbalance in resource acquisition which is manifested as a resource limitation. Acclimation to the resource stresses resulted in the secretion of reduced byproducts in a behavior analogous to classic overflow metabolism in heterotrophs. The byproduct-secreting phenotypes represent a competitive and economical response to the stress [1]. It is worth noting that photobioreactor observations and computational predictions for BP-1 are in general agreement with the transcriptional patterns and physiological trends observed in the closely related *Synechococcus* sp. PCC 7002 [83]. The predicted byproduct secretion profiles furthermore control nutrient niches for proximal heterotrophic partners (Figure 7). In some cases, heterotrophic consumption of the byproducts represents a mutually beneficial interaction in that byproduct removal prevents accumulation of byproducts to a degree that represents an additional stress. This mutually beneficial interaction template likely plays a significant role in the many reported occurrences of photoautotrophic-heterotrophic consortia [84–86]. In fact, cross-feeding between BP-1 and the aerobic heterotroph *M. ruber* strain A has been both predicted by genome-scale modeling and observed in a laboratory setting [14,33].

The computational analyses investigated several metabolic acclimations to photosynthesis-associated stresses that apply broadly to photoautotrophs, including photosystem utilization and photorespiration strategies, the nature of reduced carbon byproducts, and the severity of O_2/CO_2 competition at RuBisCO. PSII was predicted to increase in photon absorption relative to PSI as irradiance increased (Figure 3), supported by transcriptomic data [33] (Tables 3 and A2). Increased relative photon absorption of PSII under higher irradiance is also reported in the literature from studies with the mesophilic cyanobacterium *Synechocystis* sp. PCC 6803 and is hypothesized to aid in reducing overall electron transport [68]. Additionally, increased utilization of the C2 photorespiration cycle at high irradiances may intersect with byproduct secretion strategies and contribute to amino acid synthesis as a resource-efficient strategy at high irradiances. Photorespiration permits salvage of carbon from unusable RuBisCO oxygenation byproducts; this carbon may be directed toward other byproduct pathways. The C2 photorespiration cycle requires more enzymatic steps and thus more biosynthetic resources (e.g., nitrogen) than the glycerate pathway, but links into glycine-serine interconversion and amino acid synthesis pathways.

Formate is the least reduced organic byproduct considered in the model (Table 1). It is predicted to be a more competitive byproduct secretion strategy at intermediate irradiance-induced stress and O_2/CO_2 competition, releasing a minimal quantity of electrons in the form of reduced carbon byproducts and retaining the remaining electrons for biomass (Figure 3). Alanine is predicted to be a competitive byproduct at high stress levels due to its high degree of reduction (Table 1). At high electron load (supported by high rates of oxygenic photosynthesis) and high O_2/CO_2 competition, alanine synthesis consumes more electrons per Cmol, resulting in a more efficient redox sink (Figure 3). Alanine was selected in this study as a representative amino acid; however, amino acids with higher nitrogen content, such as arginine, histidine, or lysine, would serve as even more effective electron sinks when nitrate is the nitrogen source. Genes involved in synthesis pathways for several amino acids beside alanine (Table A3 in Appendix C) were identified as upregulated under high irradiance conditions in the transcriptomic data [33]. Additionally, qualitative measurements from BP-1 steady state cultures have previously identified a variety of amino acids in the extracellular environment, including glutamate, isoleucine, leucine, lysine, phenylalanine, serine, threonine, and valine [33].

Experimental assessment of O_2/CO_2 competition and actual concentrations of O_2 and CO_2 at the active site of RuBisCO *in vivo* is challenging. The specificity factor, a kinetic constant describing the relative affinity of RuBisCO for CO_2 versus O_2 ($v_c/v_o = \mathrm{SF}[CO_2]/[O_2]$) [87], has been measured for a variety of phototrophs and is typically obtained from enzyme extracts. Falkowski and Raven [88] compiled a list of specificity factors from a variety of organisms, including cyanobacteria, algae, and plants, and estimated v_o/v_c ratios under assumptions of air equilibrium at 25 °C. These experimental estimates were compared with the predicted v_o/v_c values from the BP-1 model irradiance tradeoff curve (Figure 8a(A)). The variation in values within and among different types of organisms highlights the diversity of RuBisCO enzyme properties, which organisms are thought to have optimized over time based on different selective pressures [89,90]. However, the experimental estimates do not account for the optimal temperature environment of the organism, the confounding influence of photosynthetic O_2 evolution, or effects of the carbon-concentrating mechanism, which may also be influenced by pH [91].

The specificity factor can be used to convert predicted O_2/CO_2 competition values to local relative O_2/CO_2 concentrations around RuBisCO [87], thereby permitting extension of stoichiometric modeling into the kinetic realm. Equivalent relative O_2/CO_2 competition values convert to different relative concentrations depending upon the magnitude of the specificity factor. Values for mesophilic cyanobacteria range from 45 to 70, and higher plants have an average value around 100 [87]. Figure 8a(B–D) shows the effect of varying the specificity factor on the O_2/CO_2 concentrations necessary to achieve the predicted v_o/v_c values along the irradiance tradeoff curve in Figure 3a; relative concentrations are lowered with a smaller specificity factor and raised with a higher specificity factor. A higher specificity factor indicates a greater tolerance to stress from O_2/CO_2 competition. *In vivo* measurements of oxygenation and carboxylation rates are sparse in the literature, particularly for microbial species; Taffs et al. [20] calculated a range of 3–7% oxygenation based on measurements of extracellularly secreted glycolate [92], but these values are likely an underestimate considering glycolate may be salvaged through the complete photorespiration pathway rather than excreted (Figure 3). Isotopic labeling studies of cyanobacteria also provide experimental data, but extrapolation of v_c/v_o ratios should be exercised with caution. Studies have shown operation of photorespiration even under high CO_2 (5%) conditions [93]. Another study has presented both modeling and experimental validation of the necessity of photorespiration even under saturating CO_2 conditions, positing that high CO_2 stimulates high photosynthetic rates to provide adequate energy for carbon fixation, which thereby leads to increased O_2 production levels [94]. Additionally, elevated temperatures have been shown to enhance oxygenation due to both changes in the specificity of RuBisCO and the different solubilities of O_2 and CO_2 [95] (Figure 8b). Experimental data on v_o/v_c. values is variable and dependent on the conditions under which the measurements were made. However, an environmental scenario with low O_2/CO_2 ratios may indicate that greater priority is placed on minimizing O_2/CO_2 competition than on minimizing photon absorption cost particularly under high irradiance conditions, e.g., O_2/CO_2 competition is a stronger driver of stress acclimation. Byproduct production and existence of heterotrophic partners is observed in experimental cyanobacterial systems, suggesting that byproduct production is an effective strategy for managing electrons from excess photon absorption. Instead, the cell may be simultaneously optimizing for other stresses such as biosynthetic nutrient investment like nitrogen or iron (Figure A2).

The systems-level analysis provided by this study indicated that the suite of metabolic carbon and electron sinks (i.e., secreted byproducts and biomass) is dependent upon environmental stressors. Pathway utilization and resource investments were co-dependent upon irradiance, O_2/CO_2 competition at RuBisCO, and DIC, nitrogen, and iron levels. These results provided novel insight into ecologically competitive metabolic strategies that cyanobacteria use to acclimate to environmental conditions. Physiological and transcriptomic [33] data paralleled the predictions, providing an additional level of support to the stoichiometric modeling predictions. It is noted that the stoichiometric model does not account for kinetic constraints, regulatory effects, or other aspects of thermodynamics beside reaction reversibilities [97,98], which may account for some of the differences between predictions and data and represents an avenue for further development. Finally, analysis of predicted optimal growth phenotypes was extended to make inferences about the nature of photoautotrophic-heterotrophic interactions and provide a theoretical basis for examining community composition. Taken holistically, this work presents a synergistic experimental and theoretical approach for understanding metabolic acclimation and provides a new level of insight into how different cyanobacterial systems, such as the photosynthetic apparatus and central carbon metabolism, coordinate and respond to environmental stresses that influence resource allocation.

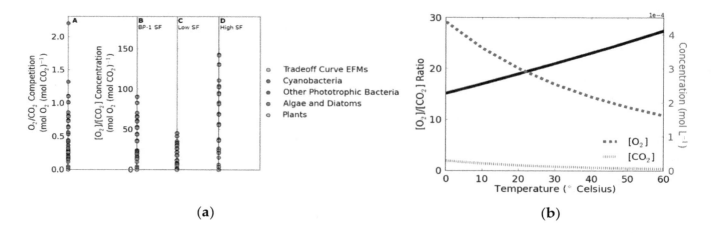

(a)   (b)

Figure 8. Comparison of computational and experimental O_2/CO_2 competition and concentrations at RuBisCO. (a) Predicted O_2/CO_2 competition values (mol O_2 (mol CO_2)$^{-1}$ consumed at RuBisCO) from the irradiance tradeoff curve in Figure 3a are shown in gray (**A**). Experimental O_2/CO_2 values for a variety of organisms calculated at air equilibrium and 25 °C [88] are overlaid in color. Predicted O_2/CO_2 competition values (gray points) were converted to relative O_2/CO_2 concentrations around RuBisCO by multiplying by the specificity factor (SF). The experimentally measured SF for BP-1 of 82 [87] was used for conversion in (**B**). Experimental data points from Falkowski and Raven (colored points) were converted to O_2/CO_2 concentrations via the respective SF of each organism [88]. Comparison of the BP-1 SF with lower and higher SF values is visualized using a lower SF of 41 (representative of *Synechococcus* sp.) (**C**) and a higher SF of 129 (representative of a red alga) (**D**). A lower SF indicates that lower relative O_2/CO_2 concentrations result in higher O_2/CO_2 competition ratios, whereas a higher SF indicates that an organism is more tolerant of higher relative O_2/CO_2 concentrations. (b) Temperature affects the relative propensity of RuBisCO for oxygenation. Dashed and dotted blue curves represent O_2 and CO_2 concentrations in aqueous phase at equilibrium with atmospheric concentrations, calculated using Henry's law constants from Sander [96]. The black curve represents the ratio of the $[O_2]$ to $[CO_2]$ curves, showing that the relative proportion of O_2 increases with elevated temperature. Calculations are provided in File S6.

Acknowledgments: This work is a contribution of the PNNL Foundational Scientific Focus Area (Principles of Microbial Community Design) subcontracted to Montana State University. Ashley E. Beck was supported by the Office of the Provost at Montana State University through the Molecular Biosciences Program and NSF (DMS-1361240). Hans C. Bernstein was supported by the Linus Pauling Distinguished Postdoctoral Fellowship which is a Laboratory Directed Research and Development program at PNNL, operated for the DOE by Battelle Memorial Institute under Contract DE-AC05-76RLO 1830. A portion of the research was performed using EMSL, a DOE Office of Science User Facility sponsored by BER under user proposal number 49356. The authors would like to acknowledge and thank Lye Meng Markillie, Eric Hill, Ryan McClure, and Margaret Romine for assisting with the culturing, RNA sequencing, and genome annotation activities that helped support this study and Reed Taffs for primary MATLAB algorithm development. The authors would also like to thank Kristopher Hunt, Zackary Jay, Heidi Schoen, Lee McGill, and William Moore for critical reading of the manuscript.

Author Contributions: A.E.B. and R.P.C. conceived and designed the experiments; A.E.B. and H.C.B. performed the experiments; A.E.B. analyzed the data; H.C.B. contributed reagents/materials/analysis tools; A.E.B. and R.P.C. wrote the manuscript.

Appendix A. Biomass Composition Analytical Methods

Appendix A.1. DNA, After Downs and Wilfinger [38]

50 μL of frozen cell pellet equivalent to approximately 1.5 mg of biomass (dry weight) was re-suspended in 50 μL of alkali extraction solution (1 N NH₄OH, 0.2% Triton X-100 with nuclease-free

water) in 2-mL Eppendorf tubes. Tubes were incubated at 37 °C for 10 min in a block heater. After 10 min, samples were diluted to 2 mL total volume with assay buffer (100 mM NaCl, 10 mM EDTA, 10 mM Tris, pH 7.0 with HCl, nuclease-free water) and transferred to 15-mL Falcon tubes for centrifugation ($2500 \times g$, 30 min, 4 °C). Calf thymus DNA standards were prepared by making a DNA stock solution in nuclease-free water about 300 μg/mL (stored at 4 °C). Exact concentration was measured with a NanoDrop 1000 spectrophotometer.

The standard solution was diluted to a working stock of 100 μg/mL with standard buffer (assay buffer containing the same concentration of alkali extraction solution as the diluted samples (100 mM NaCl, 10 mM EDTA, 10 mM Tris, pH 7.0 with HCl, 0.025 N NH_4OH, 0.005% Triton X-100)). The DNA working stock was then diluted into a standard series with standard buffer (1–5 μg/mL). 50 μL of sample or standard were added to a black polystyrene 96-well plate with clear bottom (Corning 3603). 295 μL of Hoechst working reagent was added to each well.

Hoechst working reagent was prepared fresh daily from an intermediate stock of 200 μg/mL by diluting to 1 μg/mL with assay buffer. The intermediate stock was prepared from a 10 mg/mL stock solution by diluting to 200 μg/mL with nuclease-free water.

Stock solutions and working stocks were stored at 4 °C wrapped in aluminum foil to protect from light. The wells were then read in a Synergy fluorescent plate reader using the following settings: (plate type) 96 well plate; (set temperature) setpoint 30 °C, preheat before moving to next step; (shake) double orbital 30 s, frequency 180 cpm;(read) fluorescence endpoint, 352 nm excitation, 461 nm emission, bottom optics, gain 100, Xenon flash light source, high lamp energy, normal read speed, 100 ms delay, 10 measurements/data point.

Three reaction wells of sample or standard were performed for each sample or standard. The concentration of the samples was determined based on the average of the three standard calibration curves.

Appendix A.2. Glycogen, After Del Don et al. [39]

Anthrone reagent was prepared fresh daily according to Herbert et al. [99] and stored at 4 °C. Frozen cell pellet (-80 °C) was thawed and divided into three equal parts by mass in 2-mL Eppendorf tubes, approximately 0.5 mg dry weight. Each aliquot was re-suspended in 200 μL 2% sodium sulfate (w/v). Eppendorf tubes were sealed with parafilm and heated for 10 min at 70 °C in a block heater. After heating, 1 mL methanol was added to each tube and vortexed to co-precipitate glycogen and sodium sulfate. The precipitate was pelleted by centrifuging for 15 s at 10,000 rpm. The precipitate was washed with 1 mL methanol, until the pellet was white, to remove impurities. Pellets were then re-suspended in 1 mL reverse osmosis water and transferred to clean glass test tubes and placed on ice to chill. 5 mL of ice-cold anthrone reagent was added to each test tube. After adding reagent, tubes were chilled on ice for 5 min, vortexed gently to homogenize the solution, and transferred to a boiling water bath for 10 min. Tubes were then returned to ice for 5–10 min until cool, vortexed gently to mix contents, and absorbance at 625 nm was read with a Genysys spectrophotometer using a reagent blank. A glucose standard curve (10–190 μg/mL) was treated identically with anthrone reagent.

For total carbohydrate quantitation, the cell pellet aliquot was re-suspended in 1 mL reverse osmosis water and transferred to a clean glass test tube, and the anthrone procedure detailed above was followed. For quantitation of other cellular carbohydrates, the residual methanol from the extraction and washings were collected in an aluminum pan and evaporated, re-suspended in 1 mL reverse osmosis water, and the anthrone procedure detailed above was followed.

Appendix A.3. Lipid, After Bligh and Dyer [40]

Frozen cell pellet (10 mg) was re-suspended to 0.6 mL using Milli-Q water in a 15-mL polypropylene centrifuge tube. Chloroform (0.75 mL) and methanol (1.5 mL) were sequentially added, adhering to the 1:2:0.8 chloroform:methanol:water volume ratio recommended by Bligh and Dyer. The mixture was vortexed 15 min at speed setting 3 using a VWR vortex mixer. Chloroform (0.75 mL) and Milli-Q water (0.75 mL) were sequentially added, vortexing 10–15 seconds at speed setting 7 after each addition. Upon centrifugation (4000 rpm, 15 min, 20 °C), the lower chloroform phase, containing lipids, chlorophyll, and pigments, was transferred via micropipette to an aluminum pan that had been pre-dried at room temperature and pre-weighed. The liquid was evaporated in a fume hood and weighed at three different time intervals following evaporation. Weights were measured with a Mettler Toledo MT5 microbalance with accuracy to 0.001 mg and recorded as an average of three measurements. It was noted that chloroform may leach compounds from polypropylene materials; thus a blank reaction using 0.6 mL Milli-Q water was used and its weight was subtracted from the biological sample weight.

Appendix A.4. Protein and Amino Acid Distribution, After Henderson et al. [39]

Amount approximately equivalent to 3 mg of frozen cell pellet was transferred to borosilicate HPLC vials with PTFE/silicone caps. 50 µL 6 M HCl per mg biomass was added to each vial. The vials were tightly capped and hydrolyzed at 105 °C for 24 h using a block heater. After 24 h, the samples were then neutralized with 6 M NaOH to pH 7.0 and filtered with 0.22 µm PES spin filter in microfuge for 5 min at 10,000 rpm. Samples were then placed at -80 °C to freeze before lyophilizing for 24 h (VirTis benchtop lyophilizer). After lyophilization samples were placed at -80 °C until HPLC analysis. HPLC analysis was performed according to the following protocol validated and published by Agilent Technologies [41] using an Agilent 1100 HPLC equipped with fluorescence detector. Borate buffer was 0.4 N borate, pH 10.2 with NaOH; *o*-phthalaldehyde (OPA) reagent, 9-fluorenylmethylchloroformate (FMOC) reagent, and amino acid standards were obtained from Agilent. OPA and FMOC reagents were replaced daily in amber vials. Upon opening a vial of reagent, analyses were performed within 10 days. Solvent A was 40 mM sodium phosphate buffer (using 1:1 ratio of NaH_2PO_4 and Na_2HPO_4), pH 7.8 with NaOH, 0.2 µm filtered. Solvent B was 45:45:10 acetonitrile:methanol:water ($v/v/v$), 0.2 µm filtered. The pump rate was 1 mL/min, 47 min per injection, with gradient settings as follows:

Time (Min)	% Solvent B
0	0
3.8	0
36.2	57
37.2	100
44.6	100
46.4	0
47	0

The flow rate was halved and the timing was doubled from the procedure reported in the Agilent technical note to improve resolution and reduce wear on equipment. The column thermostat was set at 40 °C, and the autosampler thermostat was set at 4 °C. The fluorescence detector settings were as followed, to switch from OPA- to FMOC-derivatized amino acids:

Time (Min)	Ex/Em (nm)	PMT Gain
0	340/450	10
30	266/305	9

The injection program was as follows:

Step	Instruction
Step 1	Draw 2.5 µL from vial 1 (borate buffer)
Step 2	Draw 0.5 µL from sample
Step 3	Mix 3 µL in air, max speed, 2×
Step 4	Wait 0.5 min
Step 5	Draw 0 µL from vial 2 (needle wash)
Step 6	Draw 0.5 µL from vial 3 (OPA)
Step 7	Mix 3.5 µL in air, max speed, 6×
Step 8	Draw 0 µL from vial 2 (needle wash)
Step 9	Draw 0.5 µL from vial 4 (FMOC)
Step 10	Mix 4 µL in air, max speed, 6×
Step 11	Draw 32 µL from vial 5 (water)
Step 12	Mix 18 µL in air, max speed, 2×
Step 13	Inject
Auxiliary settings	Drawspeed = 200 µL/min
	Ejectspeed = 600 µL/min
	Draw position = 0.0 mm

The integration parameters for collecting the data were set according to the following parameters.

Parameter	Value
Slope Sensitivity	1
Peak Width	0.04
Area Reject	1
Height Reject	0.4
Shoulders	OFF

Appendix A.5. RNA, After Benthin et al. *[43]*

Samples were thawed and washed three times with 3 mL 0.7 M $HClO_4$ for degradation of cell walls, vortexing to re-suspend in between washing and centrifuging at 4000 rpm for 10 min at 4 °C. The pellet was then re-suspended in 3 mL 0.3 M KOH to lyse the cells and was incubated in a 37 °C water bath for 1 h, shaking at 15-min intervals. After 1 h, samples were cooled and 1 mL 3 M $HClO_4$ was added for neutralization. The solution was centrifuged at the same specifications as before, and the supernatant was poured off into a new centrifuge tube. The pellet was washed twice with 4 mL 0.5 M $HClO_4$, centrifuged, and supernatant added to the new tube. 0.5 M $HClO_4$ extracts the RNA, while DNA, which is stable even in strong alkali, and protein, which does not solubilize in the alkali, remain in the precipitate. The collection of extracts was made up to a volume of 15 mL by adding 3 mL 0.5 M $HClO_4$ and was centrifuged once more to remove any non-visible precipitates of $KClO_4$. Upon final centrifugation, absorbance was measured at 260 nm against a 0.5 M $HClO_4$ blank using disposable UV cuvettes rated to 220 nm. Linearity of the spectrophotometer was confirmed within that range by successively diluting the sample twice with 0.5 M $HClO_4$ and confirming a linear fit to the three measured absorbances at 260 nm. Calculation of RNA quantity was performed by assuming 1 unit of absorbance at 260 nm corresponds to 38 µg/mL RNA on average [100].

Appendix B. Supplemental Figures

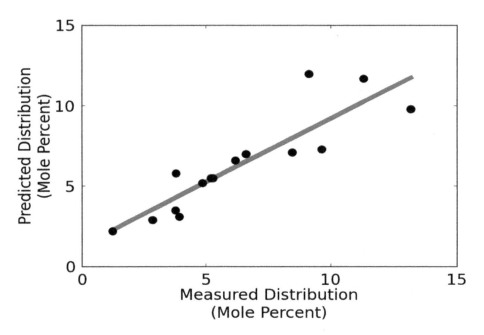

Figure A1. BP-1 amino acid distribution. Correlation between the predicted amino acid distribution (based on protein-coding gene sequences) and the experimentally measured distribution. Cysteine, methionine, and tryptophan are excluded from the correlation due to degradation during hydrolysis of the biomass samples to extract the protein. The equation of the best-fit linear trendline is $y = 0.79x + 1.29$ with $R^2 = 0.79$.

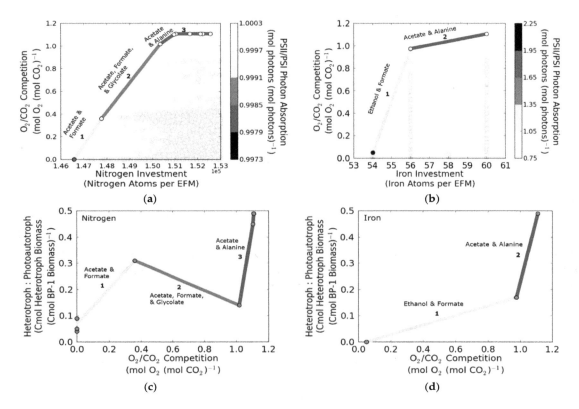

Figure A2. *Cont.*

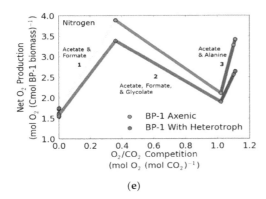
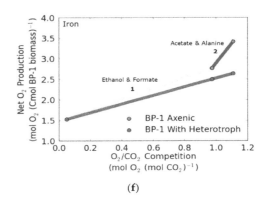

(e)   (f)

Figure A2. Computational analysis of nutrient availability and O_2/CO_2 competition at RuBisCO in cyanobacterium BP-1. (a) Nitrogen availability. O_2/CO_2 competition (mol O_2 (mol CO_2)$^{-1}$ consumed at RuBisCO) as a function of nitrogen investment (nitrogen atoms per EFM) for biomass-producing EFMs. The tradeoff curve defining competitive strategies between O_2/CO_2 competition and nitrogen limitation was divided into three distinct phenotypic regions based on byproduct secretion, labeled accordingly (intensity of shading increases with increasing stress levels). Points in the plot area are representative of 1,430,252 EFMs. **(b)** Iron availability. O_2/CO_2 competition (mol O_2 (mol CO_2)$^{-1}$ consumed at RuBisCO) as a function of iron investment (iron atoms per EFM, considering only photosynthetic and central metabolism reactions) for biomass-producing EFMs. The tradeoff curve defining competitive strategies between O_2/CO_2 competition and iron limitation was divided into two distinct phenotypic regions based on byproduct secretion, labeled accordingly (intensity of shading increases with increasing stress levels). Color scale represents the photon absorption at PSII relative to PSI for EFMs on the tradeoff curve (mol photons absorbed at PSII (mol photons absorbed at PSI)$^{-1}$). Each point represents a unique EFM. Modeled biomass production did not include maintenance energy requirements. Points in the plot area are representative of 4,615,500 EFMs. **(c,d)** Heterotrophic biomass yield per BP-1 biomass (Cmol Cmol^{-1}) is presented as a function of O_2/CO_2 competition at RuBisCO (mol O_2 (mol CO_2)$^{-1}$ consumed) for the EFMs forming the optimal tradeoffs with nitrogen and iron availability, respectively. **(e,f)** Presence of a heterotroph lowers net O_2 production per Cmol BP-1 biomass as a function of O_2/CO_2 competition (mol O_2 (mol CO_2)$^{-1}$ consumed) for the EFMs forming the optimal tradeoffs with nitrogen and iron availability, respectively, which reduces O_2 inhibition. The distinct phenotypic regions defined by the tradeoff between O_2/CO_2 competition and nutrient availability stress are labeled according to byproduct secretion patterns as in panels **(a,b)**.

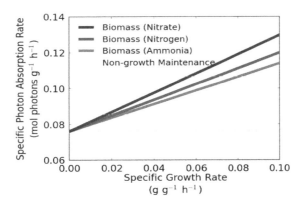

Figure A3. Analysis of influence of nitrogen source on photon requirement. Comparison of the impact of various nitrogen sources on the theoretical specific photon absorption rate necessary for biomass production. Photon absorption rates (mol photons absorbed (g CDW)$^{-1}$ h^{-1}) were calculated using the BP-1 experimentally measured biomass composition. Ammonia is a completely reduced form of nitrogen, whereas and molecular nitrogen and nitrate are less reduced forms and require successively more energy for reduction to be assimilated into biomass, causing an increase in the specific photon absorption rate. Nitrate serves as the most effective sink for excess electrons from the photosynthetic electron transport chain.

Appendix C. Supplemental Tables

Table A1. Experimentally measured amino acid distribution from OPA/FMOC derivatization and HPLC fluorescence detection. Amino acids are abbreviated according to IUPAC 1-letter convention; average mole percent of two separately hydrolyzed samples and percent relative standard deviation are reported.

Amino Acid	Mole %	% RSD
Q/E	13.1	0.16
N/D	10.1	0.07
L	9.4	0.08
A	9.2	0.24
R	8.5	0.96
V	6.1	0.97
I	5.4	0.29
G	5.0	0.02
F	4.9	0.15
T	4.9	0.07
K	4.5	1.96
Y	4.1	0.01
S	4.0	0.08
P	3.4	1.21
H	1.5	0.14

Table A2. Upregulated BP-1 genes under high versus low irradiance conditions (2000 versus 200 µmol photons $m^{-2} s^{-1}$) [33] involved in photosystem II (PSII), carbon-concentrating mechanism (CCM), Calvin cycle carbon fixation, glycolysis, TCA cycle, oxidative phosphorylation (OP), and nitrate and sulfate uptake and assimilation. For each pathway, upregulated genes are listed in the left column with corresponding fold change in the right column.

PSII		CCM		Calvin		Glycolysis	
psbA3	42.0	ccmK1	2.5	gap2	5.3	eno	4.6
psbP	4.3	ccmK2	2.1	gapA	4.5	gap2	5.3
psbX	3.3	ccmK3	2.6	glpX	6.6	gapA	4.5
psbQ	2.7	ccmM	2.5	pgk	2.1	gmpA	6.4
psb29	2.6	tlr0311	2.4	prkB	13.4	gpmI	5.4
psb32	4.1			rpiA	16.2	pfkA	2.3
				tpiA	5.2	pgk	2.1
						pyk	3.5
						tpiA	5.2

TCA		OP		Nitrate		Sulfate	
acnB	2.2	atpC	2.2	narM	2.3	cysA	2.5
fumC	3.2	atpD	5.2	nirA	3.4	cysC1	3.9
gltA	4.7	atpE	2.2	ntcB	4.9	cysH	4.6
idh3	6.3	atpF	2.9	nrtA	2.2	cysQ	3.1
sdhC	2.0	atpG	3.2	nrtD	2.5	met3	8.5
		atpH	2.8	tll1357	3.1		
		cydB	3.6				
		ndh	2.0				
		ndhA	6.3				
		ndhB	4.4				
		ndhC	6.1				
		ndhD3	13.2				
		ndhE	4.7				
		ndhF3	2.4				
		ndhG	3.0				
		ndhH	2.2				
		ndhI	4.9				
		ndhJ	6.6				
		ndhK	3.1				
		ndhL	4.9				
		ndhM	13.1				
		ndhN	9.8				
		ppa	3.6				
		sdhC	2.0				

Table A3. Genes involved in amino acid synthesis pathways found to be upregulated under high irradiance conditions (2000 versus 200 μmol photons m^{-2} s^{-1}) [33]. Amino acids are abbreviated according to IUPAC 1-letter convention and number of nitrogen atoms follows in parentheses. For each amino acid, upregulated genes are listed in the left column with corresponding fold change in the right column.

R (4)		N/D (2/1)		C (1)		Q (2)		H (3)		I/L/V (1/1/1)		K (2)		M (1)		F/W/Y (1/1/1)		P (1)		S (1)		T (1)		Transporters	
argB	3.0	asnB	2.5	cysE	4.0	glnA	3.6	hisA	3.4	avtA	2.8	dapA	2.9	metE	12.9	aroA	9.2	proB	6.7	glyA	4.5	thrA	2.6	slr1046	2.2
argC	4.8	asnA2	3.0					hisB	3.9	cimA	2.5	dapB	4.5			aroB	4.3			serA	7.8	thrB1	3.7	slr1067	3.4
argD	2.5	aspB	2.7					hisIE	6.7	ilvE	5.6	dapF	4.1			aroC	4.3					thrCII	6.3	sll1105	2.4
argG	14.0							hisG	2.3	leuB	6.0	dapL1	4.0			aroE	2.4							slr1120	16.5
argH	4.4							hisF	2.0	leuC	2.9	lysA	3.3			trpA	2.6							slr1121	3.6
argJ	4.0											lysC	2.1			trpB	2.5							slr1151	2.3
																trpC	2.2							sll1318	2.7
																trpE	4.1							slr1592	2.1
																trpF	2.3								
																tyrAa	2.4								

Table A4. Byproduct yields with respect to BP-1 biomass for pathways that secreted byproducts along the tradeoff curve for irradiance-induced stress and O_2/CO_2 competition and the corresponding yield of heterotroph biomass. Heterotroph yield was obtained via substrate consumption costs estimated from the literature for the respective byproducts (Cmol per Cmol heterotroph biomass): 4.53 for formate [77], 2.66 for alanine [78], and 2.61 for acetate [76]. Photons absorbed per BP-1 are reported in mol Cmol^{-1}, O_2/CO_2 competition is reported in mol mol^{-1}, and yields are reported in Cmol Cmol^{-1}

Photons Absorbed Per BP-1 Biomass	O$_2$/CO$_2$ Competition	Formate Per BP-1 Biomass	Alanine Per BP-1 Biomass	Acetate Per BP-1 Biomass	Heterotroph Biomass Per BP-1 Biomass
33.26	0.86	0.13	NA	NA	0.03
51.36	1.01	NA	0.30	NA	0.11
52.44	1.02	NA	0.31	0.01	0.12
100.05	1.10	NA	0.84	0.36	0.45
105.55	1.11	NA	0.90	0.40	0.49

Appendix D. Nitrogen and Iron Limitation

Nitrogen is a major component of protein but is often scarce in the environment [101]. Ecologically competitive acclimation to increased O_2/CO_2 competition as a function of pathway nitrogen requirement is shown in Figure A2a in Appendix B; the tradeoff surface defines three phenotypic regions according to byproduct secretion. The EFM at the lower left corner of the plot represents nitrogen-limited cyanobacterial growth under low O_2/CO_2 competition; byproduct secretion is predicted even at the lowest nitrogen stress. With increasing O_2/CO_2 competition and nitrogen investment, BP-1 is predicted to secrete a number of reduced carbon compounds along the tradeoff surface, including acetate, formate, glycolate, and under the highest stress, the amino acid alanine. Production of byproducts is predicted to achieve the most efficient nitrogen utilization while simultaneously minimizing O_2/CO_2 competition. EFMs on the nitrogen tradeoff surface exclusively use the C2 photorespiration cycle whereas the glycerate pathway is not used (Figure A2a in Appendix B), similar to the result for irradiance-induced stress in Figure 3a. Also similar to the result for irradiance-induced stress, relative PSII/PSI photon absorption increases along the tradeoff surface, with greater relative photon absorption at PSII at higher O_2/CO_2 competition and higher nitrogen investment (Figure A2a in Appendix B).

Biologically available iron is often limiting in microbial habitats due to low solubility, which is exacerbated at elevated pH [102]. Figure A2b in Appendix B predicts acclimation to increased O_2/CO_2 competition as a function of pathway iron investment. Two phenotypic regions of byproduct secretion were defined by the tradeoff surface, including combinations of ethanol, formate, and acetate, and finally alanine under the highest stress. Under low iron, relative PSII/PSI photon absorption decreases along the tradeoff surface as O_2/CO_2 competition and iron investment increase, showing a reversed trend compared with irradiance-induced stress and nitrogen investment. Analogous to the result for irradiance-induced stress in Figure 3a, the C2 cycle is the predominant photorespiration strategy as O_2/CO_2 competition increases. At low O_2/CO_2 competition, the relative PSII/PSI photon absorption is nearly twice that at high O_2/CO_2 competition (the scale shows greater variability than the scale for light stress or nitrogen investment in Figures 3a and A3a) and again indicates higher gross O_2 and ATP production. Additionally, as compared to the tradeoff surface in Figure 3a, the responses to nitrogen and iron limitation are less robust; fewer suboptimal pathways exist in close proximity to the tradeoff surface.

Appendix E. Biomass Yield Comparison

The physiological light response experiments allowed for comparison of photon costs for synthesizing BP-1 biomass with results reported in previously published studies for *Cyanothece* sp. ATCC 51142 and the green alga *Chlamydomonas reinhardtii* [30,103]. This comparison is of interest because experimental photon requirement values are relatively uncommon. The photon cost per biomass for BP-1 was about three times higher than the costs for these two organisms. This result may be due to the thermophilic nature of the organism and/or higher maintenance costs incurred by alkaline habitats. The nonlinear increase observed in the overall experimental photon absorption rate as growth rate increases (Figure 5) may correspond to increased cellular stress with higher maintenance energy requirements or greater thermal dissipation at higher irradiances. For example, higher irradiance may necessitate increased repair of photosystem proteins or a greater proportion of light may be lost to inefficiency [88]. These experiments and simulations demonstrate the wide range of irradiances under which BP-1 is capable of growing, stimulating interest in the metabolic strategies microbes such as this thermophilic cyanobacterium use to manage the daily fluctuations in irradiance and the accompanying stresses.

References

1. Folsom, J.P.; Carlson, R.P. Physiological, biomass elemental composition and proteomic analyses of *Escherichia coli* ammonium-limited chemostat growth, and comparison with iron- and glucose-limited chemostat growth. *Microbiology* **2015**, *161*, 1659–1670. [CrossRef] [PubMed]

2. Schimel, J.; Balser, T.C.; Wallenstein, M. Microbial stress-response physiology and its implications for ecosystem function. *Ecology* **2007**, *88*, 1386–1394. [CrossRef] [PubMed]

3. Rousseaux, C.S.; Gregg, W.W. Interannual variation in phytoplankton primary production at a global scale. *Remote Sens.* **2014**, *6*, 1–19. [CrossRef]

4. Bullerjahn, G.S.; Post, A.F. Physiology and molecular biology of aquatic cyanobacteria. *Front. Microbiol.* **2014**, *5*. [CrossRef] [PubMed]

5. Abed, R.M.M.; Dobretsov, S.; Sudesh, K. Applications of cyanobacteria in biotechnology. *J. Appl. Microbiol.* **2009**, *106*, 1–12. [CrossRef] [PubMed]

6. Singh, S.; Kate, B.N.; Banerjee, U.C. Bioactive compounds from cyanobacteria and microalgae: An overview. *Crit. Rev. Biotechnol.* **2005**, *25*, 73–95. [CrossRef] [PubMed]

7. Raven, J.A. The cost of photoinhibition. *Physiol. Plant.* **2011**, *142*, 87–104. [CrossRef] [PubMed]

8. Bailey, S.; Grossman, A. Photoprotection in cyanobacteria: Regulation of light harvesting. *Photochem. Photobiol.* **2008**, *84*, 1410–1420. [CrossRef] [PubMed]

9. Bernstein, H.C.; Kesaano, M.; Moll, K.; Smith, T.; Gerlach, R.; Carlson, R.P.; Miller, C.D.; Peyton, B.M.; Cooksey, K.E.; Gardner, R.D.; et al. Direct measurement and characterization of active photosynthesis zones inside wastewater remediating and biofuel producing microalgal biofilms. *Bioresour. Technol.* **2014**, *156*, 206–215. [CrossRef] [PubMed]

10. Ward, D.M.; Bateson, M.M.; Ferris, M.J.; Kuhl, M.; Wieland, A.; Koeppel, A.; Cohan, F.M. Cyanobacterial ecotypes in the microbial mat community of Mushroom Spring (Yellowstone National Park, Wyoming) as species-like units linking microbial community composition, structure and function. *Philos. Trans. R. Soc. B* **2006**, *361*, 1997–2008. [CrossRef] [PubMed]

11. Eisenhut, M.; Ruth, W.; Haimovich, M.; Bauwe, H.; Kaplan, A.; Hagemann, M. The photorespiratory glycolate metabolism is essential for cyanobacteria and might have been conveyed endosymbiontically to plants. *Proc. Natl. Acad. Sci. USA* **2008**, *105*, 17199–17204. [CrossRef] [PubMed]

12. Espie, G.S.; Kimber, M.S. Carboxysomes: Cyanobacterial rubisco comes in small packages. *Photosynth. Res.* **2011**, *109*, 7–20. [CrossRef] [PubMed]

13. Rae, B.D.; Long, B.M.; Whitehead, L.F.; Forster, B.; Badger, M.R.; Price, G.D. Cyanobacterial carboxysomes: Microcompartments that facilitate CO_2 fixation. *J. Mol. Microbiol. Biotechnol.* **2013**, *23*, 300–307. [CrossRef] [PubMed]

14. Henry, C.S.; Bernstein, H.C.; Weisenhorn, P.; Taylor, R.C.; Lee, J.Y.; Zucker, J.; Song, H.S. Microbial community metabolic modeling: A community data-driven network reconstruction. *J. Cell. Physiol.* **2016**, *231*, 2339–2345. [CrossRef] [PubMed]

15. Llaneras, F.; Pico, J. Stoichiometric modelling of cell metabolism. *J. Biosci. Bioeng.* **2008**, *105*, 1–11. [CrossRef] [PubMed]

16. Maarleveld, T.R.; Khandelwal, R.A.; Olivier, B.G.; Teusink, B.; Bruggeman, F.J. Basic concepts and principles of stoichiometric modeling of metabolic networks. *Biotechnol. J.* **2013**, *8*, 997–1008. [CrossRef] [PubMed]

17. Orth, J.D.; Thiele, I.; Palsson, B.O. What is flux balance analysis? *Nat. Biotechnol.* **2010**, *28*, 245–248. [CrossRef] [PubMed]

18. Perez-Garcia, O.; Lear, G.; Singhal, N. Metabolic network modeling of microbial interactions in natural and engineered environmental systems. *Front. Microbiol.* **2016**, *7*. [CrossRef] [PubMed]

19. Trinh, C.T.; Wlaschin, A.; Srienc, F. Elementary mode analysis: A useful metabolic pathway analysis tool for characterizing cellular metabolism. *Appl. Microbiol. Biotechnol.* **2009**, *81*, 813–826. [CrossRef] [PubMed]

20. Taffs, R.; Aston, J.E.; Brileya, K.; Jay, Z.; Klatt, C.G.; McGlynn, S.; Mallette, N.; Montross, S.; Gerlach, R.; Inskeep, W.P.; et al. In silico approaches to study mass and energy flows in microbial consortia: A syntrophic case study. *BMC Syst. Biol.* **2009**, *3*, 114. [CrossRef] [PubMed]

21. Schuster, S.; Hilgetag, C. On elementary flux modes in biochemical reaction systems at steady state. *J. Biol. Syst.* **1994**, *2*, 165–182. [CrossRef]

22. Beck, A.E.; Hunt, K.A.; Bernstein, H.C.; Carlson, R.P. Interpreting and designing microbial communities for bioprocess applications, from components to interactions to emergent properties. In *Biotechnology for Biofuel Production and Optimization*; Eckert, C.E., Trinh, C.T., Eds.; Elsevier: Amsterdam, The Netherlands, 2016; pp. 407–432.

23. Carlson, R.P. Decomposition of complex microbial behaviors into resource-based stress responses. *Bioinformatics* **2009**, *25*, 90–97. [CrossRef] [PubMed]

24. Carlson, R.P.; Oshota, O.J.; Taffs, R.L. Systems analysis of microbial adaptations to simultaneous stresses. In *Reprogramming Microbial Metabolic Pathways*; Wang, X., Chen, J., Quinn, P., Eds.; Springer: Dordrecht, The Netherlands, 2012; pp. 139–157.

25. Molenaar, D.; van Berlo, R.; de Ridder, D.; Teusink, B. Shifts in growth strategies reflect tradeoffs in cellular economics. *Mol. Syst. Biol.* **2009**, *5*. [CrossRef] [PubMed]

26. Mori, M.; Hwa, T.; Martin, O.C.; De Martino, A.; Marinari, E. Constrained allocation flux balance analysis. *PLoS Comput. Biol.* **2016**, *12*, e1004913. [CrossRef] [PubMed]

27. Knoop, H.; Grundel, M.; Zilliges, Y.; Lehmann, R.; Hoffmann, S.; Lockau, W.; Steuer, R. Flux balance analysis of cyanobacterial metabolism: The metabolic network of *Synechocystis* sp. PCC 6803. *PLoS Comput. Biol.* **2013**, *9*, e1003081. [CrossRef] [PubMed]

28. Nogales, J.; Gudmundsson, S.; Knight, E.M.; Palsson, B.O.; Thiele, I. Detailing the optimality of photosynthesis in cyanobacteria through systems biology analysis. *Proc. Natl. Acad. Sci. USA* **2012**, *109*, 2678–2683. [CrossRef] [PubMed]

29. Vu, T.T.; Hill, E.A.; Kucek, L.A.; Konopka, A.E.; Beliaev, A.S.; Reed, J.L. Computational evaluation of *Synechococcus* sp. PCC 7002 metabolism for chemical production. *Biotechnol. J.* **2013**, *8*, 619–630. [CrossRef] [PubMed]

30. Vu, T.T.; Stolyar, S.M.; Pinchuk, G.E.; Hill, E.A.; Kucek, L.A.; Brown, R.N.; Lipton, M.S.; Osterman, A.; Fredrickson, J.K.; Konopka, A.E.; et al. Genome-scale modeling of light-driven reductant partitioning and carbon fluxes in diazotrophic unicellular cyanobacterium *Cyanothece* sp. ATCC 51142. *PLoS Comput. Biol.* **2012**, *8*, e1002460. [CrossRef] [PubMed]

31. Everroad, R.C.; Otaki, H.; Matsuura, K.; Haruta, S. Diversification of bacterial community composition along a temperature gradient at a thermal spring. *Microbes Environ.* **2012**, *27*, 374–381. [CrossRef] [PubMed]

32. Yamaoka, T.; Satoh, K.; Katoh, S. Photosynthetic activities of a thermophilic blue-green-alga. *Plant Cell Physiol.* **1978**, *19*, 943–954. [CrossRef]

33. Bernstein, H.C.; McClure, R.S.; Thiel, V.; Sadler, N.C.; Kim, Y.M.; Chrisler, W.B.; Hill, E.A.; Bryant, D.A.; Romine, M.F.; Jansson, J.K.; et al. Indirect interspecies regulation; transcriptional and physiological responses of a cyanobacterium to heterotrophic partnership. *mSystems* **2017**, *2*. [CrossRef] [PubMed]

34. Stanier, R.Y.; Kunisawa, R.; Mandel, M.; Cohen-Bazire, G. Purification and properties of unicellular blue-green algae (order Chroococcales). *Bacteriol. Rev.* **1971**, *35*, 171–205. [PubMed]

35. Bernstein, H.C.; Konopka, A.; Melnicki, M.R.; Hill, E.A.; Kucek, L.A.; Zhang, S.Y.; Shen, G.Z.; Bryant, D.A.; Beliaev, A.S. Effect of mono- and dichromatic light quality on growth rates and photosynthetic performance of *Synechococcus* sp. PCC 7002. *Front. Microbiol.* **2014**, *5*. [CrossRef] [PubMed]

36. Melnicki, M.R.; Pinchuk, G.E.; Hill, E.A.; Kucek, L.A.; Stolyar, S.M.; Fredrickson, J.K.; Konopka, A.E.; Beliaev, A.S. Feedback-controlled led photobioreactor for photophysiological studies of cyanobacteria. *Bioresour. Technol.* **2013**, *134*, 127–133. [CrossRef] [PubMed]

37. Shibata, K.; Benson, A.A.; Calvin, M. The absorption spectra of suspensions of living micro-organisms. *Biochim. Biophys. Acta* **1954**, *15*, 461–470. [CrossRef]

38. Downs, T.R.; Wilfinger, W.W. Fluorometric quantification of DNA in cells and tissue. *Anal. Biochem.* **1983**, *131*, 538–547. [CrossRef]

39. Del Don, C.; Hanselmann, K.W.; Peduzzi, R.; Bachofen, R. Biomass composition and methods for the determination of metabolic reserve polymers in phototrophic sulfur bacteria. *Aquat. Sci.* **1994**, *56*, 1–15. [CrossRef]

40. Bligh, E.G.; Dyer, W.J. A rapid method of total lipid extraction and purification. *Can. J. Biochem. Physiol.* **1959**, *37*, 911–917. [CrossRef] [PubMed]

41. Henderson, J.W.; Ricker, R.D.; Bidlingmeyer, B.A.; Woodward, C. *Rapid, Accurate, Sensitive, and Reproducible HPLC Analysis of Amino Acids*; Agilent Technologies: Santa Clara, CA, USA, 2000.

42. Fountoulakis, M.; Lahm, H.W. Hydrolysis and amino acid composition analysis of proteins. *J. Chromatogr. A* **1998**, *826*, 109–134. [CrossRef]

43. Benthin, S.; Nielsen, J.; Villadsen, J. A simple and reliable method for the determination of cellular RNA-content. *Biotechnol. Tech.* **1991**, *5*, 39–42. [CrossRef]

44. Klamt, S.; Saez-Rodriguez, J.; Gilles, E.D. Structural and functional analysis of cellular networks with CellNetAnalyzer. *BMC Syst. Biol.* **2007**, *1*. [CrossRef] [PubMed]

45. Klamt, S.; von Kamp, A. An application programming interface for CellNetAnalyzer. *Biosystems* **2011**, *105*, 162–168. [CrossRef] [PubMed]

46. Nakamura, Y.; Kaneko, T.; Sato, S.; Ikeuchi, M.; Katoh, H.; Sasamoto, S.; Watanabe, A.; Iriguchi, M.; Kawashima, K.; Kimura, T.; et al. Complete genome structure of the thermophilic cyanobacterium *Thermosynechococcus elongatus* BP-1. *DNA Res.* **2002**, *9*, 123–130. [CrossRef] [PubMed]

47. Caspi, R.; Billington, R.; Ferrer, L.; Foerster, H.; Fulcher, C.A.; Keseler, I.M.; Kothari, A.; Krummenacker, M.; Latendresse, M.; Mueller, L.A.; et al. The MetaCyc database of metabolic pathways and enzymes and the BioCyc collection of pathway/genome databases. *Nucleic Acids Res.* **2016**, *44*, D471–D480. [CrossRef] [PubMed]

48. Kanehisa, M.; Goto, S.; Sato, Y.; Furumichi, M.; Tanabe, M. KEGG for integration and interpretation of large scale molecular data sets. *Nucleic Acids Res.* **2012**, *40*, D109–D114. [CrossRef] [PubMed]

49. Schomburg, I.; Chang, A.; Placzek, S.; Sohngen, C.; Rother, M.; Lang, M.; Munaretto, C.; Ulas, S.; Stelzer, M.; Grote, A.; et al. BRENDA in 2013: Integrated reactions, kinetic data, enzyme function data, improved disease classification: New options and contents in BRENDA. *Nucleic Acids Res.* **2013**, *41*, D764–D772. [CrossRef] [PubMed]

50. Beck, C.; Knoop, H.; Axmann, I.M.; Steuer, R. The diversity of cyanobacterial metabolism: Genome analysis of multiple phototrophic microorganisms. *BMC Genom.* **2012**, *13*, 56. [CrossRef] [PubMed]

51. Neidhardt, F.C.; Ingraham, J.L.; Schaechter, M. *Physiology of the Bacterial Cell: A Molecular Approach*; Sinauer Associates: Sunderland, MA, USA, 1990.

52. Maslova, I.P.; Mouradyan, E.A.; Lapina, S.S.; Klyachko-Gurvich, G.L.; Los, D.A. Lipid fatty acid composition and thermophilicity of cyanobacteria. *Russ. J. Plant Physiol.* **2004**, *51*, 353–360. [CrossRef]

53. Miyairi, S. CO_2 assimilation in a thermophilic cyanobacterium. *Energy Convers. Manag.* **1995**, *6–9*, 763–766. [CrossRef]

54. Petroutsos, D.; Amiar, S.; Abida, H.; Dolch, L.J.; Bastien, O.; Rebeille, F.; Jouhet, J.; Falconet, D.; Block, M.A.; McFadden, G.I.; et al. Evolution of galactoglycerolipid biosynthetic pathways—From cyanobacteria to primary plastids and from primary to secondary plastids. *Prog. Lipid Res.* **2014**, *54*, 68–85. [CrossRef] [PubMed]

55. Roels, J.A. *Energetics and Kinetics in Biotechnology*; Elsevier Biomedical Press: Amsterdam, The Netherlands, 1983.

56. Oliver, J.W.K.; Atsumi, S. Metabolic design for cyanobacterial chemical synthesis. *Photosynth. Res.* **2014**, *120*, 249–261. [CrossRef] [PubMed]

57. Shikanai, T.; Munekage, Y.; Kimura, K. Regulation of proton-to-electron stoichiometry in photosynthetic electron transport: Physiological function in photoprotection. *J. Plant Res.* **2002**, *115*, 3–10. [CrossRef] [PubMed]

58. Flamholz, A.; Noor, E.; Bar-Even, A.; Milo, R. eQuilibrator-the biochemical thermodynamics calculator. *Nucleic Acids Res.* **2012**, *40*, D770–D775. [CrossRef] [PubMed]

59. Noor, E.; Bar-Even, A.; Flamholz, A.; Lubling, Y.; Davidi, D.; Milo, R. An integrated open framework for thermodynamics of reactions that combines accuracy and coverage. *Bioinformatics* **2012**, *28*, 2037–2044. [CrossRef] [PubMed]

60. Raven, J.A.; Evans, M.C.W.; Korb, R.E. The role of trace metals in photosynthetic electron transport in O_2-evolving organisms. *Photosynth. Res.* **1999**, *60*, 111–149. [CrossRef]

61. Terzer, M.; Stelling, J. Large-scale computation of elementary flux modes with bit pattern trees. *Bioinformatics* **2008**, *24*, 2229–2235. [CrossRef] [PubMed]

62. Klahn, S.; Hagemann, M. Compatible solute biosynthesis in cyanobacteria. *Environ. Microbiol.* **2011**, *13*, 551–562. [CrossRef] [PubMed]

63. Beliaev, A.S.; Romine, M.F.; Serres, M.; Bernstein, H.C.; Linggi, B.E.; Markillie, L.M.; Isern, N.G.; Chrisler, W.B.; Kucek, L.A.; Hill, E.A.; et al. Inference of interactions in cyanobacterial-heterotrophic co-cultures via transcriptome sequencing. *ISME J.* **2014**, *8*, 2243–2255. [CrossRef] [PubMed]

64. Kim, Y.M.; Nowack, S.; Olsen, M.T.; Becraft, E.D.; Wood, J.M.; Thiel, V.; Klapper, I.; Kuhl, M.; Fredrickson, J.K.; Bryant, D.A.; et al. Diel metabolomics analysis of a hot spring chlorophototrophic microbial mat leads to new hypotheses of community member metabolisms. *Front. Microbiol.* **2015**, *6*. [CrossRef] [PubMed]

65. Senger, R.S. Biofuel production improvement with genome-scale models: The role of cell composition. *Biotechnol. J.* **2010**, *5*, 671–685. [CrossRef] [PubMed]

66. Hill, J.F.; Govindjee. The controversy over the minimum quantum requirement for oxygen evolution. *Photosynth. Res.* **2014**, *122*, 97–112. [CrossRef] [PubMed]

67. Kawamura, M.; Mimuro, M.; Fujita, Y. Quantitative relationship between two reaction centers in the photosynthetic system of blue-green algae. *Plant Cell Physiol.* **1979**, *20*, 697–705.

68. Sonoike, K.; Hihara, Y.; Ikeuchi, M. Physiological significance of the regulation of photosystem stoichiometry upon high light acclimation of *Synechocystis* sp. PCC 6803. *Plant Cell Physiol.* **2001**, *42*, 379–384. [CrossRef] [PubMed]

69. Vasilikiotis, C.; Melis, A. Photosystem-II reaction-center damage and repair cycle—Chloroplast acclimation strategy to irradiance stress. *Proc. Natl. Acad. Sci. USA* **1994**, *91*, 7222–7226. [CrossRef] [PubMed]

70. Folsom, J.P.; Parker, A.E.; Carlson, R.P. Physiological and proteomic analysis of *Escherichia coli* iron-limited chemostat growth. *J. Bacteriol.* **2014**, *196*, 2748–2761. [CrossRef] [PubMed]

71. Muramatsu, M.; Hihara, Y. Acclimation to high-light conditions in cyanobacteria: From gene expression to physiological responses. *J. Plant Res.* **2012**, *125*, 11–39. [CrossRef] [PubMed]

72. Rochaix, J.D. Regulation of photosynthetic electron transport. *Biochim. Biophys. Acta* **2011**, *1807*, 375–383. [CrossRef] [PubMed]

73. Kirilovsky, D. Photoprotection in cyanobacteria: The orange carotenoid protein (OCP)-related non-photochemical-quenching mechanism. *Photosynth. Res.* **2007**, *93*, 7–16. [CrossRef] [PubMed]

74. Li, G.; Brown, C.M.; Jeans, J.A.; Donaher, N.A.; McCarthy, A.; Campbell, D.A. The nitrogen costs of photosynthesis in a diatom under current and future pCO_2. *New Phytol.* **2015**, *205*, 533–543. [CrossRef] [PubMed]

75. Murphy, C.D.; Roodvoets, M.S.; Austen, E.J.; Dolan, A.; Barnett, A.; Campbell, D.A. Photoinactivation of photosystem II in *Prochlorococcus* and *Synechococcus*. *PLoS ONE* **2017**, *12*, e0168991. [CrossRef] [PubMed]

76. Edwards, J.S.; Ibarra, R.U.; Palsson, B.O. In silico predictions of *Escherichia coli* metabolic capabilities are consistent with experimental data. *Nat. Biotechnol.* **2001**, *19*, 125–130. [CrossRef] [PubMed]

77. Goldberg, I.; Rock, J.S.; Ben-Bassat, A.; Mateles, R.I. Bacterial yields on methanol, methylamine, formaldehyde, and formate. *Biotechnol. Bioeng.* **1976**, *18*, 1657–1668. [CrossRef] [PubMed]

78. Hunt, K.A.; Jennings, R.D.; Inskeep, W.P.; Carlson, R.P. Stoichiometric modeling of assimilatory and dissimilatory biomass utilization in a microbial community. *Environ. Microbiol.* **2016**, *18*, 4946–4960. [CrossRef] [PubMed]

79. Janssen, P.H.; Hugenholtz, P. Fermentation of glycolate by a pure culture of a strictly anaerobic gram-positive bacterium belonging to the family *Lachnospiraceae*. *Arch. Microbiol.* **2003**, *179*, 321–328. [CrossRef] [PubMed]

80. Luttik, M.A.H.; Van Spanning, R.; Schipper, D.; Van Dijken, J.P.; Pronk, J.T. The low biomass yields of the acetic acid bacterium *Acetobacter pasteurianus* are due to a low stoichiometry of respiration-coupled proton translocation. *Appl. Environ. Microbiol.* **1997**, *63*, 3345–3351. [PubMed]

81. Nagpal, S.; Chuichulcherm, S.; Livingston, A.; Peeva, L. Ethanol utilization by sulfate-reducing bacteria: An experimental and modeling study. *Biotechnol. Bioeng.* **2000**, *70*, 533–543. [CrossRef]

82. Seifritz, C.; Frostl, J.M.; Drake, H.L.; Daniel, S.L. Glycolate as a metabolic substrate for the acetogen *Moorella thermoacetica*. *FEMS Microbiol. Lett.* **1999**, *170*, 399–405. [CrossRef]

83. Bernstein, H.C.; McClure, R.S.; Hill, E.A.; Markillie, L.M.; Chrisler, W.B.; Romine, M.F.; McDermott, J.E.; Posewitz, M.C.; Bryant, D.A.; Konopka, A.E.; et al. Unlocking the constraints of cyanobacterial productivity: Acclimations enabling ultrafast growth. *mBio* **2016**, *7*. [CrossRef] [PubMed]

84. Cole, J.K.; Hutchison, J.R.; Renslow, R.S.; Kim, Y.M.; Chrisler, W.B.; Engelmann, H.E.; Dohnalkova, A.C.; Hu, D.H.; Metz, T.O.; Fredrickson, J.K.; et al. Phototrophic biofilm assembly in microbial-mat-derived unicyanobacterial consortia: Model systems for the study of autotroph-heterotroph interactions. *Front. Microbiol.* **2014**, *5*. [CrossRef] [PubMed]

85. Paerl, H.W.; Pinckney, J.L. A mini-review of microbial consortia: Their roles in aquatic production and biogeochemical cycling. *Microb. Ecol.* **1996**, *31*, 225–247. [CrossRef] [PubMed]

86. Paerl, H.W.; Pinckney, J.L.; Steppe, T.F. Cyanobacterial-bacterial mat consortia: Examining the functional unit of microbial survival and growth in extreme environments. *Environ. Microbiol.* **2000**, *2*, 11–26. [CrossRef] [PubMed]

87. Gubernator, B.; Bartoszewski, R.; Kroliczewski, J.; Wildner, G.; Szczepaniak, A. Ribulose-1,5-bisphosphate carboxylase/oxygenase from thermophilic cyanobacterium *Thermosynechococcus elongatus*. *Photosynth. Res.* **2008**, *95*, 101–109. [CrossRef] [PubMed]

88. Falkowski, P.G.; Raven, J.A. *Aquatic Photosynthesis*, 2nd ed.; Princeton University Press: Princeton, NJ, USA, 2007.

89. Savir, Y.; Noor, E.; Milo, R.; Tlusty, T. Cross-species analysis traces adaptation of rubisco toward optimality in a low-dimensional landscape. *Proc. Natl. Acad. Sci. USA* **2010**, *107*, 3475–3480. [CrossRef] [PubMed]

90. Tcherkez, G.G.B.; Farquhar, G.D.; Andrews, T.J. Despite slow catalysis and confused substrate specificity, all ribulose bisphosphate carboxylases may be nearly perfectly optimized. *Proc. Natl. Acad. Sci. USA* **2006**, *103*, 7246–7251. [CrossRef] [PubMed]

91. Mangan, N.M.; Flamholz, A.; Hood, R.D.; Milo, R.; Savage, D.F. pH determines the energetic efficiency of the cyanobacterial CO_2 concentrating mechanism. *Proc. Natl. Acad. Sci. USA* **2016**, *113*, E5354–E5362. [CrossRef] [PubMed]

92. Bateson, M.M.; Ward, D.M. Photoexcretion and fate of glycolate in a hot spring cyanobacterial mat. *Appl. Environ. Microbiol.* **1988**, *54*, 1738–1743. [PubMed]

93. Huege, J.; Goetze, J.; Schwarz, D.; Bauwe, H.; Hagemann, M.; Kopka, J. Modulation of the major paths of carbon in photorespiratory mutants of *Synechocystis*. *PLoS ONE* **2011**, *6*, e16278. [CrossRef] [PubMed]

94. Yokota, A.; Iwaki, T.; Miura, K.; Wadano, A.; Kitaoka, S. Model for the relationships between CO_2-concentrating mechanism, CO_2 fixation, and glycolate synthesis during photosynthesis in *Chlamydomonas-reinhardtii*. *Plant Cell Physiol.* **1987**, *28*, 1363–1376.

95. Jordan, D.B.; Ogren, W.L. The CO_2/O_2 specificity of ribulose 1,5-bisphosphate carboxylase oxygenase—Dependence on ribulosebisphosphate concentration, pH and temperature. *Planta* **1984**, *161*, 308–313. [CrossRef] [PubMed]

96. Sander, R. Compilation of Henry's law constants (version 4.0) for water as solvent. *Atmos. Chem. Phys.* **2015**, *15*, 4399–4981. [CrossRef]

97. Gerstl, M.P.; Jungreuthmayer, C.; Zanghellini, J. tEFMA: Computing thermodynamically feasible elementary flux modes in metabolic networks. *Bioinformatics* **2015**, *31*, 2232–2234. [CrossRef] [PubMed]

98. Peres, S.; Jolicoeur, M.; Moulin, C.; Dague, P.; Schuster, S. How important is thermodynamics for identifying elementary flux modes? *PLoS ONE* **2017**, *12*, e0171440. [CrossRef] [PubMed]

99. Herbert, D.; Phipps, P.J.; Strange, R.E. Chemical analysis of microbial cells. In *Methods in Microbiology*; Norris, J.R., Ribbons, D.W., Eds.; Academic Press: New York, NY, USA, 1971; pp. 209–344.

100. Carnicer, M.; Baumann, K.; Toplitz, I.; Sanchez-Ferrando, F.; Mattanovich, D.; Ferrer, P.; Albiol, J. Macromolecular and elemental composition analysis and extracellular metabolite balances of *Pichia pastoris* growing at different oxygen levels. *Microb. Cell Fact.* **2009**, *8*. [CrossRef] [PubMed]

101. Elser, J.J.; Bracken, M.E.S.; Cleland, E.E.; Gruner, D.S.; Harpole, W.S.; Hillebrand, H.; Ngai, J.T.; Seabloom, E.W.; Shurin, J.B.; Smith, J.E. Global analysis of nitrogen and phosphorus limitation of primary producers in freshwater, marine and terrestrial ecosystems. *Ecol. Lett.* **2007**, *10*, 1135–1142. [CrossRef] [PubMed]

102. Johnson, D.B.; Kanao, T.; Hedrich, S. Redox transformations of iron at extremely low pH: Fundamental and applied aspects. *Front. Microbiol.* **2012**, *3*, 96. [CrossRef] [PubMed]

103. Kliphuis, A.M.J.; Klok, A.J.; Martens, D.E.; Lamers, P.P.; Janssen, M.; Wijffels, R.H. Metabolic modeling of *Chlamydomonas reinhardtii*: Energy requirements for photoautotrophic growth and maintenance. *J. Appl. Phycol.* **2012**, *24*, 253–266. [CrossRef] [PubMed]

6

In Silico Identification of Microbial Partners to Form Consortia with Anaerobic Fungi

St. Elmo Wilken [1], **Mohan Saxena** [1], **Linda R. Petzold** [2] **and Michelle A. O'Malley** [1,*]

[1] Department of Chemical Engineering, University of California, Santa Barbara, CA 93106, USA; stelmo@ucsb.edu (S.E.W.); mohan_saxena@umail.ucsb.edu (M.S.)

[2] Department of Computer Science, University of California, Santa Barbara, CA 93106, USA; petzold@engineering.ucsb.edu

* Correspondence: momalley@engineering.ucsb.edu

Abstract: Lignocellulose is an abundant and renewable resource that holds great promise for sustainable bioprocessing. However, unpretreated lignocellulose is recalcitrant to direct utilization by most microbes. Current methods to overcome this barrier include expensive pretreatment steps to liberate cellulose and hemicellulose from lignin. Anaerobic gut fungi possess complex cellulolytic machinery specifically evolved to decompose crude lignocellulose, but they are not yet genetically tractable and have not been employed in industrial bioprocesses. Here, we aim to exploit the biomass-degrading abilities of anaerobic fungi by pairing them with another organism that can convert the fermentable sugars generated from hydrolysis into bioproducts. By combining experiments measuring the amount of excess fermentable sugars released by the fungal enzymes acting on crude lignocellulose, and a novel dynamic flux balance analysis algorithm, we screened potential consortia partners by qualitative suitability. Microbial growth simulations reveal that the fungus *Anaeromyces robustus* is most suited to pair with either the bacterium *Clostridia ljungdahlii* or the methanogen *Methanosarcina barkeri*—both organisms also found in the rumen microbiome. By capitalizing on simulations to screen six alternative organisms, valuable experimental time is saved towards identifying stable consortium members. This approach is also readily generalizable to larger systems and allows one to rationally select partner microbes for formation of stable consortia with non-model microbes like anaerobic fungi.

Keywords: anaerobic fungi; in silico modeling; microbial consortia; dynamic flux balance analysis; non-model organism; lignocellulose

1. Introduction

Modern biotechnology is well poised to take advantage of the current shift towards a more sustainable chemical industry [1]. Harnessing the estimated 1.3 billion tons of energy rich, lignocellulosic agricultural waste generated world wide each year is a promising avenue towards this goal [2]. However, extracting cellulose (40–50%) and hemicellulose (20–40%) from raw plant biomass has proven to be challenging due to the high lignin content of the substrate [3]. Current industrial techniques used to overcome this barrier include physical, chemical and biological treatment (e.g., milling, acid hydrolysis and enzyme treatment, respectively) [4].

Biological conversion attempts to exploit natural mechanisms to produce chemicals from lignocellulose. Currently, two competing alternatives are being investigated: consolidated bioprocessing and microbial consortia approaches [5]. The former seeks to engineer a single organism to both degrade biomass and produce a high value commodity chemical [6]. The latter seeks to leverage specialist organisms to split the associated metabolic burden between them [7]. Exploiting the natural degradation powers of non-model fungi could prove beneficial in this endeavor.

Currently, fungal enzymes from a handful of organisms, e.g., *Trichoderma reesei* or *Aspergillus sp.*, are utilized on an industrial scale to break down plant biomass [8]. A recent report illustrates the utility of developing consortia between a cellulose degrader like *T. reesei* and the model bacterium *Escherichia coli* [9]. A potential drawback of this pairing is that *T. reesei* encodes for the smallest diversity of cellulolytic enzymes of any fungus capable of plant cell wall degradation [10]. This could necessitate the addition of (expensive) beta-glucosidases, to convert cellobiose to glucose, in some applications. It is hypothesized that under-explored fungal clades, like *Neocallimastigomycota*, coud offer substantial benefits in this regard [11].

Anaerobic gut fungi, in the phylum *Neocallimastigomycota*, found in the gastrointestinal tract of ruminants, have been shown to be prodigious degraders of plant biomass [12]. Moreover, they possess the highest diversity of lignocellulolytic enzymes, largely untapped, within the fungal kingdom [13]. These organisms play a pivotal role in the digestion of plant biomass in herbivores, due to the physical and chemical way in which they degrade plant biomass [14]. Recent work highlights the bounty of biotechnological applications of these fungi [15]. Given that these organisms typically thrive in consortia, it is desirable to emulate nature to unlock their potential for bioconversion of unpretreated lignocellulose.

However, these organisms are under-studied, and the mechanisms that promote the formation of stable microbial consortia with anaerobic fungi are unknown. Given the wealth of omics-related data available, we speculate that model driven design could elucidate some of these questions [11]. Indeed, model driven analysis has successfully been used to study anaerobic organisms [16]. Necessary components for such analyses are accurate genome-scale models of anaerobic gut fungi and their consortia partners. While a full genome-scale model of the gut fungi is still under active development, it is possible to narrow the field in search of potential consortia partners by making use of extant high quality genome-scale models to highlight mechanisms of interaction that would promote microbial partnership and consortium stability.

In this work, we present a marriage of experimental and computational tools used to identify suitable consortia partners for anaerobic gut fungi. Given the vast number of potential candidates, it is infeasible to experimentally test all combinations. Instead, we filter microbes by simulation to test their compatibility in silico. As a first approximation, we assume no interaction between the organisms in consortia: the excess fermentable sugars released by fungal hydrolysis of plant biomass, measured experimentally, is available for consumption regardless of the identity of the partner microbe. By predicting the growth rate and waste production of the partner, we can rank order microbes by the likelihood that they would stably co-exist with the gut fungi over the course of active fungal growth in a batch bioreactor. This is a valuable tool to reduce the number of costly and time-consuming wet-lab experiments necessary to identify suitable partners for anaerobic gut fungal-based consortia. Finally, we introduce a novel dynamic flux balance analysis algorithm specifically developed for this task.

2. Materials and Methods

2.1. Strains and Culture Conditions

Three isolated anaerobic gut fungi were investigated in this work: *Neocallimastix californiae*, *Anaeromyces robustus* and a previously uncharacterized fungus *Neocallimastix* sp. S1 (confirmed by ITS sequencing, see the Supplementary Materials). Anaerobic conditions, as described in [17], were maintained for all experiments. Starter cultures for each experiment were grown on complex media [17], with reed canary grass used as a substrate, in 75 mL serum bottles. After four days of growth, these cultures were used to start experiments by inoculating 4 mL from them into the experiment serum bottles. Gas accumulation in the head space of the starter cultures was vented daily. All experiments were conducted in triplicate using 40 mL of M2 media [18] loaded with 2 g of corn stover grass, (4 mm particle size) supplied by the USDA-ARS research center (Madison, WI, USA), in 75 mL serum bottles.

2.2. Growth and Metabolite Measurements

Fungal growth was monitored by measuring pressure in the head space of the serum bottles twice daily, approximately 12 h apart [19]. Cultures that accumulated significantly more pressure than a control set, without the carbon source corn stover, were deemed to be growing. The gaseous product is primarily composed of hydrogen and carbon dioxide. After the pressure was measured, and prior to venting, 0.2 mL of media was sampled for sugar concentration analysis on a high performance liquid chromatography (HPLC) device. Samples were stored at $-20\,^\circ$C for batch-wise analysis. After thawing the samples at room temperature, they were centrifuged for 5 min at 21,000 $\times$ g. By avoiding the pellet, 100 µL was transfered to HPLC vials containing 100 µL de-ionized, 0.45 µm filtered water (1:1 dilution). Subsequently, 20 µL of each sample was run on an Agilent 1260 Infinity HPLC (Agilent, Santa Clara, CA, USA) using a Bio-Rad Aminex HPX-87P column (Part No. 1250098, Bio-Rad, Hercules, CA, USA) with inline filter (Part No. 5067-1551, Agilent, Santa Clara, CA, USA), Bio-rad Micro-Guard De-Ashing column (Part No. 1250118, Bio-Rad, Hercules, CA, USA), and Bio-Rad Micro-Guard CarboP column (Part No. 1250119, Bio-Rad, Hercules, CA, USA) in the following orientation: inline filter $\rightarrow$ De-Ashing $\rightarrow$ CarboP $\rightarrow$ HPX-87P columns. Samples were run with water acting as the mobile phase at a flow rate of 0.6 mL/min and column temperature of 60 $^\circ$C. Signals were detected using a refractive index detector (RID) with a temperature set point of 40 $^\circ$C. HPLC standards were created in triplicate for cellobiose, glucose, fructose, xylose and arabinose at 5 g/L, 1 g/L, and 0.1 g/L concentrations in M2. The concentration of each sugar was measured by subtracting the RID signal from a blank M2 sample.

2.3. Evaluation and Selection of Model Organisms

The BIGG database is an online repository of curated genome-scale metabolic models [20]. Currently (Accessed December 2017) the database consists of 84 models from a wide diversity of organisms. We hypothesized that the higher level of understanding implied by these models may be leveraged into the formation of stable consortia with the relatively understudied anaerobic fungi. The first step in identifying possible consortia partners is to screen the modeled organisms by three criteria: (1) is the organism an obligate aerobe, (2) is the organism pathogenic and (3) is the organism obviously incompatible with the anaerobic fungi? If any of these criteria were positive, the model was discarded. For example, *Helicobacter pylori* is a modeled pathogen and is therefore excluded. In addition, *Thermotoga maritima* is a modeled hyperthermophilic bacterium; it cannot be co-cultured with the anaerobic fungi and is immediately discarded as a potential consortia partner. By filtering all 84 potential models, we are left with six possible partners, shown in Table 1.

Table 1. Genome-scale models of potential consortia partners for the un-modeled anaerobic gut fungi used in this work.

Organism	Notes	Reference
Clostridium ljungdahlii str. 13528	Bacterium, obligate anaerobe, acetogen	[21]
Escherichia coli str. K-12 substr. MG1655	Bacterium, facultative anaerobe	[22]
Escherichia coli str. ZSC113	Bacterium, facultative anaerobe, glucose deficient	[23]
Lactococcus lactis subsp. cremoris MG1363	Bacterium, facultative anaerobe	[24]
Methanosarcina barkeri str. Fusaro	Methanogen, obligate anaerobe	[25]
Saccharomyces cerevisiae S288C	Fungus, facultative anaerobe	[26]

2.4. Dynamic Flux Balance Analysis Formulation

Flux balance analysis (FBA) is a widely used computational tool that simplifies and recasts the metabolic reaction network of a cell into a linear program by making use of a genome-scale model [27]. Central to FBA is the assumption of metabolic steady state, $\frac{dx}{dt} = \mathbf{S}\mathbf{v} = \mathbf{0}$. The space of fluxes, $\mathbf{v}$, that satisfy the mass balance implied by the stoichiometric matrix, $\mathbf{S}$, is reduced by assuming that the cell strives to maximize an empirically defined biomass objective function, $\mu(\mathbf{v})$, subject to additional flux constraints, $\mathbf{v}_{\min} \leq \mathbf{v} \leq \mathbf{v}_{\max}$. Typically, FBA is applied to systems in a steady state; this poses a problem for modeling anaerobic gut fungi because no continuous reactor has been developed for them yet.

Dynamic flux balance analysis (dFBA) is a well-established tool used to extend FBA to dynamic settings [28]. It relies on the assumption that intra-cellular dynamics are much faster than extra-cellular dynamics. This allows one to discretize time and apply classical FBA at each time step. The resultant fluxes are then used to update the biomass (X), external substrate ($\mathbf{s}$), and product ($\mathbf{p}$), concentrations by integrating

$$
\begin{aligned}
\frac{dX}{dt} &= \mu X, \\
\frac{d\mathbf{s}}{dt} &= \mathbf{v}_s X, \\
\frac{d\mathbf{p}}{dt} &= \mathbf{v}_p X,
\end{aligned}
\tag{1}
$$

where μ, $\mathbf{v}_s$ and $\mathbf{v}_p$ are the growth rate, substrate and product fluxes, respectively. These are then used to update the flux constraints,

$$
\mathbf{v}_{\min}(\mathbf{s}, \mathbf{P}) \leq \mathbf{v} \leq \mathbf{v}_{\max}(\mathbf{s}, \mathbf{P}),
\tag{2}
$$

used in the FBA algorithm for the next time step [29]. dFBA has been successfully applied to mono-culture [30,31] and community [32,33] modeling.

An inherent weakness of FBA, and by extension dFBA, is the non-uniquess of the fluxes that maximize the cellular growth rate [34]. Sampling from the space of optimal fluxes is feasible for FBA applications because the computational cost is paid only once (typically a mixed integer linear program needs to be solved [35]). For dFBA applications, this is prohibitively expensive due to the iterative nature of the algorithm. However, it is well recognized that non-uniqueness of the fluxes can pose problems when integrating Equation (1).

Techniques developed to deal with this problem typically involve hierarchal optimization, subsequent to the biomass maximization, to constrain the fluxes further. One possibility is to maximize the growth rate and then sequentially optimize each external flux using the previous optimization problem as a constraint in the current one [36,37]. This method effectively deals with the non-uniqueness problem but requires additional assumptions per external flux. These assumptions can dramatically affect the results of the simulation but seem to be a problem only when modeling multiple species [37].

An alternative method is to perform only a single secondary optimization subsequent to the biomass maximization, in the hope that this constrains the fluxes sufficiently to ameliorate the non-uniqueness issue when performing the integration of Equation (1). An example of this approach is to minimize the absolute fluxes, based on the principle of maximum enzyme efficiency [38]. The drawback with this approach is that it requires the solution of a quadratic program (QP) at each time step. For larger models, this can be computationally expensive.

We chose to keep the imposition of additional assumptions on the modeled systems to a minimum because the work is exploratory in nature. Therefore, we combine aspects of [37] with the single secondary optimization approach. In our case, the secondary optimization seeks to ensure that the derivative change of each modeled flux is minimized between each time step. The rationale for this is that over small time steps the flux is unlikely to jump suddenly. Therefore, at each time step, the following procedure is followed:

1. The flux bounds, Equation (2), are updated. Typically, Michaelis–Menten kinetics are assumed [39]. Since detailed expression for glucose and xylose uptake rates are not known for all the organisms, we assumed, for comparative fairness,

$$v_{\text{min, glucose}} = \max\left(v_{\text{Glc}}^{\max}, -\frac{G + \Delta t f_G^{\text{produced}}}{\Delta t X m_{\text{glucose}}}\right),$$

$$v_{\text{max, glucose}} = 0,$$

$$v_{\text{min, xylose}} = \max\left(v_{\text{Xyl}}^{\max}, -\frac{Z + \Delta t f_Z^{\text{produced}}}{\Delta t X m_{\text{xylose}}}\frac{1}{1 + \frac{G}{0.005}}\right),$$ (3)

$$v_{\text{max, xylose}} = 0,$$

where f_G^{produced}, f_Z^{produced} are the fluxes of glucose and xylose produced by the extracellular enzymes, G, Z are the current concentrations of glucose and xylose, and m_{glucose}, m_{xylose} are the respective molar masses. The glucose inhibition term ensures that glucose is preferentially metabolized before xylose [32]. The maximum flux constants, $v_{\text{Glc}}^{\max}$ and $v_{\text{Xyl}}^{\max}$, were taken from literature and are supplied in Section 2.5. See the Supplement for motivation of the derivation of Equation (3).

2. A linear program feasibility problem,

$$\min_{\mathbf{s}_1, \mathbf{s}_2} \sum_{i=1}^{N} s_{1,i} + s_{2,i} \text{ (where } N \text{ is the number of fluxes)},$$

$$\text{s.t.} \quad \mathbf{S}\mathbf{v} + \mathbf{s}_1 - \mathbf{s}_2 = \mathbf{b} \text{ (where } \mathbf{b} \text{ is typically the zero vector in this context)},$$ (4)

$$\mathbf{v}_{\min} \leq \mathbf{v} \leq \mathbf{v}_{\max},$$

$$0 \leq s_{1,i}, s_{2,i} \; \forall i \in [1, \ldots, N],$$

is solved to ensure that the genome-scale model is feasible for steps 3 and 4. This problem is solved for the "relaxation variables" $\mathbf{s}_1$ and $\mathbf{s}_2$ (see [36] for justification).

3. A standard FBA linear program (LP) is solved to determine the optimal growth rate of the organism given the constraints of step 1. This problem,

$$\max_{\mathbf{v}} \quad \mu(\mathbf{v}),$$

$$\text{s.t.} \quad \mathbf{S}\mathbf{v} + \mathbf{s}_1 - \mathbf{s}_2 = \mathbf{b},$$ (5)

$$\mathbf{v}_{\min} \leq \mathbf{v} \leq \mathbf{v}_{\max},$$

is solved for the unique optimal growth rate μ^*. Given μ^* from Equation (5), it is possible to solve for the organism biomass concentration by using $\frac{dX}{dt} = \mu^* X$ for at least one time step into the future.

4. A secondary LP,

$$\min_{\mathbf{v}} \quad \sum_i \gamma_i \text{ for } i \in \mathcal{M},$$

$$\text{s.t.} \quad \mathbf{S}\mathbf{v} + \mathbf{s}_1 - \mathbf{s}_2 = \mathbf{b},$$

$$\mathbf{v}_{\min} \leq \mathbf{v} \leq \mathbf{v}_{\max},$$ (6)

$$\mu(\mathbf{v}) = \mu^*,$$

$$-\gamma_i \leq 1 - \frac{v_{t-1,i}}{v_{t-1,i} - v_{t-2,i}} - \frac{v_{t,i}}{v_{t-1,i} - v_{t-2,i}} \leq \gamma_i \text{ for } i \in \mathcal{M},$$

is solved to ensure that the resultant fluxes used to integrate Equation (1) are sufficiently smooth. Here, $\mathcal{M}$ is the index set of all modeled substrates and products. A full derivation of Equation (6) is given in the Supplement. Briefly, the objective function asserts that $\sum_i \left| 1 - \frac{dv_i}{dt}_t / \frac{dv_i}{dt}_{t-1} \right| \, \forall \, i \, \in \, \mathcal{M}$ is minimized, where the flux derivative at time t, $\frac{dv_i}{dt}_t$, is approximated to first order.

5. Using an integration scheme of choice, e.g., backward Euler, the full dynamic profile of the system may be iteratively simulated. If products are being generated at each time step, Equation (1) needs to include those fluxes as well.

The primary benefit of Equation (6) is that there is only a single secondary LP imposed on the system. From a computational point of view, this is very desirable compared to the other existing algorithms that solve either a QP or multiple sequential LPs.

2.5. Simulation Parameters

All simulations restricted the oxygen flux into the system to zero. It was assumed that the gas produced by the fungi is 90% carbon dioxide and 10% hydrogen on a mole basis. This is in line with previous experimental observations. The maximum glucose and xylose uptake flux constraints, shown in Equation (3), were taken from the papers introducing the models (see Table 1 for the references). These are summarized in Table 2.

Table 2. Glucose and xylose maximum uptake rates.

Organism	$v_{Glc}\left[\frac{mmol}{g_{DW}h}\right]$	$v_{Xyl}\left[\frac{mmol}{g_{DW}h}\right]$
Clostridium ljungdahlii str. 13528	5	5
Escherichia coli str. K-12 substr. MG1655	10.5	6
Escherichia coli str. ZSC113	0	6
Lactococcus lactis subsp. cremoris MG1363	14.5	0
Methanosarcina barkeri str. Fusaro	0	0
Saccharomyces cerevisiae S288C	6.44	0

Note that *M. barkeri* does not consume glucose or xylose. Instead, it autotrophically metabolizes hydrogen and carbon dioxide into methane. The maximum hydrogen uptake rate was set at $v_{H_2} = 41.5 \left[\frac{mmol}{g_{DW}h}\right]$, and the maximum carbon dioxide uptake rate was unbounded [25]. All products P produced by the fungi, e.g., sugar and gas (in the form of pressure accumulation) were assumed to follow the logistic function,

$$P(t) = \frac{k_1}{1 + e^{-k_2(t-k_3)}},\tag{7}$$

where the constants were fit to experimental data. Henry's law was used to model the concentration of dissolved gases (hydrogen, carbon dioxide and methane) in the liquid fraction given the gas pressure. A backward Euler scheme was used to integrate Equation (1) with a time step of 0.1 h. The initial conditions for all the substrates and products consumed and produced by the partner microbes were assumed to be zero. The initial biomass concentration was assumed to be 1 mg/L.

3. Results and Discussion

Both experimental and computational data were gathered to evaluate the organisms listed in Table 1 for their ability to form stable consortia with anaerobic gut fungi. Batch growth experiments were used to model the rate of sugar release from the raw plant biomass during fungal digestion, as well as the gas accumulation profile. This sheds light on the ability of the fungi to accommodate another organism, likely through nutritional linkage of primary metabolites. Computational experiments were then used to predict growth rates and waste generation of a model partner microbe, given the excess fermentable products determined via the batch experiments.

3.1. Anaerobic Fungi Release an Assortment of Products to Enable Consortia Formation

Figure 1 shows the experimentally observed sugar release and gas production profiles over time of the three anaerobic fungi we investigated. It can be seen that *A. robustus* produced the highest concentration of soluble sugars and the next to highest accumulated pressure. In accordance with the variance between culture replicates, *N. californiae* displayed more erratic growth. This behavior is uncharacteristic of the fungus when cultured in complex media. We speculate that the M2 defined minimal media was a contributing factor to this phenomenon. *Neocallimastix* sp. S1 performed between the other two fungi in terms of stability and sugar/gas production.

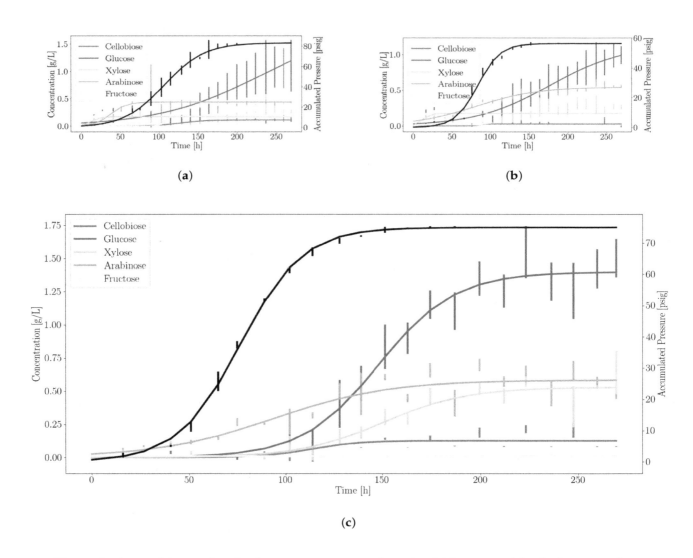

Figure 1. Anaerobic gut fungi release excess sugars for microbial partnership during growth on corn stover. The solid black line denotes the profile of the accumulated pressure. Other colors represent distinct fermentable sugars generated during growth, as indicated. The vertical bars are standard deviations of errors for each triplicate measurement. (**a**) *N. californiae*; (**b**) *Neocallimastix* sp. S1; (**c**) *A. robustus*.

Based on these data, we selected *A. robustus* as the best candidate for consortia experiments that combine anaerobic fungi with model microbes due to the more stable sugar and gas production rates. Constants used to model substrate production rates for glucose, xylose and pressure accumulation were fit to Equation (7) using *A. robustus* data, as shown in Table 3.

Table 3. Glucose, xylose and gas production rate constants fit to Equation (7) for *A. robustus*.

Product	k_1 (g/L/h or psi/h)	k_2 (1/h)	k_3 (h)
Glucose	1.39	0.05	148.17
Xylose	0.53	0.05	150.41
Pressure	75.04	0.06	76.51

For completeness, we compare the measured gut fungal net specific growth rates found in M2 defined media, used here, with that of complex media (see Table 4). Predictably, the growth rates are lower in minimal defined media. *A. robustus* consistently outperforms the other fungi when grown on corn stover. The superior growth characteristics of *A. robustus* further motivate its selection as the gut fungus to investigate in greater depth.

Table 4. Anaerobic gut fungi growth rates in defined media compared to rich media.

Organism	Growth Rate in M2 (1/h)	Growth rate in MC [15] (1/h)
N. californiae	0.029	0.046
A. robustus	0.033	0.065
Neocallimastix sp. S1	0.027	No data

3.2. Dynamic Simulations Predict Consortia Partner Feasibility

By making use of the dFBA algorithm introduced in Section 2.4, and using the experimental data of *A. robustus* to fit Equation (7) for both glucose and xylose separately, we can simulate the growth of the co-cultured partner organisms listed in Table 1 dynamically. We chose to focus only on glucose and xylose utilization at this stage of modeling because more is known about the relative preference of each sugar in microbial metabolism [40]. The two classes, fermentable sugar consuming heterotrophs, and hydrogen/carbon dioxide consuming autotrophs, of possible consortia partners were treated separately.

3.2.1. Heterotroph Partnership with Anaerobic Fungi

As suggested by Equation (3), we assumed, for simplicity, that only glucose and xylose are capable of being fermented by each organism under analysis. Furthermore, we assumed that glucose would be consumed preferentially to xylose whenever possible. Figure 2 illustrates the output of the dFBA algorithm when pairing the anaerobic bacterium *C. ljungdahlii* with the gut fungus *A. robustus*. Similar results are available for the other organisms of Table 1 in the Supplement.

C. ljungdahlii can metabolize both glucose and xylose; this is reflected in the sequential utilization of the substrates in the simulated time course. To determine the effective average growth rate, we fit $\frac{dX}{dt} = e^{\mu t}$ to the simulated biomass output. The fit indicated that $\mu \approx 0.08$ 1/h. The growth rate is the primary criterion we used to determine suitability for consortia with the gut fungi. We hypothesized that an optimal pairing would occur if the growth rates of the organisms are similar. This would reduce the risk of them out-competing each other. Inter-cellular communication, another pivotal component of consortia, is neglected at this stage of analysis, as it requires detailed experimental data to model.

Each modeled organism is also capable of producing metabolic by-products, e.g., ethanol, acetate and formate, that are known to inhibit microbial growth. We also recorded the final concentration of each compound as a secondary criteria to ascertain compatibility with the fungi. The summarized characteristics of each organism, simulated to pair with *A. robustus*, are shown in Table 5.

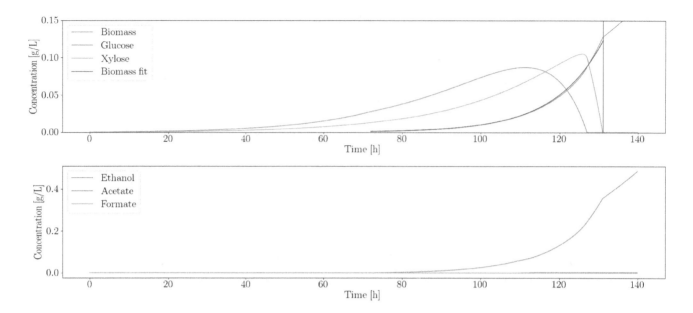

Figure 2. Dynamic simulation of *C. ljungdahlii* shows that it consumes all the excess sugars released by *A. robustus*. The vertical red line indicates the point where both sugars were depleted. Even though the fungal enzymes continuously release sugars, the rate at which they release them is exactly equal to the consumption rate beyond the vertical red line. Simulation artifacts cause the growth to continue linearly beyond this point. All the simulations assume an inoculation time at 72 h into the experiment. This allows the slower-growing gut fungi to establish themselves and produce fermentable products prior to the start of the co-culture.

Table 5. Growth rate and end point metabolic by-product concentrations produced by each partner microbe assuming inoculation after 72 h of fungal growth. The end point concentrations are taken when the fermentable substrates were depleted for each organism.

Organism	Growth Rate (1/h)	Ethanol (g/L)	Acetate (g/L)	Formate (g/L)
C. ljungdahlii	0.08	0	0.35	0
E. coli MG1655	0.17	0.02	0.02	0.03
E. coli ZSC113	0.04	0.01	0.02	0.03
L. lactis	0.04	0.13	0.32	0.51
S. cerevisiae	0.12	0.02	0	0

The models predicted that both *S. cerevisiae* and *E. coli* MG1655 have a significantly higher growth rate than *A. robustus*. This suggests that maintaining population stability could be difficult for these co-cultures if paired with anaerobic fungi [41]. While *L. lactis* has a comparable growth rate to *A. robustus*, it is unable to metabolize xylose; therefore, it would directly compete for glucose. Additionally, *L. lactis* produces a wide spectrum of metabolic by-products (ethanol, acetate and formate) at relatively high concentrations; this lessens its attractiveness as a consortia partner. The glucose deficient *E. coli* strain ZSC113 also has a comparable growth rate but produces less metabolic waste products. Additionally, it is genetically amenable to engineering [42]—this suggests that it could be a favorable organism for consortia formation. Finally, *C. ljungdahlii* is also a competitive choice for consortia. While its growth rate is higher than *A. robustus*, it is not in the range of *S. cerevisiae* and *E. coli* MG1655. *C. ljungdahlii* can ferment a wide range of sugars as well as autotrophically consume hydrogen (not modeled); this suggests that the organism can take full advantage of the fungal products. Recently, genetic engineering tools have become available for *C. ljungdahlii*, further increasing its viability as a consortia partner.

3.2.2. Autotroph Partnership with Anaerobic Fungi

While the organisms shown in Section 3.2.1 utilized the fermentable sugars released by the gut fungal enzymes as their carbon source (or preferred carbon source in the case of *C. ljungdahlii*), *M. barkeri*, a methanogen, metabolizes carbon dioxide and hydrogen. It is well known that methanogens are natural consortia partners of gut fungi due to their symbiotic relationship [43]. Methanogens consume the hydrogen gas, a likely growth inhibitor, produced by an intracellular organelle of the fungi called the hydrogenosome [44]. Furthermore, it has been shown that methanogens co-cultured with gut fungi significantly increase their cellulolytic efficiency [45].

Figure 3 illustrates the simulated growth profile of *M. barkeri*. Negligible quantities of ethanol, acetate and formate are produced, while hydrogen is almost completely consumed. The effective growth rate is 0.03 1/h. Since the gas produced by the fungi drive their growth, it is not surprising that their growth rates are similar.

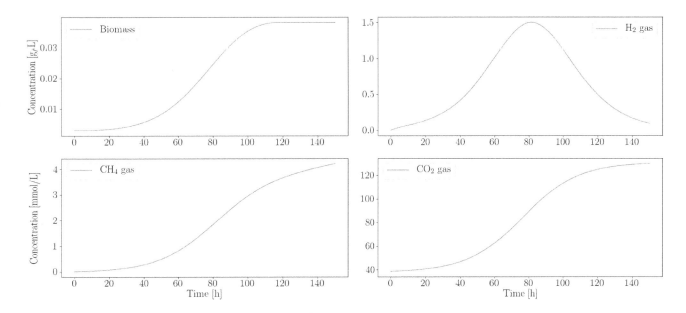

Figure 3. Computationally predicted growth profile of *M. barkeri* biomass accumulation over time shows a strong dependence on the fungal metabolic by-products. Hydrogen and carbon dioxide, produced by the fungi, are consumed by the methanogen. Simultaneous inoculation is assumed because the microbes do not compete for their preferred carbon source. All gas concentrations are in mmol/L.

M. barkeri is also an attractive candidate for synthetic gut fungal consortia due to the mutualism exhibited by the pairing of fungi and methanogens in nature [46]. The recent development of genetic technology to manipulate *Methanosarcina* suggests that the pairing is also feasible for bioproduction [47]. Finally, given the low levels of by-products generated by *M. barkeri*, it is plausible to consider tri-cultures of *A. robustus*, *M. barkeri* and another microbe, like *C. ljungdahlii*. Such a system would be, theoretically, minimally negatively interactive due to the reduced substrate competition. This is a desirable property for community stability.

The benefit of using the dFBA, to screen for consortia partners, is that it is readily generalizable to higher order systems. Known interactions can easily be accounted for, and quantitative predictions of by-product generation can be used to evaluate partner suitability (cf. qualitative literature surveys). The simulation approach is particularly useful for non-model organisms, like anaerobic fungi, because growth rate predictions in their unique culture conditions are not often readily available.

Experimental validation of these predictions will take the form of community composition tracking and by-product generation monitoring. The latter technique is particularly applicable to the anaerobic fungi because it is one of the few non-invasive methods that can be used to measure growth in gut fungal systems [19]. For example, in the case of the *A. robustus* and *M. barkeri* pairing, the methane, carbon dioxide and hydrogen production over time, compared to the mono-cultures, will indicate the success of the co-culture. Similar indirect measurements could be used to validate the other predictions. However, these detailed experiments are beyond the scope of the current work.

4. Conclusions

To assess the suitability of each organism in Table 1 to form stable microbial consortia with anaerobic fungi, the identities and contributions of both the gut fungus and partner microbe need to be justified. In this work, experiments were used to select an anaerobic fungus and simulations, making the least number of assumptions, were used to screen possible consortia partners.

The experimental results of Section 3.1 indicate that *A. robustus* is a more desirable building block for consortia (or even mono-cultures) compared to other strains tested here—both in terms of higher growth rates on corn stover (see Table 4) as well as enzyme effectiveness at releasing fermentable sugars (see Figure 1). Barring the generation of unknown inhibitory agents, it should be prioritized for further experimentation.

M. barkeri, a methanogen, is a natural consortia partner for gut fungi [45]. This is clear from the similar growth rates to *A. robustus* and consumption of hydrogen, a known inhibitor of fungal growth. Additionally, it produces minimal by-products that could retard fungal growth. *C. ljungdahlii* and *E. coli* ZSC113 are also potentially suitable consortia partners. On the other hand, *L. lactis*, *S. cerevisiae* and *E. coli* MG1655 were all ruled out due to their by-product generation or significantly higher growth rates. We introduced a novel dFBA algorithm that is computationally efficient and that does not impose many extra assumptions on the system. Making use of computational tools, such as this, to reduce the number of costly and time-consuming experiments is a boon to developing and designing scalable synthetic biosystems [48].

Moreover, building predictive models of consortia systems can be critical to fully leveraging the inherent capabilities of micro-organisms as it allows engineers additional insight into the mechanics of these complex systems [49]. Fully unlocking the inherent capabilities of non-model organisms, like anaerobic gut fungi, will require novel tools to inexpensively generate and test hypotheses. Current consortia analysis techniques typically assume that the identities of the partner microbes are known and that they are modeled. This work provides a framework that can be used to rationally select the them even if some of the microbes are not modeled.

Acknowledgments: The authors gratefully acknowledge funding support from the Office of Science (BER), U.S. Department of Energy (DE-SC0010352), the Institute for Collaborative Biotechnologies through grant W911NF-09-0001 from the U.S. Army Research Office, the National Science Foundation (MCB-1553721), and the Dow Discovery Fellowship (to SW). The authors also thank John K. Henske for isolation and ITS characterization of *Neocallimastix* sp. S1, and Paul Weimer from the United States Department of Agriculture (USDA) for providing freshly milled biomass substrates.

Author Contributions: S.E.W., L.R.P. and M.A.O. conceived and designed the experiments; S.E.W. and M.S. performed the experiments; S.E.W. and M.S. analyzed the data; S.E.W. and L.R.P. developed the dFBA algorithm; S.E.W., L.R.P. and M.A.O. wrote the paper.

References

1.   Otero, J.M.; Nielsen, J. Industrial systems biology. *Biotechnol. Bioeng.* **2010**, *105*, 439–460.
2.   Saini, J.K.; Saini, R.; Tewari, L. Lignocellulosic agriculture wastes as biomass feedstocks for second-generation bioethanol production: Concepts and recent developments. *3 Biotech* **2015**, *5*, 337–353.

3. Liao, J.C.; Mi, L.; Pontrelli, S.; Luo, S. Fuelling the future: Microbial engineering for the production of sustainable biofuels. *Nat. Rev. Microbiol.* **2016**, *14*, 288–304.

4. Sindhu, R.; Binod, P.; Pandey, A. Biological pretreatment of lignocellulosic biomass—An overview. *Bioresour. Technol.* **2016**, *199*, 76–82.

5. Alper, H.; Stephanopoulos, G. Engineering for biofuels: Exploiting innate microbial capacity or importing biosynthetic potential? *Nat. Rev. Microbiol.* **2009**, *7*, 715–723.

6. Lynd, L.R.; Van Zyl, W.H.; McBride, J.E.; Laser, M. Consolidated bioprocessing of cellulosic biomass: An update. *Curr. Opin. Biotechnol.* **2005**, *16*, 577–583.

7. Brenner, K.; You, L.; Arnold, F.H. Engineering microbial consortia: A new frontier in synthetic biology. *Trends Biotechnol.* **2008**, *26*, 483–489.

8. Paloheimo, M.; Haarmann, T.; Mäkinen, S.; Vehmaanperä, J. Production of Industrial Enzymes in Trichoderma reesei. In *Gene Expression Systems in Fungi: Advancements and Applications*; Schmoll, M., Dattenböck, C., Eds.; Springer International Publishing: Cham, Switzerland, 2016; pp. 23–57.

9. Minty, J.J.; Singer, M.E.; Scholz, S.A.; Bae, C.H.; Ahn, J.H.; Foster, C.E.; Liao, J.C.; Lin, X.N. Design and characterization of synthetic fungal-bacterial consortia for direct production of isobutanol from cellulosic biomass. *Proc. Natl. Acad. Sci. USA* **2013**, *110*, 14592–14597.

10. Martinez, D.; Berka, R.M.; Henrissat, B.; Saloheimo, M.; Arvas, M.; Baker, S.E.; Chapman, J.; Chertkov, O.; Coutinho, P.M.; Cullen, D.; et al. Genome sequencing and analysis of the biomass-degrading fungus Trichoderma reesei (syn. Hypocrea jecorina). *Nat. Biotechnol.* **2008**, *26*, 553–560.

11. Seppälä, S.; Elmo Wilken, S.; Knop, D.; Solomon, K.V.; O'Malley, M.A. The importance of sourcing enzymes from non-conventional fungi for metabolic engineering & biomass breakdown. *Metab. Eng.* **2017**, *44*, 45–59.

12. Resch, M.G.; Donohoe, B.S.; Baker, J.O.; Decker, S.R.; Bayer, E.A.; Beckham, G.T.; Himmel, M.E. Fungal cellulases and complexed cellulosomal enzymes exhibit synergistic mechanisms in cellulose deconstruction. *Energy Environ. Sci.* **2013**, *6*, 1858–1867.

13. Solomon, K.V.; Haitjema, C.H.; Henske, J.K.; Gilmore, S.P.; Borges-Rivera, D.; Lipzen, A.; Brewer, H.M.; Purvine, S.O.; Wright, A.T.; Theodorou, M.K.; et al. Early-branching gut fungi possess a large, comprehensive array of biomass-degrading enzymes. *Science* **2016**, *1431*, 1192–1195.

14. Gruninger, R.J.; Puniya, A.K.; Callaghan, T.M.; Edwards, J.E.; Youssef, N.; Dagar, S.S.; Fliegerova, K.; Griffith, G.W.; Forster, R.; Tsang, A.; et al. Anaerobic fungi (phylum Neocallimastigomycota): Advances in understanding their taxonomy, life cycle, ecology, role and biotechnological potential. *FEMS Microbiol. Ecol.* **2014**, *90*, 1–17.

15. Henske, J.K.; Wilken, S.E.; Solomon, K.V.; Smallwood, C.R.; Shutthanandan, V.; Evans, J.E.; Theodorou, M.K.; O'Malley, M.A. Metabolic characterization of anaerobic fungi provides a path forward for two-stage bioprocessing of crude lignocellulose. *Biotechnol. Bioeng.* **2017**, doi:10.1002/bit.26515.

16. Senger, R.S.; Yen, J.Y.; Fong, S.S. A review of genome-scale metabolic flux modeling of anaerobiosis in biotechnology. *Curr. Opin. Chem. Eng.* **2014**, *6*, 33–42.

17. Theodorou, M.K.; Brookman, J.L.; Trinci, A.P. Anaerobic fungi. In *Methods in Gut Microbial Ecology for Ruminants*, 1st ed.; Makkar, H.P., McSweeney, C.S., Eds.; Springer: Dordrecht, The Netherlands, 2005; Chapter 2.4, pp. 55–67.

18. Teunissen, M.J.; Op den Camp, H.J.M.; Orpin, C.G.; Huls In 't Veld, J.II.J., Vogels, G.D. Comparison of growth characteristics of anaerobic fungi isolated from ruminant and non-ruminant herbivores during cultivation in a defined medium. *J. Gen. Microbiol.* **1991**, *137*, 1401–1408.

19. Theodorou, M.K.; Davies, D.R.; Nielsen, B.B.; Lawrence, M.I.; Trinci, A.P. Determination of growth of anaerobic fungi on soluble and cellulosic substrates using a pressure transducer. *Microbiology* **1995**, *141*, 671–678.

20. King, Z.A.; Lu, J.; Dräger, A.; Miller, P.; Federowicz, S.; Lerman, J.A.; Ebrahim, A.; Palsson, B.O.; Lewis, N.E. BiGG Models: A platform for integrating, standardizing and sharing genome-scale models. *Nucleic Acids Res.* **2016**, *44*, D515–D522.

21. Nagarajan, H.; Sahin, M.; Nogales, J.; Latif, H.; Lovley, D.R.; Ebrahim, A.; Zengler, K. Characterizing acetogenic metabolism using a genome-scale metabolic reconstruction of Clostridium ljungdahlii. *Microb. Cell Fact.* **2013**, *12*, 118,

22. Monk, J.M.; Lloyd, C.J.; Brunk, E.; Mih, N.; Sastry, A.; King, Z.; Takeuchi, R.; Nomura, W.; Zhang, Z.; Mori, H.; et al. iML1515, a knowledgebase that computes *Escherichia coli* traits. *Nat. Biotechnol.* **2017**, *35*, 904–908.

23. Curtis, S.J.; Epstein, W.   Phosphorylation of D-glucose in *Escherichia coli* mutants defective in glucosephosphotransferase, mannosephosphotransferase, and glucokinase. *J. Bacteriol.* **1975**, *122*, 1189–1199.

24. Flahaut, N.A.; Wiersma, A.; van de Bunt, B.; Martens, D.E.; Schaap, P.J.; Sijtsma, L.; Dos Santos, V.A.; De Vos, W.M. Genome-scale metabolic model for Lactococcus lactis MG1363 and its application to the analysis of flavor formation. *Appl. Microbiol. Biotechnol.* **2013**, *97*, 8729–8739.

25. Feist, A.M.; Scholten, J.C.; Palsson, B.; Brockman, F.J.; Ideker, T.   Modeling methanogenesis with a genome-scale metabolic reconstruction of Methanosarcina barkeri. *Mol. Syst. Biol.* **2006**, *2*, 1–14.

26. Mo, M.L.; Palsson, B.Ø.; Herrgård, M.J. Connecting extracellular metabolomic measurements to intracellular flux states in yeast. *BMC Syst. Biol.* **2009**, *3*, 37.

27. Orth, J.D.; Thiele, I.; Palsson, B.Ø. What is flux balance analysis? *Nat. Comput. Biol.* **2010**, *28*, 245–248.

28. Varma, A.; Palsson, B.O.; Varma, A.; Palsson, B.O. Stoichiometric Flux Balance Models Quantitatively Predict Growth and Metabolic By-Product Secretion in Wild-Type *Escherichia coli* W3110. *Appl. Environ. Microbiol.* **1994**, *60*, 3724–3731.

29. Henson, M.A.; Hanly, T.J. Dynamic flux balance analysis for synthetic microbial communities. *IET Syst. Biol.* **2013**, *8*, 214–229.

30. Henson, J.L.; Hjersted, M.A. Steady-state and dynamic flux balance analysis of ethanol production by Saccharomyces cerevisiae. *IET Syst. Biol.* **2009**, *3*, 167–179.

31. Mahadevan, R.; Edwards, J.S.; Doyle, F.J. Dynamic Flux Balance Analysis of Diauxic Growth in *Escherichia coli*. *Biophys. J.* **2002**, *83*, 1331–1340.

32. Hanly, T.J.; Henson, M.A.   Dynamic flux balance modeling of microbial co-cultures for efficient batch fermentation of glucose and xylose mixtures. *Biotechnol. Bioeng.* **2011**, *108*, 376–385.

33. Hanly, T.J.; Urello, M.; Henson, M.A. Dynamic flux balance modeling of S. cerevisiae and *E. coli* co-cultures for efficient consumption of glucose/xylose mixtures. *Appl. Microbiol. Biotechnol.* **2012**, *93*, 2529–2541.

34. Mahadevan, R.; Schilling, C.H. The effects of alternate optimal solutions in constraint-based genome-scale metabolic models. *Metab. Eng.* **2003**, *5*, 264–276.

35. Saa, P.; Nielsen, L.K. ll-ACHRB: A scalable algorithm for sampling the feasible solution space of metabolic networks. *Bioinformatics* **2016**, *32 15*, 2330–2337.

36. Hoffner, K.; Harwood, S.M.; Barton, P.I.   A Reliable Simulator for Dynamic Flux Balance Analysis. *Biotechnol. Bioeng.* **2013**, *110*, 792–802.

37. Gomez, J.A.; Höffner, K.; Barton, P.I. DFBAlab: A fast and reliable MATLAB code for dynamic flux balance analysis. *BMC Bioinform.* **2014**, *15*, 409.

38. Sánchez, B.J.; Pérez-Correa, J.R.; Agosin, E. Construction of robust dynamic genome-scale metabolic model structures of Saccharomyces cerevisiae through iterative re-parameterization. *Metab. Eng.* **2014**, *25*, 159–173.

39. Hanly, T.J.; Henson, M.A. Unstructured modeling of a synthetic microbial consortium for consolidated production of ethanol. *IFAC Proc. Vol.* **2013**, *12*, 157–162.

40. Eiteman, M.A.; Lee, S.A.; Altman, E. A co-fermentation strategy to consume sugar mixtures effectively. *J. Biol. Eng.* **2008**, *2*, 3.

41. Goers, L.; Freemont, P.; Polizzi, K.M. Co-culture systems and technologies: Taking synthetic biology to the next level. *J. R. Soc. Interface* **2014**, *11*, doi:10.1098/rsif.2014.0065.

42. Bokinsky, G.; Peralta-Yahya, P.P.; George, A.; Holmes, B.M.; Steen, E.J.; Dietrich, J.; Soon Lee, T.; Tullman-Ercek, D.; Voigt, C.A.; Simmons, B.A.; et al. Synthesis of three advanced biofuels from ionic liquid-pretreated switchgrass using engineered *Escherichia coli*. *Proc. Natl. Acad. Sci. USA* **2011**, *108*, 19949–19954.

43. Haitjema, C.H.; Solomon, K.V.; Henske, J.K.; Theodorou, M.K.; O'Malley, M.A. Anaerobic gut fungi: Advances in isolation, culture, and cellulolytic enzyme discovery for biofuel production. *Biotechnol. Bioeng.* **2014**, *111*, 1471–1482.

44. Müller, M. Review article: The hydrogenosome. *J. Gen. Microbiol.* **1993**, *139*, 2879–2889.

45. Marvin-Sikkema, F.D.; Pedro Gomes, T.M.; Grivet, J.P.; Gottsehal, J.C.; Prins, R.A. Characterization of hydrogenosomes and their role in glucose metabolism of Neocallimastix sp. L2. *Arch. Microbiol.* **1993**, *160*, 388–396.

46. Peng, X.N.; Gilmore, S.P.; O'Malley, M.A. Microbial communities for bioprocessing: Lessons learned from nature. *Curr. Opin. Chem. Eng.* **2016**, *14*, 103–109.

47. Kohler, P.R.; Metcalf, W.W. Genetic manipulation of Methanosarcina spp. *Front. Microbiol.* **2012**, *3*, 1–9.

48. Höffner, K.; Barton, P.I. Design of microbial consortia for industrial biotechnology. *Comput. Aided Chem. Eng.* **2014**, *34*, 65–74.

49. Mahadevan, R.; Henson, M.A. Genome-Based Modeling and Design of Metabolic Interactions in Microbial Communities. *Comput. Struct. Biotechnol. J.* **2012**, *3*, e201210008.

7

Dispersal-Based Microbial Community Assembly Decreases Biogeochemical Function

Emily B. Graham * and James C. Stegen *

Pacific Northwest National Laboratory, P.O. Box 999, Richland, WA 99352, USA
* Correspondence: emily.graham@pnnl.gov (E.B.G.); james.stegen@pnnl.gov (J.C.S.)

Abstract: Ecological mechanisms influence relationships among microbial communities, which in turn impact biogeochemistry. In particular, microbial communities are assembled by deterministic (e.g., selection) and stochastic (e.g., dispersal) processes, and the relative balance of these two process types is hypothesized to alter the influence of microbial communities over biogeochemical function. We used an ecological simulation model to evaluate this hypothesis, defining biogeochemical function generically to represent any biogeochemical reaction of interest. We assembled receiving communities under different levels of dispersal from a source community that was assembled purely by selection. The dispersal scenarios ranged from no dispersal (i.e., selection-only) to dispersal rates high enough to overwhelm selection (i.e., homogenizing dispersal). We used an aggregate measure of community fitness to infer a given community's biogeochemical function relative to other communities. We also used ecological null models to further link the relative influence of deterministic assembly to function. We found that increasing rates of dispersal decrease biogeochemical function by increasing the proportion of maladapted taxa in a local community. Niche breadth was also a key determinant of biogeochemical function, suggesting a tradeoff between the function of generalist and specialist species. Finally, we show that microbial assembly processes exert greater influence over biogeochemical function when there is variation in the relative contributions of dispersal and selection among communities. Taken together, our results highlight the influence of spatial processes on biogeochemical function and indicate the need to account for such effects in models that aim to predict biogeochemical function under future environmental scenarios.

Keywords: stochastic; deterministic; microbial ecology; simulation; null model; ecosystem function

1. Introduction

Recent attempts to link microbial communities and environmental biogeochemistry have yielded mixed results [1–6], leading researchers to propose the inclusion of community assembly mechanisms such as dispersal and selection in our understanding of biogeochemistry [2,7–9]. Although much work has examined how assembly processes influence the maintenance of diversity and other ecosystem-level processes in macrobial systems [10–13], our comprehension of how these processes influence microbially-mediated biogeochemical cycles is still nascent [2,8,14]. Thus, there is a need to discern the circumstances under which knowledge on assembly processes is valuable for predicting biogeochemical function.

Community assembly processes collectively operate through space and time to determine microbial community composition [3,7,14,15]. They fall into two predominate categories that can be summarized as influenced (i.e., deterministic) or uninfluenced (i.e., stochastic) by biotic and abiotic environmental conditions. Stochastic processes can be further classified into dispersal, evolutionary diversification, and ecological drift, while determinism is largely dictated by selection [7,16]. We refer readers to a recent review article for a more nuanced understanding of deterministic influences on dispersal and of stochastic influences on selection, which are not discussed here [17]. Experimental research has shown unpredictable relationships between microbial diversity and biogeochemical function (generically defined here to represent any biogeochemical reaction of interest), leading to the hypothesis that differences in community assembly history—and thus the relative contributions of stochastic and deterministic processes—drives relationships between microbial community structure and biogeochemical function [8,9].

Dispersal in particular may vary the relationship between community structure and biogeochemical function [7]. Both positive and negative associations between dispersal and community function have been hypothesized (reviewed in [18]).

The 'portfolio effect' argues for enhanced community functioning under high levels of dispersal, proposing that high diversity communities are more likely to contain more beneficial species properties on average than lower diversity communities [19,20]. Additionally, if dispersal increases biodiversity, there should be a greater chance that the community can occupy more niche space (i.e., niche complementarity), reducing direct competition and increasing function [21].

Alternatively, dispersal may decrease community-level biogeochemical function [7,22]. High rates of dispersal can add organisms to a microbial community that are not well-suited to local environmental conditions (i.e., mass effect or source-sink dynamics [23,24]). Maladapted individuals may invest more in cell maintenance to survive as opposed to investing in cellular machinery associated with biogeochemical reactions needed to obtain energy for growth and reproduction. In this case, the community's ability to drive biogeochemical reactions may be depressed.

For instance, pH [25] and salinity [26,27] are widely considered strong regulators of microbial community structure. If microorganisms are well adapted to and disperse from a moderate pH or salinity environment to a more extreme environment, they may be maladapted and have to expend energy to express traits that maintain neutral internal pH (e.g., H+ pumps) or maintain cellular water content (e.g., osmotic stress factors).

These cell maintenance activities detract from the energy available to transform biogeochemical constituents and may suppress overall community rates of biogeochemical function. In contrast, locally adapted species would putatively have more efficient mechanisms for cell maintenance in the local environment and be able to allocate more energy for catalyzing biogeochemical reactions.

These dispersal effects also interact with local selective pressures and the physiological ability of organisms to function across a range of environments to collectively influence biogeochemical function in uncertain ways.

Here, we propose that (1) communities more influenced by dispersal are composed of species that are less well adapted to the local environment and, in turn, that (2) dispersal-based assembly processes decrease biogeochemical function (Figure 1).

Our aim is to formalize these hypotheses and provide a simulation-based demonstration of how dispersal-based assembly can influence function. To do so, we employ an ecological simulation model to explicitly represent dispersal and selection-based processes, and we leverage ecological null models that have a long history of use in inferring assembly processes [28]. We link the resulting communities to biogeochemical function through organismal fitness. In our conceptualization, biogeochemical function is a generic representation, and thus, our results can be applied to any process of interest.

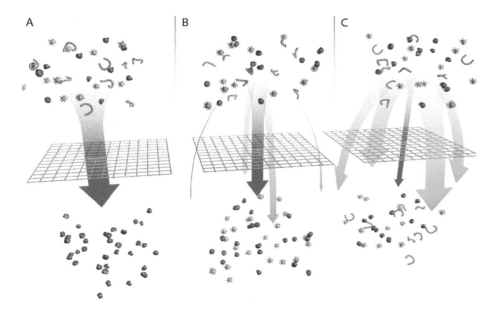

Figure 1. We propose a conceptual model in which dispersal-based assembly processes decrease biogeochemical function. Purple organisms in this figure represent all species that are well-adapted to and are thus good competitors in a given environment. Yellow and green organisms represent all species that are less adapted to the environment than purple organisms. While not displayed for simplicity, we conceptualize multiple species within each color. We acknowledge that the environment influences microbial community composition through effects of both abiotic (e.g., resource availability) and biotic (e.g., competition and predator-prey interactions) factors. We use the term 'selective filter' to indicate influences of both factors on an organism's fitness [29]. (**A**) In a community structured entirely by determinism, selective filtering restricts community composition to species that are well-adapted to prevailing conditions, resulting in enhanced biogeochemical function. (**B**) In communities with moderate stochasticity (here, moderate rates of dispersal), there is an increase in the abundance of maladapted organisms in the community. In turn, the community is less efficient and exhibits lower biogeochemical function. (**C**) Under high levels of stochasticity (here, high rates of dispersal), a large portion of community members are maladapted, resulting in the lowest rates of biogeochemical function.

2. Materials and Methods

All simulations, null models, statistical analyses, and graphics were completed in *R* software (https://cran.r-project.org/). The simulation model consisted of two parts and was followed by statistical analysis. The model builds upon previous work by Stegen, Hulbert, Graham, and others [2,14,15,30–33]. Relative to this previous work, the model used here is unique in connecting evolutionary diversification, variation in the relative influences of dispersal and selection, null models to infer those influences, and biogeochemical function. Previous models have addressed some subset of those features (e.g., connecting evolutionary processes with stochastic and deterministic ecology), but as far as we are aware, previous studies have not integrated all features examined here.

A central purpose of the simulation model was to vary the influences of community assembly processes. Previously developed null models (see below) were used to identify parameter combinations that provided a range of scenarios across which the relative balance among community assembly processes varied. As such, parameter values were selected to generate conceptual outcomes needed to evaluate the relationship between assembly processes and biogeochemical function. Specific parameter values do not, therefore, reflect conditions in any particular ecosystem. Likewise, the model reflects a general timescale across which there are (1) large numbers of birth/death events such that community composition closely tracks environmentally-imposed differences in organismal fitness, and/or (2) opportunities for significant immigration into local communities via dispersal. The model's spatial scale is also a general representation that depends on the rate at which individual cells can move through space in a given environment. Therefore, the model's spatial and temporal scales depend on the environment of interest and may be short (e.g., microaggregates in unsaturated soils or communities with fast generation times) or long (e.g., stream biofilms influenced by hydrologic transport across long distances or communities with long generation times). One hundred replicates were run for each parameter combination in the simulation model.

2.1. Regional Species Pool Simulation

First, a regional species was constructed following the protocol outlined in Stegen et al. [15]. Regional species pools were constructed by simulating diversification in which entirely new species arise through mutations in the environmental optima of ancestral organisms. Environmental optima evolve along an arbitrary continuum from 0 to 1, following a Brownian process. Regional species pools reach equilibria according to the constraints described by Stegen et al. [15] and Hurlbert and Stegen [30] and summarized here: (1) we define a maximum number of total individuals in the pool (2 million) such that the population size of a given species declines with an increasing number of species, and (2) the probability of extinction for a given species increases as its population size decreases according to a negative exponential function [population extinction probability $\propto \exp(-0.001 \times$ population size$)$].

The evolution of a regional species pool was initiated from a single ancestor with a randomly chosen environmental optimum (initially comprising all two million individuals in the population). Mutation probability was set as 1.00×10^{-5}. A descendant's environmental optimum deviated from its ancestor by a quantity selected from a Gaussian distribution with mean 0 and standard deviation 0.2. Following mutation, population sizes were reduced evenly such that the total number of individuals remained at two million. The simulation was run for 250 time steps, which was sufficient to reach equilibrium species richness.

2.2. Community Assembly

The model's second component assembled four local communities from the regional species pool according to scenarios conceptualized to test our hypotheses. In the model, both selection and dispersal are probabilistic. Selection is based on the difference between species environmental optima and local environmental conditions, while dispersal is unrelated to environmental conditions.

Species were drawn from the regional species pool to generate a source community under weak selection and three receiving communities with no dispersal, moderate dispersal, and high dispersal in which organismal niche breadth (n, 0.0075 to 0.175) and environmental conditions (E, 0.05 to 0.95) were allowed to vary across simulations. A simplifying assumption of the model was that all organisms in a simulation had equivalent niche breadth. The purpose of this assumption was to examine tradeoffs between communities comprised of high-functioning, specialist organisms vs. those comprised of lower-functioning, generalist species. Our intent was to simulate communities across a gradient in the degree of specialization (i.e., niche breadth). This allowed for an evaluation of the influence of niche breadth on the relationship between assembly processes and biogeochemical function. All communities had 100 species and 10,000 individuals, drawn probabilistically from the regional species pool. To define species presence/absence in each community, we drew 100 species without replacement from the regional species pool based on selection probabilities described below. In turn, we drew 10,000 individuals with replacement into those 100 species using the same selection probabilities. Selection probabilities (P) of each species from the regional pool were set by a Gaussian function with variance n (reflecting niche breadth) and the deviation (d) of each species environmental optimum from the local environment per the following equation:

$$P = \frac{1}{\sqrt{2n\pi}}e^{-\frac{d^2}{2n}} \tag{1}$$

This equation represents the probability of an individual from a given species surviving in a given environment—and thus the strength of selection for or against it—as directly related to three factors: (1) its own environmental optimum, (2) the simulated environment in which it finds itself, and (3) its niche breadth.

For assembly of the source community, we used one niche breadth (n) for all simulations, which was the maximum value used for receiving communities (0.175). This value represents generalist organisms, which allows for assembly of species representing a broader range of environmental optima than when niches are narrow. The environmental conditions in the source community were also set to a single value using the following procedure: we generated 10 regional species pools and combined species abundances and environmental optima from these pools to generate one aggregate pool representative of the probable distributions of environmental optima yielded by our simulations. We set the environmental optimum of the source community to one end of this spectrum (5th percentile) to allow for comparisons with receiving communities that had the same or larger environmental values. This allowed us to study emergent behavior across a broad range of environmental differences between the source and receiving communities.

For receiving communities, we allowed the environmental conditions and niche breadth to vary across simulations. Environmental conditions ranged from 0.05 to 0.95 by intervals of 0.04736842 to yield 20 conditions. Environmental conditions were static within each simulation. Niche breadth ranged from 0.0075 to 0.175 by 0.008815789 to yield 20 conditions. Receiving communities were assembled under all possible combinations of environmental conditions and niche breadths. Communities for the selection-only case (i.e., no dispersal from the source community) were assembled based only on the selection probabilities as defined by Equation (1), using the same approach as for the assembly of the source community. For moderate and homogenizing dispersal, we modified selection probabilities to incorporate species dispersing from the source community as defined by the following equation:

$$P_{disp} = P + 0.05(S_{source}{}^{D}) \tag{2}$$

where P_{disp} is the selection probability of a given species accounting for dispersal, S_{source} is the abundance of that species in the source community, and D a parameter reflecting dispersal rate. This equation alters the selection probability without dispersal (Equation (1) with an exponential modifier that enhances the probability of selection for species that are abundant in the source community. Parameter D was set to 1 for moderate dispersal and 2 for homogenizing dispersal. All possible communities were simulated with 100 replicate regional species pools such that all possible combination of parameters were used once with each regional species pool.

Equation (2) simplifies dispersal as a probabilistic function without regard to phylogeny, although we acknowledge that the ability of organisms to disperse is not phylogenetically random in natural settings [17]. In our view of community assembly (and in our simulation model), both selection and dispersal are probabilistic. Selection is based on the difference between species environmental optima and local environmental conditions, while dispersal is unrelated to environmental conditions. In this view, the word 'deterministic' indicates that the environment determines the probability of drawing a given species into a local community, even though assembly is still probabilistic. Likewise, the word 'stochastic' indicates that the random movement of organisms is the only factor influencing local community assembly. Future studies should build upon this work to examine the influence of phylogenetically-structured dispersal probabilities in affecting biogeochemical function.

Our estimation of biogeochemical function is meant to be illustrative and is not associated with any specific reaction. Given this perspective, we make a simplifying conceptual assumption that individuals well-fit to their environment generate higher rates of biogeochemical function than maladapted individuals. The motivation for this assumption is that individuals that are maladapted to a given environmental condition will have to use a larger portion of available energy to maintain their physiological state than well-adapted organisms. In turn, maladapted organisms can invest less in the production of enzymes needed to carry out biogeochemical reactions, thereby leading to lower biogeochemical rates.

In our model, selection probability of a given species in a given environment (Equation (1)) defines how adapted an individual of that species is to its local environment. This leads to another simplifying assumption: the contribution of an individual to the overall biogeochemical rate (B) is directly proportional to how well adapted it is to the local environment such that the contribution of each individual is a linear function of its selection probability within a given environment. The biogeochemical contribution of each species is therefore found by multiplying its selection probability by its abundance. To find the total biogeochemical rate for each community, we then summed across all species in a community. Biogeochemical function for each community was thus calculated as:

$$B = \sum_{i=1}^{100} a_i P_i \qquad (3)$$

where B is the biogeochemical function for a given community and a_i and P_i are the abundance and probability of selection for species i, respectively (note there were 100 species within each community). An inherent result of this calculation is that simulations with smaller niche breadth have higher maximum selection probabilities (see Equation (1)), which can lead to higher biogeochemical function, relative to simulations with larger niche breadth. Our formulation therefore assumes higher biogeochemical function for specialist organisms, but only if they are well adapted to their local environment. This assumption reflects a tradeoff between the breadth of environments an individual can persist in and the maximum fitness of an individual in any one environment (discussed in [34]).

2.3. Ecological Inferences Using Null Models

Following the assembly of communities, the relative influences of stochasticity (i.e., dispersal-based) and determinism (i.e., selection) in structuring communities were estimated using a null modeling approach previous described in Stegen et al. [15,31]. We refer the reader to these earlier publications for full details and provide only a summary of the major elements of the null modeling approach here. The composition of each receiving community was compared to an associated source community that was assembled from the same regional species pool. We first estimated pairwise phylogenetic turnover between a given pair of communities. This was done by calculating the abundance-weighted β-mean-nearest-taxon-distance (βMNTD) [35,36]. A null model was then run 999 times. In each iteration of the null model, species names were moved randomly across the tips of the regional pool phylogeny. This breaks phylogenetic relationships among taxa observed in each community. Using the resulting (randomized) phylogenetic relationships, we re-calculated phylogenetic turnover between the pair of communities and refer to this as $\beta\text{MNTD}_{\text{null}}$. Running the null model 999 times generated a distribution of $\beta\text{MNTD}_{\text{null}}$ values. We then compared the observed βMNTD to the mean of the $\beta\text{MNTD}_{\text{null}}$ distribution and normalized this difference by the standard deviation of the $\beta\text{MNTD}_{\text{null}}$ distribution. The difference between βMNTD and the $\beta\text{MNTD}_{\text{null}}$ distribution was therefore measured in units of standard deviations and is referred to as the β-nearest taxon index (βNTI) [32]. Values of βNTI that are <-2 or $>+2$ are deemed significant in the sense that observed βMNTD deviated significantly from the $\beta\text{MNTD}_{\text{null}}$ distribution. The $\beta\text{MNTD}_{\text{null}}$ distribution is what's expected when community assembly is not strongly influenced by deterministic ecological selection. Significant deviation from this distribution therefore indicates that selective pressures are very similar (βNTI < -2) or very different (βNTI $> +2$) between the two communities being compared. Following the convention of Dini-Andreote et al. [37] we refer to βNTI <-2 as indicating homogeneous selection (i.e., significantly less turnover than expected due to consistent selective pressures) and βNTI $> +2$ as indicating variable selection (i.e., significantly more turnover than expected due to divergent selective pressures). Inferences from βNTI have previously been shown to be robust [15]. This method has also been used extensively across a broad range of systems (e.g., [2,14,38–40]) and is described in detail in previous work [32].

2.4. Statistical Analysis

We analyzed differences in model outputs using standard statistical approaches. We calculated the alpha diversity of each source and receiving community using the Inverse Simpson Index [41,42] in the *R* package 'vegan' [43]. Differences in alpha diversity across communities were evaluated with one-way ANOVA followed by post-hoc Tukey's HSD tests. We used pairwise Kolmogorov-Smirnov tests to compare distributions of species optima between simulations (distributions were non-normal). To compare biogeochemical function of the three dispersal cases, we used one-way ANOVA followed by post-hoc Tukey's HSD tests. We also analyzed how biogeochemical function changed as the environmental difference between source and receiving communities increased; this was done using quadratic regressions due to non-linearity in the relationships. We also compared the influence of dispersal on biogeochemical function across different niche breadths.

This was done by first finding the ratio of function in selection-only communities to function in associated homogenizing dispersal communities. Ratios were calculated by comparing communities assembled from the same regional species pool and with identical environmental condition and niche breadth. The resulting distributions of ratios were compared across different niche breadths using one-way ANOVA followed by post-hoc Tukey's HSD tests.

To evaluate the relationship between the relative influence of dispersal-based assembly (inferred from the value of βNTI) and biogeochemical function, correlations between βNTI and biogeochemical function were assessed with linear regression. In most studies βNTI values are not independent of each other such that statistical significance requires a permutation-based method such as a Mantel test. Here, each βNTI estimate is independent whereby standard statistical methods that assume independence are appropriate.

3. Results and Discussion

As ecosystem process models become more sophisticated (e.g., [44–46]), there is a need to improve these models by better understanding the linkages among community assembly processes and ecosystem function. Here, we used an ecological simulation model to highlight the importance of dispersal-based microbial community assembly for biogeochemical function. Our results suggest that incorporating assembly processes into ecosystem models may improve model predictions of biogeochemical function under future environmental conditions.

3.1. Microbial Community Composition in Response to Dispersal

We found that diversity was highest when both dispersal and selection influenced community structure (Figure 2). In communities assembled with moderate to broad niches, intermediate amounts of dispersal led to the highest diversity (Figure 2A,B). These moderate-dispersal communities were characterized by distributions of environmental optima (across species and individuals) that did not match the source or selection-only distributions, and instead reflect an influence of both dispersal from the source and local selection (Figure 3B,D,F). Both moderate- and homogenizing-dispersal cases exhibited higher diversity than source or selection-only communities (Figure 2A,B). We note that the slight differences in diversity between source and selection-only communities were due to environmental conditions in source communities being defined at one end of the environmental spectrum. This edge-effect truncated its distribution of species environmental optima, causing the distribution to be right skewed (Figure 3A,C,E). Our results suggest a conceptual parallel to Connell's [47] Intermediate Disturbance Hypothesis, whereby intermediate levels of dispersal lead to the highest overall diversity, but only when niche breadth is broad enough to allow for strong contributions from both dispersal and selection (Figure 2).

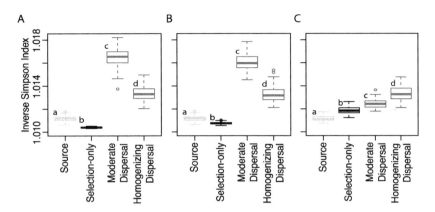

Figure 2. Alpha diversity (inverse Simpson Index) of communities assembled under wide ((**A**) niche breadth = 0.175), moderate ((**B**) niche breadth = 0.086842105), and narrow ((**C**) niche breadth = 0.0075), niches in the mid-point environment (0.476315789). Upper and lower hinges of the box plots represent the 75th and 25th percentiles and whiskers represent 1.5 times the 75th and 25th percentiles, respectively. Colors coincide with labels on the x-axis.

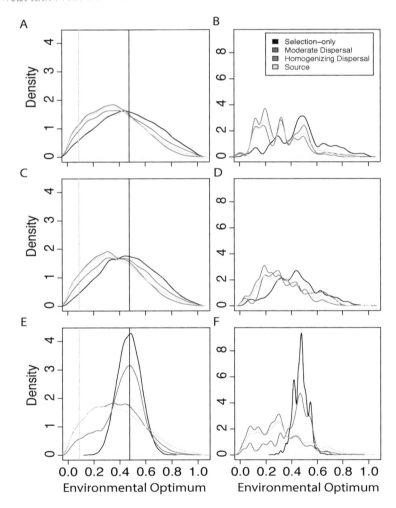

Figure 3. Kernel density of species optima are shown under wide ((**A,B**) niche breadth = 0.175), moderate ((**C,D**) niche breadth = 0.086842105), and narrow ((**E,F**) niche breadth = 0.0075) niches in the mid-point environment (0.476315789, vertical black line). Column 1 displays distributions of species optima without accounting for abundances. Column 2 displays distributions of individuals' optima. Distributions for the source community and its environment condition (vertical line) are displayed in gray. The same communities were selected as examples to generate Figures 2 and 3.

With the narrowest niche breadths (Figure 2C), we observed a distinct pattern of diversity relative to broader niche breadths (Figure 2A,B). Diversity in moderate-dispersal cases decreased substantially as niche breadth narrowed, indicating that moderate levels of dispersal can be overwhelmed when local selective pressures are strong. In contrast, homogenizing dispersal cases maintained consistent levels of diversity across niche breadths and displayed distributions of environmental optima that tracked those of the source community (Figure 3). Diversity in selection-only cases was greatest under the narrowest niche breadth. This was due to only very well-adapted species being part of the community, which led to high abundance across all species in those communities (Figure 3F). For selection-only communities, broader niche breadths resulted in more species with low abundances, and thus lower diversity (cf. black lines in Figure 3B,D,F).

3.2. Dispersal, Microbial Community Composition, and Biogeochemical Function

We found that microbial community assembly history altered the degree to which organisms within a community were adapted to their local environment. Given our assumption of the connection between the degree of adaptation and biogeochemical function (see Methods), assembly history was therefore found to have an indirect effect on biogeochemical function. The environmental optima of taxa in selection-only communities more closely matched their simulated environmental conditions compared to communities assembled with dispersal (Figure 3, 1st column, $p < 0.001$). When niche breadth was broad (Figure 3A), species' environmental optima were distributed around the simulated environment under all dispersal cases. However, as niche breadth decreased (Figure 3C,E), the species distribution of selection-only cases tightened around the simulated environment, with moderate and homogenizing dispersal cases having a wider distribution than the selection-only case. These disparities were maintained when accounting for species abundances (Figure 3B,D,F), in which selection-only communities had unimodal distributions separate from the source community, while moderate and homogenizing dispersal communities had distributions ranging from unimodal to multi-modal, depending on niche breadth. Dispersal from the source therefore resulted in significant numbers of individuals having large deviations between their environmental optima and the local environmental condition. The large number of maladapted individuals in communities experiencing dispersal from the source resulted in selection-only communities having the highest rates of biogeochemical function, on average, regardless of the simulated environment (Figure 4, $p < 0.0001$).

In natural systems, microbial community compositional differences can be due to competitive dynamics that select for organisms based on their niche optima [48,49] and to immigration of new taxa from the regional species pool [7,32,50]. Strong local selective pressures can lead to more fit species and enhanced biogeochemistry [7]. Due to the lack of immigrating maladapted species in the selection-only simulations, biogeochemical rates were maintained regardless of the difference between source and receiving community environments. This indicates that biogeochemical function can be enhanced by species adaptation to local conditions. Indeed, a plethora of literature demonstrates that environmental features such as pH [25], nutrients [51], and salinity [26,27] impact microbial community structure and biogeochemical function, and our results indicate that the linkage between community structure and function is due to microbial adaptation to local conditions.

Our results also indicate that when immigrating microorganisms are derived from environments that differ from the receiving community (e.g., dispersal across steep geochemical gradients), biogeochemical function may be suppressed. When we included dispersal from a source community, greater differences between the source and receiving communities led to decreases in biogeochemical function in the receiving communities (Figure 4B, $p < 0.0001$), and this effect became more pronounced as the rate of dispersal increased. Natural systems are influenced by some combination of dispersal and selection and our results indicate that function is maximized when dispersal is minimized and selection is maximized.

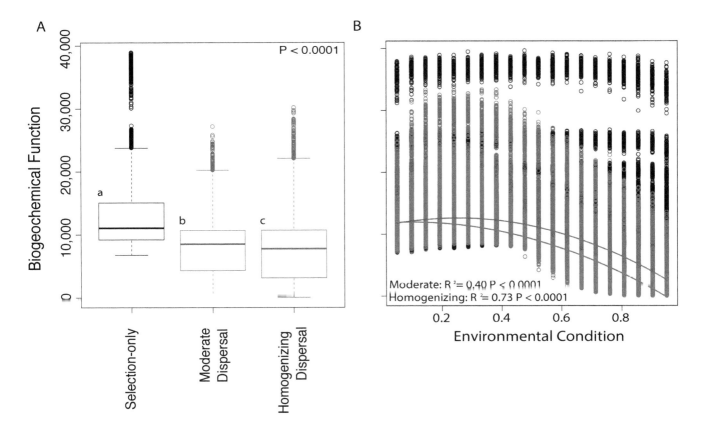

Figure 4. (**A**) Biogeochemical function across dispersal cases. Upper and lower hinges of the box plot represent the 75th and 25th percentiles and whiskers represent 1.5 times the 75th and 25th percentiles, respectively. Different letters indicate statistically significant differences in mean values. (**B**) Biogeochemical function across environmental conditions in receiving communities (vertical axis is the same as in panel A). In the selection-only case (black), biogeochemical function did not vary with environmental condition such that no regression line is drawn. With moderate (blue) and homogenizing (red) dispersal, biogeochemical maxima occurred when the receiving community's environmental condition aligned with the environmental optima of species in the source community (compare to Figure 3). For these two cases, quadratic regression was used and resulting models are shown as solid lines (statistics provided).

Dispersal had the greatest influence on biogeochemical function when niche breadth was narrow (Figure 5). The biogeochemical function of selection-only communities in comparison to homogenizing-dispersal communities was greatest under the narrowest niche breadth (0.0075) and rapidly decreased when transitioning to broader niche breadths. Selection-only communities simulated with narrow niches are comprised of specialist species that can generate high biogeochemical rates and that are well adapted to their local environment. Increasing niche breadth results in the assembly of species with a broader range of environmental optima and that generate lower biogeochemical rates even if their environmental optimum matches the environmental condition (see Methods for a discussion of this assumed trade-off). Thus, high rates of dispersal combined with narrow niche breadth causes replacement of high-functioning specialist organisms with maladapted taxa, thereby significantly reducing community-level biogeochemical function. When niche breadth is broader, immigrating organisms replace lower-functioning organisms (i.e., generalists), resulting in a smaller decreased in community biogeochemical function. We note that our model does not represent dispersal-competition tradeoffs [19], nor does it explicitly represent organismal interactions; exploring the influence of these features would be an interesting extension of the model presented here.

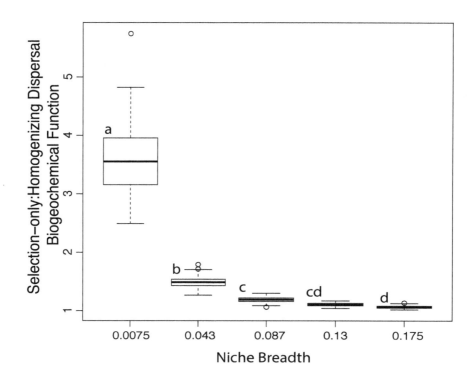

Figure 5. The ratio of biogeochemical function in selection-only cases to homogenizing dispersal cases across five niche breadths that span the entire parameter range (0.0075 to 0.175). Each column represents all replicates across all environments for a given niche breadth. Different letters indicate statistically significant differences in mean values. Upper and lower hinges of the box plots represent the 75th and 25th percentiles and whiskers represent 1.5 times the 75th and 25th percentiles, respectively.

Regardless of dispersal, simulations with broader niche breadth led to lower rates of biogeochemical function, supporting a tradeoff between communities comprised of specialist vs. generalist species [52–54]. Previous work in microbial systems has posited life-history tradeoffs between specialist vs. generalist species, whereby specialists expend more energy to establish their niches but function at higher levels once established [55]. Specialist species have also been found to be more sensitive to changes in the environment due to strong adaptation to their local environment, with generalists being more resilient to change [56–59]. While we do not address temporal dynamics in our model, the separation of biogeochemical function based on niche breadth indicates a central role for the balance of specialist vs. generalist microorganisms within a community in determining function, regardless of prevailing environmental conditions.

3.3. Impact of Assembly Processes on Biogeochemical Function

We also observed that niche breadth within the receiving community was a key parameter in dictating biogeochemical function when environmental conditions (and thus selective pressures) differed between source and receiving communities. In cases without dispersal, biogeochemical function was dictated entirely by niche breadth regardless of differences in selective environments (as inferred from βNTI) between source and receiving communities (Figure 6A,D). Selective pressures in the selection-only receiving communities were most dissimilar to the source community (βNTI > 2) in simulations with both narrow niche breadth and environmental conditions that were very different from the source community (Figure 6A). This relationship was also apparent (but weaker) in simulations with an intermediate amount of dispersal (Figure 6B). In receiving communities with high rates of dispersal, stochasticity ($|\beta$NTI$|$ < 2) was the dominant process regardless of niche breadth or environmental condition in the receiving community (Figure 6C).

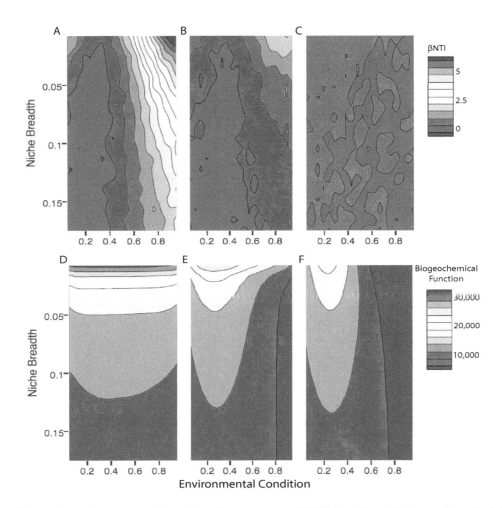

Figure 6. Interpolated contour plots showing average βNTI (**A–C**) and biogeochemical function (**D–F**) for each dispersal case across all parameter combinations. Interpolations are based on parameter combinations at each of 20 evenly spaced values across each axis. Values of βNTI that are further from 0 indicate increasing influences of deterministic assembly (and decreasing stochasticity). (**A,D**) depict selection-only communities; (**B,E**) depict moderate dispersal communities; and (**C,F**) depict homogenizing dispersal communities.

Across the full parameter space defined by niche breadth and environmental condition, cases with moderate and homogenizing dispersal were generally characterized by a dominance of stochasticity (Figure 6B,C). This increase in stochasticity relative to selection-only cases corresponded to decreased biogeochemical function. This was particularly true as the environment diverged from the source community (Figure 6D–F). Biogeochemical function in these cases was also negatively correlated to niche breadth (i.e., highest under narrow niche breadths), revealing higher functioning of specialist organisms regardless of assembly processes.

Given these apparent associations between assembly processes and biogeochemical function, we directly examined differences in relationships between βNTI and biogeochemical function across a range of environments and niche breadths (Figure 7). Our results suggest that microbial assembly processes may exert the most influence over biogeochemical function when there is significant variation in the relative contributions of deterministic and stochastic processes among communities. We found the strongest relationships between βNTI and function when environmental conditions were dissimilar to the source community, regardless of niche breadth (Figure 7G–I). βNTI had the greatest range in these scenarios, reflecting substantial variation in the contribution of stochastic and deterministic processes. By contrast, scenarios with environments more similar to the source environment had little variation in assembly processes and no relationship between βNTI and biogeochemical function (Figure 7A–F).

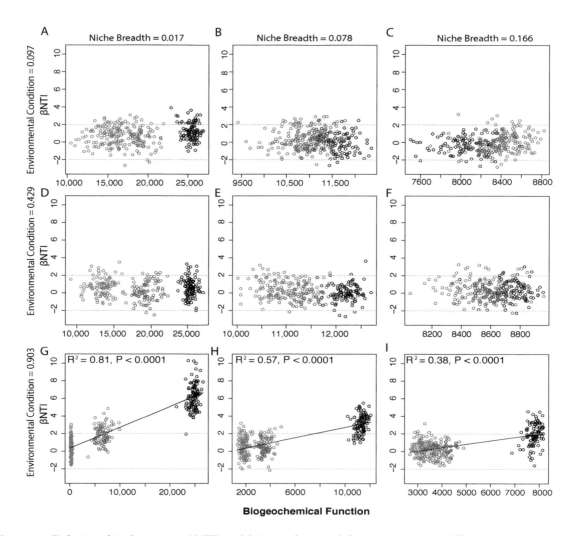

Figure 7. Relationship between βNTI and biogeochemical function across different niche breadths (columns) and different environmental conditions of the receiving communities (rows). (**A–C**), (**D–F**), and (**G–I**) depict environmental conditions similar, moderately different, and very dissimilar to that of the source community, respectively. (**A,D,G**), (**B,E,H**), and (**C,F,I**) respectively show narrow, moderate, and wide niche breadths. Values of βNTI that are further from 0 indicate increasing influences of deterministic assembly (and decreasing stochasticity). Horizontal gray lines indicate significance thresholds of −2 and +2. Relationships were evaluated with linear regression; fitted models are shown as black lines and statistics are provided on each panel. Panels without regression models had non-significant ($p > 0.05$) relationships. Note that the vertical axis is scaled the same across panels, but the horizontal axis is not. Black, blue, and red symbols indicate selection-only, moderate dispersal, and homogenizing dispersal scenarios, respectively.

Variation in the balance of stochastic and deterministic assembly processes is prevalent in natural systems [2,7,16,38], as most ecosystems experience spatially and/or temporally variable rates of dispersal. For example, hydrologic connectivity facilitates microbial dispersal and differs with physical matrix structure in soils and sediments. We therefore pose that variation in βNTI may be an effective tool for predicting biogeochemical function when biotic and abiotic conditions lead to a mixture of stochastic and deterministic assembly processes. Natural systems have repeatedly shown such a mixture, and previous field observations have revealed connections between βNTI and biogeochemical function [2,14,60]. These outcomes support our model-based inference that βNTI—as a proxy for assembly processes—offers a practical means to inform models that represent the effects of ecological processes on biogeochemical function.

While our results suggest that maladapted immigrating organisms decrease biogeochemical function, it is important to note that stochasticity may offer buffering capacity that maintains or increases biogeochemical function relative to well-adapted deterministic communities in the context of future environmental perturbations not simulated with the static environmental conditions in our model [56]. Stochastic spatial processes, such as dispersal, may lead to coexistence of species with different environmental optima resulting in a community that can rapidly adapt to changing environment conditions and maintain biogeochemical function in the face of perturbation. Researchers have long demonstrated positive relationships between biodiversity and ecosystem function in both macrobial [61,62] and microbial [63–65] systems, and new work has highlighted the role of stochasticity in maintaining this connection [66]. Conversely, a lack of stochasticity may result in species so specialized to a given environment that they are vulnerable to environmental change [56]. While these communities would putatively exhibit high rates of biogeochemical function under stable environmental conditions, their function would plummet in response to perturbation, akin to observations of a tradeoff between function and vulnerability in plant communities [21,67].

3.4. Implications for Ecosystem Models

The cumulative impacts of ecological processes through time and how they relate to ecosystem-level processes is an emerging research frontier in ecosystem science [2,44,52,68,69]. We reveal how dispersal-based community assembly can decrease adaptation to local environments and, in turn, decrease biogeochemical function. Our modelling approach demonstrates plausible outcomes of microbial assembly processes on ecosystem functioning, and integrating this knowledge with factors such as historical abiotic conditions, competitive dynamics, and life-history traits could substantially improve ecosystem model predictions.

Previous work by Hawkes and Keitt [52] laid a theoretical foundation for incorporating time-integrated ecological processes into predictions of biogeochemical function. They demonstrate that community-level microbial functions are the accretion of individual life-histories that determine population growth, composition, and fitness. However, they acknowledge their exclusion of dispersal processes from their models and do not explicitly consider dispersal in their analysis. Hawkes and Keitt [52] therefore provide a baseline for future research and call for a holistic understanding of historical processes on microbial function, with a particular emphasis on the underlying mechanisms generating these trends. Our work enhances this framework by demonstrating that community assembly processes are integral to knowledge of biogeochemical function in natural systems.

Microbially-explicit models (e.g., MIMICS, MEND) are rapidly becoming more sophisticated and are readily amenable to modules that represent ecological assembly processes [70,71]. As models begin to consider microbial ecology, there is a need to decipher linkages among spatiotemporal microbial processes and ecosystem-level biogeochemical function. We propose that new microbially-explicit models should go beyond microbial mechanisms at a given point in time or space, and building upon the foundation laid by Hawkes and Keitt [52], incorporate ecological dynamics that operate across longer time scales to influence biogeochemical function. Although there are many available avenues to merge modelling efforts in microbial ecology and ecosystem science, there is little debate that integrated models will increase the accuracy of predictions in novel future environments.

4. Conclusions

We demonstrate the influence of ecological assembly processes on biogeochemical function. Specifically, we show that dispersal can increase the abundance of maladapted organisms in a community, and in turn, decrease biogeochemical function. This impact is strongest when organismal niche breadth is narrow. We also pose that the explanatory power of microbial assembly processes on biogeochemical function is greatest when there is variation in the contributions of dispersal and selection across a collection of local communities within a broader system of interest. While our work is an encouraging advancement in understanding relationships between ecology and biogeochemistry, a key next step is incorporating assembly processes into emerging model frameworks that explicitly represent microbes and that mechanistically represent biogeochemical reactions.

Acknowledgments: We thank Andrew Pitman for help editing manuscript text and Nathan Johnson for graphical assistance. This research was supported by the US Department of Energy (DOE), Office of Biological and Environmental Research (BER), as part of Subsurface Biogeochemical Research Program's Scientific Focus Area (SFA) at the Pacific Northwest National Laboratory (PNNL). PNNL is operated for DOE by Battelle under contract DE-AC06-76RLO 1830. This research was performed using Institutional Computing at PNNL.

Author Contributions: E.B.G. and J.C.S. conceived and designed this work; E.B.G. performed the simulations, analyzed the data, and wrote the manuscript; J.C.S. contributed to manuscript revisions and code development.

References

1. Wallenstein, M.D.; Hall, E.K. A trait-based framework for predicting when and where microbial adaptation to climate change will affect ecosystem functioning. *Biogeochemistry* **2012**, *109*, 35–47. [CrossRef]
2. Graham, E.B.; Crump, A.R.; Resch, C.T.; Fansler, S.; Arntzen, E.; Kennedy, D.W.; Fredrickson, J.K.; Stegen, J.C. Coupling spatiotemporal community assembly processes to changes in microbial metabolism. *Front. Microbiol.* **2016**, *7*, 1949. [CrossRef] [PubMed]
3. Graham, E.B.; Knelman, J.E.; Schindlbacher, A.; Siciliano, S.; Breulmann, M.; Yannarell, A.; Beman, J.; Abell, G.; Philippot, L.; Prosser, J. Microbes as engines of ecosystem function: When does community structure enhance predictions of ecosystem processes? *Front. Microbiol.* **2016**, *7*, 214. [CrossRef] [PubMed]
4. Martiny, J.B.; Jones, S.E.; Lennon, J.T.; Martiny, A.C. Microbiomes in light of traits: A phylogenetic perspective. *Science* **2015**, *350*, aac9323. [CrossRef] [PubMed]
5. Bier, R.L.; Bernhardt, E.S.; Boot, C.M.; Graham, E.B.; Hall, E.K.; Lennon, J.T.; Nemergut, D.R.; Osborne, B.B.; Ruiz-González, C.; Schimel, J.P. Linking microbial community structure and microbial processes: An empirical and conceptual overview. *FEMS Microbiol. Ecol.* **2015**, *91*. [CrossRef] [PubMed]
6. Rocca, J.D.; Hall, E.K.; Lennon, J.T.; Evans, S.E.; Waldrop, M.P.; Cotner, J.B.; Nemergut, D.R.; Graham, E.B.; Wallenstein, M.D. Relationships between protein-encoding gene abundance and corresponding process are commonly assumed yet rarely observed. *ISME J.* **2015**, *9*, 1693. [CrossRef] [PubMed]
7. Nemergut, D.R.; Schmidt, S.K.; Fukami, T.; O'Neill, S.P.; Bilinski, T.M.; Stanish, L.F.; Knelman, J.E.; Darcy, J.L.; Lynch, R.C.; Wickey, P. Patterns and processes of microbial community assembly. *Microbiol. Mol. Biol. Rev.* **2013**, *77*, 342–356. [CrossRef] [PubMed]
8. Nemergut, D.R.; Shade, A.; Violle, C. When, where and how does microbial community composition matter? *Front. Microbiol.* **2014**, *5*, 497. [CrossRef] [PubMed]
9. Pholchan, M.K.; Baptista, J.D.C.; Davenport, R.J.; Sloan, W.T.; Curtis, T.P. Microbial community assembly, theory and rare functions. *Front. Microbiol.* **2013**, *4*, 68. [CrossRef] [PubMed]
10. Cardinale, B.J.; Wright, J.P.; Cadotte, M.W.; Carroll, I.T.; Hector, A.; Srivastava, D.S.; Loreau, M.; Weis, J.J. Impacts of plant diversity on biomass production increase through time because of species complementarity. *Proc. Natl. Acad. Sci. USA* **2007**, *104*, 18123–18128. [CrossRef] [PubMed]
11. Cadotte, M.W.; Carscadden, K.; Mirotchnick, N. Beyond species: Functional diversity and the maintenance of ecological processes and services. *J. Appl. Ecol.* **2011**, *48*, 1079–1087. [CrossRef]
12. Cadotte, M.W. Dispersal and species diversity: A meta-analysis. *Am. Nat.* **2006**, *167*, 913–924. [PubMed]
13. Cardinale, B.J. Biodiversity improves water quality through niche partitioning. *Nature* **2011**, *472*, 86. [CrossRef] [PubMed]
14. Graham, E.B.; Crump, A.R.; Resch, C.T.; Fansler, S.; Arntzen, E.; Kennedy, D.W.; Fredrickson, J.K.; Stegen, J.C. Deterministic influences exceed dispersal effects on hydrologically-connected microbiomes. *Environ. Microbiol.* **2017**, *19*, 1552–1567. [CrossRef] [PubMed]
15. Stegen, J.C.; Lin, X.; Fredrickson, J.K.; Konopka, A.E. Estimating and mapping ecological processes influencing microbial community assembly. *Front. Microbiol.* **2015**, *6*, 370. [CrossRef] [PubMed]
16. Vellend, M. Conceptual synthesis in community ecology. *Q. Rev. Biol.* **2010**, *85*, 183–206. [CrossRef] [PubMed]

17. Zhou, J.; Ning, D. Stochastic community assembly: Does it matter in microbial ecology? *Microbiol. Mol. Biol. Rev.* **2017**, *81*, e00002–e00017. [CrossRef] [PubMed]
18. Leibold, M.A.; Holyoak, M.; Mouquet, N.; Amarasekare, P.; Chase, J.M.; Hoopes, M.F.; Holt, R.D.; Shurin, J.B.; Law, R.; Tilman, D. The metacommunity concept: A framework for multi-scale community ecology. *Ecol. Lett.* **2004**, *7*, 601–613. [CrossRef]
19. Tilman, D.; Lehman, C.L.; Bristow, C.E. Diversity-stability relationships: Statistical inevitability or ecological consequence? *Am. Nat.* **1998**, *151*, 277–282. [PubMed]
20. Doak, D.F.; Bigger, D.; Harding, E.; Marvier, M.; O'malley, R.; Thomson, D. The statistical inevitability of stability-diversity relationships in community ecology. *Am. Nat.* **1998**, *151*, 264–276. [PubMed]
21. Tilman, D. The ecological consequences of changes in biodiversity: A search for general principles. *Ecology* **1999**, *80*, 1455–1474. [CrossRef]

22. Strickland, M.S.; Lauber, C.; Fierer, N.; Bradford, M.A. Testing the functional significance of microbial community composition. *Ecology* **2009**, *90*, 441–451. [CrossRef] [PubMed]
23. Pulliam, H.R. Sources, sinks, and population regulation. *Am. Nat.* **1988**, *132*, 652–661. [CrossRef]
24. Shmida, A.; Wilson, M.V. Biological determinants of species diversity. *J. Biogeogr.* **1985**, *12*, 1–20. [CrossRef]
25. Fierer, N.; Jackson, R.B. The diversity and biogeography of soil bacterial communities. *Proc. Natl. Acad. Sci. USA* **2006**, *103*, 626–631. [CrossRef] [PubMed]
26. Hollister, E.B.; Engledow, A.S.; Hammett, A.J.M.; Provin, T.L.; Wilkinson, H.H.; Gentry, T.J. Shifts in microbial community structure along an ecological gradient of hypersaline soils and sediments. *ISME J.* **2010**, *4*, 829. [CrossRef] [PubMed]
27. Casamayor, E.O.; Massana, R.; Benlloch, S.; Øvreås, L.; Díez, B.; Goddard, V.J.; Gasol, J.M.; Joint, I.; Rodríguez-Valera, F.; Pedrós-Alió, C. Changes in archaeal, bacterial and eukaryal assemblages along a salinity gradient by comparison of genetic fingerprinting methods in a multipond solar saltern. *Environ. Microbiol.* **2002**, *4*, 338–348. [CrossRef] [PubMed]
28. Gotelli, N.J.; Graves, G.R. Null models in ecology. *Ecology* **1996**, *78*, 189–211.
29. Cadotte, M.W.; Tucker, C.M. Should environmental filtering be abandoned? *Trends Ecol. Evol.* **2017**, *32*, 429–437. [CrossRef] [PubMed]
30. Hurlbert, A.H.; Stegen, J.C. When should species richness be energy limited, and how would we know? *Ecol. Lett.* **2014**, *17*, 401–413. [CrossRef] [PubMed]
31. Stegen, J.C.; Lin, X.; Fredrickson, J.K.; Chen, X.; Kennedy, D.W.; Murray, C.J.; Rockhold, M.L.; Konopka, A. Quantifying community assembly processes and identifying features that impose them. *ISME J.* **2013**, *7*, 2069. [CrossRef] [PubMed]
32. Stegen, J.C.; Lin, X.; Konopka, A.E.; Fredrickson, J.K. Stochastic and deterministic assembly processes in subsurface microbial communities. *ISME J.* **2012**, *6*, 1653. [CrossRef] [PubMed]
33. Stegen, J.C.; Hurlbert, A.H. Inferring ecological processes from taxonomic, phylogenetic and functional trait β-diversity. *PLoS ONE* **2011**, *6*, e20906. [CrossRef] [PubMed]

34. Chase, J.M.; Leibold, M.A. *Ecological Niches: Linking Classical and Contemporary Approaches*; University of Chicago Press: Chicago, IL, USA, 2003.
35. Webb, C.O.; Ackerly, D.D.; Kembel, S.W. Phylocom: Software for the analysis of phylogenetic community structure and trait evolution. *Bioinformatics* **2008**, *24*, 2098–2100. [CrossRef] [PubMed]

36. Fine, P.V.; Kembel, S.W. Phylogenetic community structure and phylogenetic turnover across space and edaphic gradients in western amazonian tree communities. *Ecography* **2011**, *34*, 552–565. [CrossRef]
37. Dini-Andreote, F.; Stegen, J.C.; van Elsas, J.D.; Salles, J.F. Disentangling mechanisms that mediate the balance between stochastic and deterministic processes in microbial succession. *Proc. Natl. Acad. Sci. USA* **2015**, *112*, E1326–E1332. [CrossRef] [PubMed]
38. Wang, J.; Shen, J.; Wu, Y.; Tu, C.; Soininen, J.; Stegen, J.C.; He, J.; Liu, X.; Zhang, L.; Zhang, E. Phylogenetic beta diversity in bacterial assemblages across ecosystems: Deterministic versus stochastic processes. *ISME J.* **2013**, *7*, 1310. [CrossRef] [PubMed]
39. Jurburg, S.D.; Nunes, I.; Stegen, J.C.; Le Roux, X.; Priemé, A.; Sørensen, S.J.; Salles, J.F. Autogenic succession and deterministic recovery following disturbance in soil bacterial communities. *Sci. Rep.* **2017**, *7*, 45691. [CrossRef] [PubMed]

40. Veach, A.M.; Stegen, J.C.; Brown, S.P.; Dodds, W.K.; Jumpponen, A. Spatial and successional dynamics of microbial biofilm communities in a grassland stream ecosystem. *Mol. Ecol.* **2016**, *25*, 4674–4688. [CrossRef] [PubMed]

41. Hill, M.O. Diversity and evenness: A unifying notation and its consequences. *Ecology* **1973**, *54*, 427–432. [CrossRef]

42. Morris, E.K.; Caruso, T.; Buscot, F.; Fischer, M.; Hancock, C.; Maier, T.S.; Meiners, T.; Müller, C.; Obermaier, E.; Prati, D. Choosing and using diversity indices: Insights for ecological applications from the german biodiversity exploratories. *Ecol. Evol.* **2014**, *4*, 3514–3524. [CrossRef] [PubMed]

43. Oksanen, J.; Blanchet, F.G.; Kindt, R.; Legendre, P.; Minchin, P.R.; O'hara, R.; Simpson, G.L.; Solymos, P.; Stevens, M.H.H.; Wagner, H.; et al. Package 'vegan'. Community ecology package, version 2.4-4. Available online: https://cran.r-project.org/web/packages/vegan/index.html (accessed on 24 August 2017).

44. Buchkowski, R.W.; Bradford, M.A.; Grandy, A.S.; Schmitz, O.J.; Wieder, W.R. Applying population and community ecology theory to advance understanding of belowground biogeochemistry. *Ecol. Lett.* **2017**, *20*, 231–245. [CrossRef] [PubMed]

45. Wang, G.; Jagadamma, S.; Mayes, M.A.; Schadt, C.W.; Steinweg, J.M.; Gu, L.; Post, W.M. Microbial dormancy improves development and experimental validation of ecosystem model. *ISME J.* **2015**, *9*, 226. [CrossRef] [PubMed]

46. Wieder, W.; Grandy, A.; Kallenbach, C.; Taylor, P.; Bonan, G. Representing life in the earth system with soil microbial functional traits in the mimics model. *Geosci. Model Dev.* **2015**, *8*, 1789–1808. [CrossRef]

47. Connell, J.H. Diversity in tropical rain forests and coral reefs. *Science* **1978**, *199*, 1302–1310. [CrossRef] [PubMed]

48. DeAngelis, K.M.; Silver, W.L.; Thompson, A.W.; Firestone, M.K. Microbial communities acclimate to recurring changes in soil redox potential status. *Environ. Microbiol.* **2010**, *12*, 3137–3149. [CrossRef] [PubMed]

49. Cregger, M.A.; Schadt, C.W.; McDowell, N.G.; Pockman, W.T.; Classen, A.T. Response of the soil microbial community to changes in precipitation in a semiarid ecosystem. *Appl. Environ. Microbiol.* **2012**, *78*, 8587–8594. [CrossRef] [PubMed]

50. Green, J.; Bohannan, B.J. Spatial scaling of microbial biodiversity. *Trends Ecol. Evol.* **2006**, *21*, 501–507. [CrossRef] [PubMed]

51. Schimel, D.S.; Braswell, B.; McKeown, R.; Ojima, D.S.; Parton, W.; Pulliam, W. Climate and nitrogen controls on the geography and timescales of terrestrial biogeochemical cycling. *Glob. Biogeochem. Cycles* **1996**, *10*, 677–692. [CrossRef]

52. Hawkes, C.V.; Keitt, T.H. Resilience vs. Historical contingency in microbial responses to environmental change. *Ecol. Lett.* **2015**, *18*, 612–625. [CrossRef] [PubMed]

53. Wilson, D.S.; Yoshimura, J. On the coexistence of specialists and generalists. *Am. Nat.* **1994**, *144*, 692–707. [CrossRef]

54. Kassen, R. The experimental evolution of specialists, generalists, and the maintenance of diversity. *J. Evolut. Biol.* **2002**, *15*, 173–190. [CrossRef]

55. Lennon, J.T.; Aanderud, Z.T.; Lehmkuhl, B.; Schoolmaster, D.R. Mapping the niche space of soil microorganisms using taxonomy and traits. *Ecology* **2012**, *93*, 1867–1879. [CrossRef] [PubMed]

56. Shade, A.; Peter, H.; Allison, S.D.; Baho, D.L.; Berga, M.; Bürgmann, H.; Huber, D.H.; Langenheder, S.; Lennon, J.T.; Martiny, J.B. Fundamentals of microbial community resistance and resilience. *Front. Microbiol.* **2012**, *3*, 417. [CrossRef] [PubMed]

57. Mou, X.; Sun, S.; Edwards, R.A.; Hodson, R.E.; Moran, M.A. Bacterial carbon processing by generalist species in the coastal ocean. *Nature* **2008**, *451*, 708. [CrossRef] [PubMed]

58. Langenheder, S.; Lindström, E.S.; Tranvik, L.J. Weak coupling between community composition and functioning of aquatic bacteria. *Limnol. Oceanogr.* **2005**, *50*, 957–967. [CrossRef]

59. Allison, S.D.; Martiny, J.B. Resistance, resilience, and redundancy in microbial communities. *Proc. Natl. Acad. Sci. USA* **2008**, *105*, 11512–11519. [CrossRef] [PubMed]

60. Stegen, J.C.; Fredrickson, J.K.; Wilkins, M.J.; Konopka, A.E.; Nelson, W.C.; Arntzen, E.V.; Chrisler, W.B.; Chu, R.K.; Danczak, R.E.; Fansler, S.J. Groundwater–surface water mixing shifts ecological assembly processes and stimulates organic carbon turnover. *Nat. Commun.* **2016**, *7*, 11237. [CrossRef] [PubMed]

61. Cardinale, B.J.; Matulich, K.L.; Hooper, D.U.; Byrnes, J.E.; Duffy, E.; Gamfeldt, L.; Balvanera, P.; O'Connor, M.I.; Gonzalez, A. The functional role of producer diversity in ecosystems. *Am. J. Bot.* **2011**, *98*, 572–592. [CrossRef] [PubMed]

62. Hooper, D.U.; Adair, E.C.; Cardinale, B.J.; Byrnes, J.E.; Hungate, B.A.; Matulich, K.L.; Gonzalez, A.; Duffy, J.E.; Gamfeldt, L.; O'Connor, M.I. A global synthesis reveals biodiversity loss as a major driver of ecosystem change. *Nature* **2012**, *486*, 105–108. [CrossRef] [PubMed]

63. Bell, T.; Newman, J.A.; Silverman, B.W.; Turner, S.L.; Lilley, A.K. The contribution of species richness and composition to bacterial services. *Nature* **2005**, *436*, 1157. [CrossRef] [PubMed]

64. Langenheder, S.; Bulling, M.T.; Solan, M.; Prosser, J.I. Bacterial biodiversity-ecosystem functioning relations are modified by environmental complexity. *PLoS ONE* **2010**, *5*, e10834. [CrossRef] [PubMed]

65. Levine, U.Y.; Teal, T.K.; Robertson, G.P.; Schmidt, T.M. Agriculture's impact on microbial diversity and associated fluxes of carbon dioxide and methane. *ISME J.* **2011**, *5*, 1683. [CrossRef] [PubMed]

66. Knelman, J.E.; Nemergut, D.R. Changes in community assembly may shift the relationship between biodiversity and ecosystem function. *Front. Microbiol.* **2014**, *5*, 424. [CrossRef] [PubMed]

67. Tilman, D.; Reich, P.B.; Knops, J.; Wedin, D.; Mielke, T.; Lehman, C. Diversity and productivity in a long-term grassland experiment. *Science* **2001**, *294*, 843–845. [CrossRef] [PubMed]

68. Evans, S.E.; Wallenstein, M.D. Soil microbial community response to drying and rewetting stress: Does historical precipitation regime matter? *Biogeochemistry* **2012**, *109*, 101–116. [CrossRef]

69. Fukami, T. Historical contingency in community assembly: Integrating niches, species pools, and priority effects. *Annu. Rev. Ecol. Evol. Syst.* **2015**, *46*, 1–23. [CrossRef]

70. Wang, G.; Post, W.M.; Mayes, M.A. Development of microbial-enzyme-mediated decomposition model parameters through steady-state and dynamic analyses. *Ecol. Appl.* **2013**, *23*, 255–272. [CrossRef] [PubMed]

71. Wieder, W.; Grandy, A.; Kallenbach, C.; Bonan, G. Integrating microbial physiology and physio-chemical principles in soils with the microbial-mineral carbon stabilization (mimics) model. *Biogeosciences* **2014**, *11*, 3899–3917. [CrossRef]

8

Modeling Microbial Communities: A Call for Collaboration between Experimentalists and Theorists

Marco Zaccaria *, Sandra Dedrick and Babak Momeni *

Department of Biology, Boston College, Chestnut Hill, MA 02467, USA; sandra.dedrick@bc.edu
* Correspondence: marco.zaccaria@bc.edu (M.Z.); babak.momeni@bc.edu (B.M.)

Abstract: With our growing understanding of the impact of microbial communities, understanding how such communities function has become a priority. The influence of microbial communities is widespread. Human-associated microbiota impacts health, environmental microbes determine ecosystem sustainability, and microbe-driven industrial processes are expanding. This broad range of applications has led to a wide range of approaches to analyze and describe microbial communities. In particular, theoretical work based on mathematical modeling has been a steady source of inspiration for explaining and predicting microbial community processes. Here, we survey some of the modeling approaches used in different contexts. We promote classifying different approaches using a unified platform, and encourage cataloging the findings in a database. We believe that the synergy emerging from a coherent collection facilitates a better understanding of important processes that determine microbial community functions. We emphasize the importance of close collaboration between theoreticians and experimentalists in formulating, classifying, and improving models of microbial communities.

Keywords: microbial communities; mathematical modeling; community ecology; interspecies interactions; mechanistic modeling; phenomenological modeling

1. Introduction

Biology traditionally investigates the complex, unique, case-particular phenomenology of the natural living world. This focus on exceptional instances has inadvertently limited the efforts, or perhaps the desire, compared to other scientific disciplines, to identify general and overarching principles. Throughout the history of modern science, interactions between the abstract, generalized way of Mathematics, and the detailed, case-oriented way of Biology have been of a tumultuous nature. Despite this history of mismatch in perspectives, the importance and potential impact of works merging these disciplines are broadly accepted.

Microbes are among the primary forces that have shaped life on Earth. In the context of biological research, microbiology has historically emerged as an overarching common ground among life-science disciplines. Microbes are ubiquitous; they are therefore an object of interest for research ranging from detailed organ-specific physiology to large-scale ecological issues. Microbiology also harbors a proven potential to direct and propel technical and conceptual proceedings in a variety of contexts (biomedical, agricultural, and industrial, among others) and has been central to the development of many of the most essential experimental tools in modern biology (from PCR [1] to CRISPR [2–4]), and for constructing miniature models of ecological and evolutionary processes [5–12]. Microbiology was among the first biological disciplines to embrace an interdisciplinary approach through the research by Esty and Meyer on *Clostridium botulinum*, in 1922 [13]. In their work, the dynamics of bacterial population growth were described on a semi-logarithmic scale in relation to the environmental temperature. The work is still employed to this day in the monitoring of food safety during the canning process. It is, to the best of our knowledge, among the first examples of how mathematics can effectively be an accessory to biology and produce highly impactful science. It is our belief that rapid progress is facilitated when theorists and microbiologists systematically coordinate their efforts, and we will argue our point approaching several topics that we believe to be of interest to theorists and experimentalists, both in Academia and in Industry.

This paper is divided into two main parts. The first part introduces a brief classification of mathematical models. It is meant as a brief overview of some of the models that have effectively complemented experimental research in the field of microbiology, and highlights where the efforts of theorists have focused so far. A section is also dedicated to the history of the modeling of the rumen bacterial community, which is a striking example of how complementation with mathematics can advance research in microbiology.

The second part discusses the philosophical differences that underlie how experimental research is approached by biologists and physicists, and how these differences often hamper interdisciplinary cooperation in the process of model development. We will advocate for research applicability as the focus around which experimentalists and theorists can more easily set aside their differences and effectively coordinate their efforts, especially in the process of microbial community assembly. The outstanding issue of model validation is also discussed. Finally, we speculate the potential impact of a unified catalog of modeling approaches in microbiology to make modeling more accessible to experimentalists and to inspire future research directions.

2. Background: Past Experiences in Modeling Microbes in Communities

2.1. Mathematical Modeling of Microbial Assemblies: An Overview

A mathematical model is defined as an equation, or a set of equations, that attempts to explain instances of reality in a simplified manner, utilizing only a system's most pertinent properties [14]. A scientific theory is founded when a mechanistic explanation is given for a set of observed natural phenomena. In the physical sciences, hypotheses are often converted into mathematical statements, and models are assimilated into the experimental process. In biology, however, theory is rarely the ground on which hypotheses are formulated, and mathematical models are oftentimes developed as the aftermath of a mass of data [15,16].

For instance, Pearson and colleagues developed a theoretical framework for the theory of evolution [17], Lotka & Volterra produced models for theoretical ecology that described competition and prey-predation [18], and Kermack & McKendrick created some of the first epidemiological mathematical models [19]. Such efforts, and many others of their kind, have been instrumental in advancing their respective fields [16]. Nonetheless, the development of new mathematical models in biology is often treated with skepticism. This skepticism is in part instigated by an uncertainty in the usefulness of a new modeling framework. Does the framework capture the crucial aspects of the biology? Does it address the important questions faced by researchers in the field? Is the model simple enough to inspire insights into important processes? Is the model general or is it specific to the details and nuances of a particular biological phenomenon? These questions naturally arise when studying microbial communities as well, and reflect the intrinsic trade-offs of each modeling framework as discussed below.

Thornley and France [20] outlined the basic principles of modeling, classifying models as: (1) dynamic or static; (2) deterministic or stochastic; and (3) mechanistic or empirical [15,20]. These categories are not mutually exclusive, and published models are often of a hybrid nature [20]. In the rest of this section, we will describe examples of methodologies that have been applied to microbiology. We will also illustrate the chronological progression of modeling, from basic input-output (empirical) to more comprehensive mechanistic models, by describing the advances made in modeling anaerobic fermentation by rumen bacterial communities.

2.1.1. Metabolic Models

The foundation of the metabolic model is an interconnected network of potential reactions leading to the outputs/products of interest. Genome sequencing data allow researchers to determine an organism's metabolic potential [21]. Whenever genomic data is attainable for a species of concern, annotation of the pathways can outline a comprehensive metabolic network eventually refined by the addition of biochemical data from the literature [22]. The resolution of the model can range from a single metabolic pathway [23] to the whole primary metabolism [24]. Stoichiometric data applicable to the previously established gene products/reactions are then assimilated into the model. The resulting stoichiometric matrices relate the flux rates of enzymatically-driven reactions to time derivatives of metabolic concentrations [21]. This type of model can then be used for Flux Balance Analysis (FBA) [21] and allows investigators to correlate a genotype to its phenotype (in an individual cell or in a community) through the derivation of metabolic fluxes [25–27].

Under-determination is a common issue that arises in initial metabolic models. When a model contains more reactions than metabolites, the observed outputs are not enough to fully constrain the model parameters. In order to rectify under-determination, biological constraints representing realistic cellular limitations are often imposed [28]. These constraints include, but are not limited to: physiochemical, spatial, topological, environmental, and regulatory [21].

Once stoichiometric data and related constraints are overlaid onto a metabolic network, FBA can aid in understanding how metabolic fluxes contribute to cellular physiology. FBA applies linear optimization techniques in order to determine the resulting steady-state fluxes [21]. Frequently applied objectives include: the maximum growth rate, maximum biomass production, and minimization of nutrient uptake. No single objective likely describes the flux states of a biological system in all environmental conditions. Therefore, meaningful objectives must be determined for each modeling scenario [21,29]. For example, Schuetz and colleagues tested a constraint-based stoichiometric model for *Escherichia coli* in six different environmental conditions and identified two objectives that described the fluxes in all conditions tested [29].

2.1.2. Kinetic Models

Initial descriptions of complex microbial communities utilized coarse-grained 'black-box' approaches (limited to inputs and outputs, with no intermediate mechanisms included). Black-box approaches apply empirical parameters to describe the basic kinetic function of community dynamics [30]. In general, kinetic models describe the growth of bacterial cultures through the use of empirically-derived equations that incorporate the concentration of the limiting substrate and the growth (or uptake) rates corresponding to that concentration [31]. Monod and Michaelis-Menten equations are two commonly-used kinetic equations expressing cell growth and substrate uptake, respectively, based on a single growth-limiting substrate and enzyme-catalyzed uptake [25,32]. Empirically-derived equations are useful for predicting the rate of an enzymatically-driven process when substrates are abundant and end-product concentrations are constant. However, unlike in a model, a biological system often contains low concentrations of a substrate, and end-products can accumulate, thus inhibiting the reaction; this aspect is not taken into account by Michaelis-Menten-based models which treat every reaction as irreversible.

In order to rectify the limitations of these equations, Hoh and colleagues designed a kinetic model which takes into consideration rate-limiting factors and thermodynamic theory [33]. The model requires the following assumptions: (1) a reaction that has reached equilibrium cannot proceed in any direction due to the lack of a driving force (change in Gibbs free energy); (2) a reaction that is only slightly displaced from its equilibrium will proceed at a reduced rate compared to a reaction that is further away from equilibrium; (3) the model is free of any additional empirically measured parameters, excluding the organism-specific reaction rates incorporated into the original Michaelis-Menten kinetic equations.

2.1.3. Spatial Models

Metabolic and kinetic models describe many of the major factors that drive microbial community dynamics (growth rate, substrate uptake, and metabolite production). In comparison, spatial considerations have received relatively little attention, mostly to keep the models simple. However, there are many microbial systems for which these factors are essential in defining the dynamics within a community [34]. Microbes often exist in complex, spatially structured communities such as biofilms. In this type of association, spatial features cannot be neglected.

The first efforts to develop a microbial biofilm model revolved around growth balance [35]. These types of models were initially one-dimensional and incorporated reaction-diffusion equations for nutrients and other cell-produced compounds [35,36]. In time, models have increased in dimension (2D and 3D) and have made use of individual-based modeling (IBM) [37,38] to more concisely describe the heterogeneous behavior commonly observed within a biofilm [35,39]. Although growth remains the primary focus of many biofilm models, other factors such as quorum sensing [35,40] and biofilm mechanics [35,41] have also been represented. To elucidate the features of a microbial biofilm model, the 3D simulation of a biofilm on porous media [42–44] or in unsaturated soil [45] has been considered. In both cases, the focus is on the effect that biofilms have on the hydraulic properties of soil. Graf von der Schulenburg et al. [42] modeled the velocity, pressure, nutrient concentration, and biomass distribution of a biofilm using a biofilm IBM previously established for a 2D model [46], complemented by parameters for fluid velocity, pressure, and solute concentration. Complementary to this work, Rosenzweig et al. [45] developed a channel-network model to describe the effect that biofilm spatial distribution has on soil hydraulic properties. Essential parameters that have been considered are time-dependent flow, substrate transport, and biofilm growth under various soil saturation conditions [45]. Simpler models of spatial structure have also been used to capture how the organization of cells influences range expansion [47,48], intercellular interactions [49,50], or access to environmental resources [51,52]. Even without invoking details such as biofilm mechanics, cell adhesion, or cell differentiation, these models were still useful in teaching us about how spatial structure might affect microbial communities.

2.1.4. Microbial Population Models

Population level modeling efforts have been thoroughly summarized in a recent review [34]. Here, we mention their salient traits.

Population modeling is based on one of two alternative approaches: bottom-up or top-down. In bottom-up approaches, the lower level is described in order to predict the outcome at the higher level. As an example, an IBM may characterize a microbial system using individual interactions/characterization [53,54]; these individuals can be single cells, species, or groups of microbes within a particular spatial and/or temporal context. Population level information emerges as a natural byproduct of the IBM's description [34]. IBMs are inherently more complex and case-specific, but offer highly descriptive predictions and are more suitable for modeling heterogeneity.

Conversely, top-down approaches, such as the use of Population Level Models (PLMs), describe population level changes. In contrast to IBMs, time and space are often considered continuous. PLMs can be based on either ordinary differential equations (ODEs) or partial differential equations (PDEs), depending on the spatial structure requirements of the model [55,56]. ODEs are most often applied and assume that the environmental space is homogenous. However, if spatial structure is a required aspect of the model, different ODEs can be assigned to each different 'compartment' (e.g., spatial compartment, species compartment, phenotype compartment). By assessing each compartment according to its own parameters, it allows for a more accurate assessment. In general, PLMs are simpler models with fewer input requirements leading to significantly easier analyses [34].

2.2. Empirical and Mechanistic Models: From Observations to General Principles

Computational models can also be categorized as either empirical (phenomenological) or mechanistic. Empirical models fit a set of parameters (with a presumed relationship) to the experimental data relevant to the particular system of interest [15,57]. Empirical models (also called "reverse" models [58]) thus often have narrow applicability and offer limited explanatory power outside their "training" scope. However, they are more manageable than their mechanistic counterparts and often prove useful in driving the experimental branch of studies on complex microbial systems by providing a trajectory for developing hypotheses [59]. At large, empirical modeling methodologies follow an iterative cycle of development, utilization, and refinement, which entails the continual input of experimental data followed by further regression analyses [15,17,60]. Thus, a model can evolve from a simplified to an increasingly more complex product as more data are acquired and incorporated.

A mechanistic model (also called a "forward" model [58]) is derived from assumed or known principles of nature and not from a set of experimental data [15]. A mechanistic model with well-founded principles is a powerful tool applicable to studies beyond the scope of its original dataset. In the 20th century, the advent of molecular biology lifted the curtain on the mechanisms underlying many biological processes, granting a new level of depth to phenomenological data [61–64]. Today, biologists advance into the unprecedented age of 'big data'. Many current modeling efforts have shifted to methodologies that allow for the incorporation of such data; FBA in community scale metabolic models is a good example [65].

2.3. Modeling Microbial Anaerobic Fermentation in the Rumen

Empirical and mechanistic models are distinct in many of their general characteristics. However, as a model develops over the years, this distinction blurs. Nascent models often begin as simple phenomenological descriptions of a microbial system; however, as knowledge of the system accumulates and gets refined, by incorporating more data (genetic, kinetic, etc.), the model gradually shifts towards a mechanistic semblance. A good example of this process is the mathematical modeling of anaerobic fermentation by the rumen microbial community.

The ruminal microbiota is a complex system, deeply intertwined with the health of its host [66,67]. In order to establish how an animal's diet affects its ability to produce milk, gain mass, or generate offspring, scientists must first elucidate how usable nutrients, such as volatile fatty acids (VFA), are produced within the rumen. The three main VFAs (acetate, propionate, and butyric acid), produced through the microbial fermentation of carbohydrates, are the primary sources of energy for ruminants. In order to characterize the relationship between diet/feed components and their respective fermentation products, many scientists have turned to modeling.

Early empirical models: In 1989, a publication by the National Research Council characterizing the nutrient requirements of dairy cattle, incorporated mathematical equations into the Cornell Net Carbohydrate and Protein System (CNCPS) to account for varying microbial growth yields [60,68]. The methods set forth by Murphy and colleagues applied to anaerobic fermentative communities within the rumen, and enabled investigators to directly relate fermentation products to diet composition [68]. Murphy et al. based their model on mathematical equations first established by Koong et al. [69] for sheep feeding on white clover; and like Koong, this model utilized stoichiometric measurements of major metabolic pathways in order to determine relative concentrations of VFAs, methane, and carbon dioxide in the presence of various digestible feed fractions [68,70].

Refining the assumptions on conversion efficiency: Although these initial models provided a framework for understanding the role of microbial communities in ruminant nutrition, their strictly empirical inputs left them unreliable for feeds that differed significantly from the ones used to derive the stoichiometric coefficients [70]. In response, a more dynamic model proposed by Argyle and Baldwin [71] incorporated equations allowing for the adjustment of the stoichiometric coefficients depending on the ruminant pH. This led to more reliable predictions for all energy sources and resulted in the overall improvement in model performance [60].

In parallel to the dynamic model proposed by Argyle and Baldwin [71], other modifications were being made that addressed a number of inconsistencies found between simulated and observed data due to overgeneralized metabolizable energy (ME) terms. Up to this point, modeling techniques applied to anaerobic fermentation and rumen microbial communities relied on constant efficiencies of ME, i.e., the efficiency of conversion (from catabolically-produced compounds to body/milk fat, for example) was assumed to be the same for all products. However, research has shown that the efficiency of conversion varied between individual nutrients, which leads to discrepancies between modeled and experimental outcomes [60]. For instance, the conversion efficiency for acetate is 78–80%, while the efficiency for VFAs is a significantly higher 95–97% [60].

Incorporating thermodynamic considerations: According to the second law of thermodynamics, a reaction will not proceed if the reactants are limited compared to the products [70]. More recently, thermodynamics has been assimilated into both metabolic and kinetic models of the anaerobic fermentation of microbial communities [28,72]. In the rumen, the concentration of many reactants (i.e., glucose) is oftentimes low. Thus, the incorporation of thermodynamic considerations is essential for achieving a precise characterization of low-abundance compounds. The aforementioned work by Hoh [33] made a significant contribution in this direction.

Furthermore, a dynamic model for glucose fermentation was developed by Kohn and Boston [70] in which the efficiency of glucose fermentation is established for each metabolite individually (56% efficiency for acetate, propionate, and butyrate; 70% efficiency for methane), and the initial concentrations of metabolites are set to physiologically relevant levels. This model also incorporates an ionophore effect by considering how acid production leads to increased energy expenditure by the bacteria in order to maintain internal ion concentrations. In order to determine conversion efficiencies, the Gibbs free energy maximum efficiency (threshold free energy), the point at which the reaction is as close to equilibrium as it can possibly get, is calculated for each metabolite. By considering the threshold free energy for each individual end-product, the model increased the simulation accuracy by eliminating unfavorable forward reactions at points of equilibrium. To further enhance the accuracy of the model, continual infusion of glucose is simulated into the system, while VFAs and methane are removed at a constant fractional rate to better reflect what occurs within the rumen. The result of such a model is a mechanistic explanation for previously observed conflicts between the modeling results and the experimental data.

Current diet evaluations for dairy cattle are still based on ME (i.e., net energy), and lack any consideration for VFAs and their effect on energy allocation [66]. However, Ellis and colleagues [73] demonstrated that taking a more mechanistic approach proves to be more accurate than the currently utilized energy evaluations for agricultural animals. The biggest challenge in building a model more reflective of experimental data is the implicit inaccuracy in VFA concentration predictions and how this relates to the chemical compounds within various ruminant feeds.

Incorporating meta-omic data, a step towards causality: Although the incorporation of 'omics' data (i.e., genomic, metagenomics, transcriptomic, proteomic) into rumen microbial models remains somewhat uncommon, a number of studies have queried genomic and metagenomic data to better understand the rumen microbial community [74]. For example, microbiologists have sought to unveil the microbes, and their associated enzymatic repertoires, responsible for fiber degradation in the rumen. To do this, the genomes of established fibrolytic organisms, such as *Fibrobacter succinogenes* and *Ruminococcus albus*, were screened for their fiber degradation potential [74–77]. These studies provided insight into the genetic potential of the rumen microbial community, and also facilitated the use of plant lignin manipulation techniques to improve the efficiency of fiber digestion in ruminants. Such studies are regarded as great contributions to our understanding of the mechanisms behind ruminant fiber degradation, which could be further improved through metagenomic analyses of the rumen community [74].

The concepts that stemmed from these works have some degree of universality in microbiology, and have been applied to other communities, including biotechnological systems such as wastewater treatment, bioremediation, organic acid biosynthesis, etc. In an attempt to improve upon current fermentation mixed-culture models, Rodriguez and colleagues [78] developed a mechanistic model in which product formation, thermodynamic, and pH considerations are incorporated. The authors argue that since bioreactors operate at or near thermodynamic equilibrium, the microbial diversity of the system can be neglected. Therefore, in this model, the culture is treated as a single microbe capable of catalyzing most major fermentation pathways resulting in ethanol, weak organic acid, hydrogen, biomass, and CO_2 outputs. The model is built upon a metabolic network of the major reactions for glucose fermentation, and is constrained by thermodynamic considerations (i.e., change in Gibbs free energy). The bioenergetics of the system are also considered in terms of both pH and the intracellular concentration of acidic compounds.

3. Reaching out across Disciplines

3.1. Our Message for Theorists: There May Be No Elegant Solution

In this section, we reach out to theorists who are willing to approach, or have already approached, the field of microbiology. This is a time in science when multidisciplinary efforts are encouraged, and rightly so. Biology needs the support of physicists, chemists, mathematicians, and all others willing to research the living world. This is especially true for microbiology, a discipline currently under the spotlight, on the verge of being the focus of many research projects. After all, there is a general feeling that this may well be the "Microbial Century" [79].

It has recently been stated [80] that, in modern biology, impending issues that need prompt intervention are the fragmentation of life sciences and the lack of coordination among research endeavors. In this context, we believe that, if not smoothly integrated into the research effort, the modeling of microbial communities may just add another partition to the ensemble. Our lab is made of theorists and experimentalists. To us, it is very evident how different the approaches to research can be for professionals with different backgrounds. This distance is often rooted in deep differences, almost deontological, on what is an insightful scientific question, and on what would be a satisfactory answer to that question. Nevertheless, this distance must be bridged. The contribution of theorists to microbiology is sorely needed. Biologists often cannot have the competence to critically take part in the formulation of a mathematical model, or even critically evaluate the work of those who develop mathematical models. To them, mathematics is still alienating and unfamiliar. Microbiologists are no exception: they are experimentalists by formation and, maybe more importantly, vocation. The bench biologists will often have a hard time in fully understanding a mathematical model, which to their eyes may appear non-intuitive and off-target in relation to their immediate research needs. This is not a novel issue. J.D. Murray has written about the importance of "easing" biologists into mathematics in his seminal textbook: *Mathematical Biology: An introduction* [81]: "The best models show how a process works and then predict what may follow. If these are not already obvious to the biologists and the predictions turn out to be right, then you will have the biologists' attention. (. . .) The use of esoteric mathematics arrogantly applied to biological problems by mathematicians who know little about the real biology, together with unsubstantiated claims as to how important those theories are, do little to promote the interdisciplinary involvement which is so essential". Theoretical modeling should be smoothly integrated in the process of microbiological research, lest biologists may feel their discipline is being usurped from them. Less dramatically, they may simply acknowledge and accept that certain modeling-oriented research directions, while within the microbiology field, will just be out of their area of expertise. Over time, this may lead to an extreme specialization and fragmentation of research competencies. This happens already in many branches of biology (physiology or cancer research come to mind) where experts' focus on specifics hardly leaves any room for employing mathematical models that are based on general principles. We thus run the risk of severing any exchange between experimentalists and theorists.

The living world is recalcitrant to be framed in a synthetic mathematical representation. The physicist or mathematician eager to contribute to this representation will have to resist the understandable temptation of approaching biology as they would thermodynamics, or electrical engineering. Biology is not an exact science or, at least, if there is exactness to it, our current knowledge is not yet in the condition of appreciating it (i.e., data will be noisy). Professional exchange between biologists and other groups of researchers has always had a love/hate nature; this has been especially true with physicists. In 1993, W. Daniel Hillis [82] efficiently surmised the clash between these two categories of scientists, a conflict deeply rooted in the founding principles and practices of their respective disciplines. Hillis pointed out that Biology is not endowed with the power of prediction, and even the synthesis of Darwinian evolution theory gives its best at describing phenomena, not so much at predicting them: "Biologists are annoyed when they sense that physicists blame this on biologists themselves, rather than on the inherent difficulty of the subject matter. (...) Biological systems are multi-causal, poorly partitionable and, let's face it, messy. Biological systems have a beauty of their own, but often it is a beauty of complexity and richness, rather than the stark simple reductionist elegance of physics." Indeed.

Experimentalists broadly accept that evolution is the only way through which biology makes sense (T. Dobzhansky [83]), and theorists may find that evolution has nothing to do with "stark simple reductionist elegance." It actually piles up "un-elegant" outcomes by the score. Evolution does not walk the line of extraordinary, efficient solutions. It is the progressive adaptation of fallible living systems along flickering environmental conditions. It is the struggle of the living in coping with their environment through progressive adaptation, based on and constrained by preexistent anatomical structures, in no small part driven by chance. It is a work of tinkering and make-do [84]. Photosynthesis, the pillar of many trophic chains on this planet in the last 400 million years, is a good example. The photosynthetic process, despite hundreds of millions of years of evolution, is still running on very low general efficiency rates: about 2–3% of the overall exploitable light energy [85]. Yet, Mother Nature kept her job whereas no engineer could have.

The mathematician/physicist that plans on tackling biology must keep these aspects in mind, and be ready to accept that sometimes there may be no elegance to be sought, no essentiality to be spotted. Of course, that is not to say that there are no simple general principles in Biology. Sometimes finding a simple description is a matter of perspective. Take central limit theorem as an example. A simple description may adequately represent the combined effects of many random unknown causes. However, finding a simple model that captures important features of interest is far from trivial amid the chaos of messy biological mechanisms. The history of encounters with non-intuitive, complex systems has made biologists suspicious of simple models. To say it with Hillis [82]: "Physicists have learned the lesson that a very simple theory of what is going on is often correct. Biologists have learned the opposite lesson: simple mathematical theories of biology are usually wrong."

3.2. What Experimental Microbiologists Need from Theorists: A Focus on Applications

Even though the details of specific research would be different from case to case, we believe that the following thought process, in mentioning shared features among many questions of interest, will be relevant to other researchers. As microbiologists, we often intend to employ bacteria to address a real-life issue, which could be of biomedical, environmental, or industrial concern. Essentially, we want one or more bacterial strains to employ their genetic potential for our contingent need.

This is, for instance, what currently happens in our lab: we intend to address a well-known environmental issue, specifically, the mycotoxin contamination of food commodities. Mycotoxins are fungal secondary metabolites of an unclear biological purpose [86], responsible for a vast array of pathologies (including cancer) when eaten and assimilated by mammals [87]. Mycotoxins are highly present in cereals and dairy products, exceptionally stable even at extreme environmental conditions, and very hard to denature without aggressive, chemical means. The burden of mycotoxin contamination may amount yearly to billions of dollars both for industrial and medical issues [88].

From scientific literature, and general wishful thinking, we expect that there must be a way to effectively tackle mycotoxin contamination through bioremediation by bacteria. We could identify a list of different strains up to the task. Also, from the literature, we know that different kinds of mycotoxins will often co-occur on the same substrate [89]. We are thus interested in devising a viable, efficient microbial community capable of degrading mycotoxins in the specific environmental conditions of the food production chain. In a nutshell, we are in no different predicament than most applied microbiology labs: we want to craft a community to address a specific issue. Microbiology harbors immense potential for application in all areas of biological research. Simplifying their diversity in form, these applications may oftentimes be categorized as no more than two main processes: the production or degradation of chemical compounds. In this context, what use could experimentalists have for a mathematical model?

It is our opinion that theorists need to focus on the mechanisms that will allow experimentalists to tinker with the potential of microbes, prioritizing the experimental outcome over the mechanistic insight that makes such an outcome possible. That is not to say that mechanistic insight is unnecessary. Mechanistic insight is the essence of real knowledge, but it is also a massive undertaking. In complex systems (i.e., in the real world), true mechanistic insight might be at the moment beyond our technical, or even intellectual, possibilities. This is of course not a certainty, but pursuing such an ambitious goal headlong may not be wise.

The main current challenge of modeling microbial communities is that it is unclear how much knowledge about the mechanisms is required to give us enough predictive power for functions of interest. The current trajectory of approaches is based on identifying and characterizing the activities of individual species (traditionally in monoculture assays), and then combining them to form a model of the community. Is such an approach necessary? We don't know. Is it sufficient? Unclear. A strictly mechanistic approach requires the modeler to incorporate known processes into the model, hoping that, if this is at all achievable, the formalization of such models explains how a community of different members functions. The achievement would be enormous and laudable, but could prove unrealistic and, to some extent, unnecessary. It is more pragmatic to only focus on the product we are interested in, often a specific community function or property, such as the rate of degradation of an environmental toxin, or the coexistence of community members. Modelers would be speaking the language of most biologists if they focused their efforts for the sake of experimental application. Applicability is what drives most experimentalists. In turn, experimentalists can help modelers in their search for the "proper level of abstraction" by focusing on specific communities with well-defined functions and relative characterizing traits. An understanding of the founding principles, and its mathematical synthesis, will come through the synthesis of well-characterized particular cases. But even if not, we would still be endowed with well-characterized particular cases.

3.3. What Theorists Need from Experimental Microbiologists: Data, Possibly in a Specific Form

In the general spirit of establishing a coordinated effort in microbiology, we believe it would be useful to encourage biologists to make their raw data of published work available in an open database. This would be similar to how next generation sequencing raw data are required to be available on public domains for other researchers to access them. This would give modelers the chance to find the mathematical rationale behind works that have independently achieved experimental success. In doing so, they will be able to provide opinions on what future experiments they believe would be insightful to refine the different aspects of the model. This in turn will allow coordination with biologists to further develop our understanding of the observed systems.

There may even be a specific journal dedicated to publishing mathematical modeling papers based on data coming from past experimental publications of applicative relevance and insight. In 2015, Quincey Justman wrote an editorial on Cell Systems to introduce Math | Bio [90], a novel journal founded on a very intriguing premise: to publish papers containing no data, but rather a mathematical argument. Justman is inspired by John J. Hopfield's paper on kinetic proofreading, published in

1974 in PNAS, at a time when, Justman argues, interesting ideas were enough to deserve publication. Math | Bio aims to throw ideas into the fray for biologists to pick them up and test them, if they feel they are insightful and potentially game-changing. We find this idea to be very precious and farsighted. In a much more trivial manner, it could be reversed and applied to microbiology research. In many instances, data is already available (and expanding) for researchers to be put into a mathematical framework. If modelers "adopted" a laboratory or a specific research topic, they might give new insight to published experimental results and, at the same time, provide unexpected inputs on future directions. Even though modeling is part of the current research activities in our lab, we would still love to be "adopted" to facilitate this process.

3.4. Microbial Communities Assembly: An Opportunity for Theorists and Experimentalists to Work Together

To show how research applicability could drive cooperation between theorists and experimentalists, we believe community assembly is fertile ground. In devising experimental research built around applicability, one of the first decisions to make is whether to focus on optimizing a single species for the function of interest, or employ a community of multiple species. Single species have the advantage of being easier to identify and handle in a laboratory environment. Additionally, using a single species makes processes such as artificial selection and data analysis more expedited and easily interpretable. After all, the complications of culturing communities are vast and, sometimes, hardly addressable. Cultivability is a constant issue in microbiology and, in the economy of a natural community, the loss of significant, unculturable strains can largely hamper the desired community function in controlled experimental conditions. Thus, the process of modeling itself, which often relies heavily on data acquired under controlled and monitored conditions, is made easier in in vitro conditions. Nonetheless, we believe that the successful cultivation of a community, even the most essential, is the premise for the most interesting research. From a purely speculative standpoint, the study of community-driven traits (inter-specific cross-talk, microenvironment modifications, ecological interactions) is among the most intriguing topics for present day microbiologists; also, in terms of the application potential of the findings to come, a community, once established and applied for the purpose of bioremediation/biosynthesis/biomedical needs, is likely to be more reluctant to perturbations than any species taken singularly.

To make an exemplificative argument, if we value bioremediators in terms of the genes they bear, we can consider the community as a scaffold that harbors a wide inventory of genetic potential, much wider than what a single-species bioremediator could. Being able to craft stable communities will thus grant much more potential in terms of the amplitude of applicability, and such is the general indication that comes from recent experimental findings [91]. Moreover, a more in-depth formalization of the principles underlying community assembly has been deemed essential by researchers that focus on the highly intriguing field of Synthetic Ecology. We quote from Johnson et al. [92]: "A deeper understanding of the biochemical causes of metabolic specialization could serve as a foundation for the field of synthetic ecology, where the objective would be to rationally engineer the assembly of a microbial community to perform a desired biotransformation."

3.4.1. An Intriguing First Step: Coexistence Theory

In this theoretical context, it is important to point out general concepts around which the process of mathematically describing community assembly would revolve. We believe a clear outline can be found in the principles that constitute the coexistence theory. Coexistence theory is a theoretical framework of concepts that describe and formalize principles that allow a community to retain or lose its identity. In other words, it describes the forces behind coexistence (not surprisingly) and provides insight on how to achieve a successful assembly of communities. It can thus be of great use for the general purpose of most microbiologists.

As outlined by HilleRisLambers et al. [93], the main concepts relative to community assembly are not numerous, and coexistence is described as depending on niche and fitness differences.

1. Relative fitness differences: outline the outcome of competition among species in the absence of stabilizing differences.

2. Stabilizing niche differences: when a species is self-limited by the environmental context rather than by competitors. It is a force in favor of community diversity.

3. Competitive exclusion: happens when relative fitness differences are stronger than stabilizing forces, and the relatively less fit disappears from the community over time.

4. Stable coexistence: Diversity is sustained and stable over time. Stabilizing forces play a greater role than fitness differences.

The task in hand is to experimentally examine which of these processes apply to a particular case of interest and how. Currently, there might not be enough precedence to formulate a systematic protocol for identifying and characterizing the impact of these factors. Nonetheless, as more instances are being examined, we see a hopeful perspective in a future not too far away. Minty's validated model on cross-regnum consortia for isobutanol production, or Zuroff's work on a community for ethanol production from cellulose, are examples notable in their thoroughness [94–97].

3.4.2. The Outstanding Issue of Model Verification

The task of verifying what models are suitable for representing microbial communities, while challenging, is absolutely necessary. Without this verification and refinement step, the cloud of doubt about the relevance of models will keep experimentalists suspicious of all modeling results.

There are still many open questions about the validity or relevance of common assumptions used in modeling microbial communities. As an example, consider the use of Lotka-Volterra (L-V) models for simulating microbial communities. Being the most popular platform for modeling communities, L-V models abstract all the interactions between species into pairwise fitness effects [98–101]. This is motivated by the historical precedence of community studies on prey-predation food webs [102–104] or plant-pollinator mutualisms [105–107]. The relative success of L-V models in the past to represent such communities has established this platform as the go-to model for ecological networks. Additionally, the mathematical tractability of the model gave it a central role in theoretical studies of community stability [99,100,108–110]. This further secured the position of L-V models in theoretical ecology. When simulating microbial communities, this history has been used as justification to extend the same modeling framework to represent microbial interactions. However, pairwise fitness models may not always accurately capture common situations in which multiple diverse interactions are present or when compounds mediating the interactions are shared among multiple species [111]. Identifying and recognizing such limitations allow us to use the very useful L-V modeling platform when it is applicable. We thus advocate for dedicated research to clarify the limitations and range of applicability of common modeling platforms.

Another fundamental assumption in almost all community modeling frameworks is the additivity assumption [111]. For simplicity, it is often assumed that in communities, the effects exerted on an individual or a population by different factors can be superimposed in an additive manner. There are of course many examples to the contrary. The presence of non-additive effects in fact has been widely recognized in ecological modeling under the umbrella of indirect interactions, nonlinear interactions, or higher-order interactions [101,112–114]. Researchers have even rigorously examined whether or not additivity assumption holds for examples in the utilization of resources from the environment [115,116]. Several studies on the combined effects of antibiotics have also shown synergy (or antagonism) between them, showing inhibition effects stronger (or weaker) than what is expected based on an additive model [117–124]. Nevertheless, when it comes to modeling communities, because the extent and prevalence of deviations from additivity is not established, models almost unanimously drop back to assuming additivity. Systematic work is needed in this area to clarify when and under what conditions such an assumption is acceptable.

Performing the necessary work to support and justify model assumptions requires not just the will of researchers, but also the support of the community, including peers, publishers, and funding agencies. Exploring uncharted territories and coming up with new hypotheses using a theoretical platform sounds more exciting, and is often rewarded and recognized as being innovative. This bias comes at a cost: the necessary steps of verifying the basic assumptions of such models are considered "less exciting." The unfortunate outcome of this trend is that a body of theoretical work will develop, without a clear understanding of the conditions under which those findings are relevant. In turn, when experimental data outside the range of applicability of such models deviate from predictions, it will be considered a failure of the theory, widening the divide between theoreticians and experimentalists.

We believe it is time to give the field of model-verification the attention it deserves. Groundwork verification efforts should be treated as independent research contributions on their own, rather than side-notes. There are examples in other fields, where the importance of such groundwork efforts has been recognized. A notable recent example is the reproducibility project in cancer biology to evaluate the reproducibility of previous reports [125–127]. Support from researchers, funding agencies, and publishers in this case shows an exemplary instance in which the scientific community is rallying behind a necessary groundwork. The field of microbial community modeling can certainly benefit from a similar attitude.

3.5. Compiling What Is Known, Clarifying the Assumptions, and Making Models Accessible

When experimentalists devise a novel research plan, it goes through a phase of information gathering that precedes the formalization of the details of the research. In this context, we believe the mathematical model would ideally be of assistance in between the preliminary process of information gathering and the beginning of the experimental phase itself: a good model would outline what variables of the system are more likely to be influential, which is invaluable information. Screening prior research to identify such a model, even if it existed, is not a streamlined process, and surprisingly so. After all, wouldn't it be easier for experimentalists, in deciding what model would be most appropriate for their system, to refer to previous reports and studies in related, well characterized, even if not similar, situations? Unfortunately, a database of microbial interactions and previous modeling efforts currently does not exist, to the best of our knowledge.

Models are most often based on phenomenological data pertaining to a specific biological process, and focus on a single instance within that process: they are hardly approached by researchers not already within that specific field. Proceedings in microbiology, as previously stated, are relevant to many disparate scientific disciplines. Nonetheless, at present, it is normal to assume that a hypothetical model based on a set of microbiological data relevant, for instances, to the field of Transfusion Medicine, is unlikely to be of interest to an environmental engineer interested in bacteria-mediated wastewater management. Yet, if the process under investigation is general enough (dynamics of microbial cell diffusion, for instances), one cannot decidedly rule out a fruitful cross-disciplinary cooperation. To make a more specific example, albeit outside the field of microbiology, a model that described fungal hyphal development, distilled to its essence, could more generally be viewed as a model describing the growth of apically polarized cells. As has already been observed [128,129], this implies that the mechanics relevant to hyphal growth could also be descriptive of neuronal outgrowth: neurons are *de facto* apically polarized cells. Pointing out that, in the right context, hyphal development can be representative of neuronal development is no trivial intuition. In doing so, researchers in the field have essentially observed that Mycology and Neuroscience may happen to cross paths, and we believe that mathematical analysis would be the main way to substantiate this link. Finding a mathematical synthesis for this and other kindred observations is also our best bet towards the mechanistic description of nature. At present, poor communication across disciplines is holding everyone back. If we facilitate the process of bringing together professionally distant researchers, we will probably find ourselves with similar observations. We may even find our example of the link between Transfusion Medicine and Environmental Engineering not to be so far-fetched.

If there was a public platform that collected works on modeling biology (i.e., works that attempt to distill phenomena to their essence), and that platform could intuitively be browsed by experimentalists, the situation may change. The platform would be a tool where mathematical models are uploaded, along with their respective publication. It would require the participation of both modelers and experimental biologists. The role of modelers would be to upload their work clearly stating the purpose for which the model was developed. They would also be required to provide a user-friendly graphical user interface (GUI), bearing in mind that an experimentalist with very little experience in programming and mathematical analysis is their target user. The user would be put in the condition to easily identify, through the GUI, the variables included and the entity of their effect on the model output.

The key aspect would be to make the model approachable through different research queries of interest to the broadest spectrum of researchers. Models should be sorted through the main processes they describe, the outputs for which they were devised, and with all the variables included. Examples of processes could be: diffusion, cell growth, cell-cell interaction, motility, mutation, biotransformation, and artificial or natural selection, etc. Examples of variables could be: temperature, pH, oxygen concentration, species involved, culture conditions (liquid culture or agar plate), resource availability (rich, minimal, or restrictive, medium), etc. All these elements should also be catalogued by the widest number of biologically relevant terms they could represent in a variety of biological contexts: metabolite diffusion in one context could equate to disease spread in a different context.

Resorting to this platform would be advantageous for both theorists and experimentalists.

For theorists: It would be a rare opportunity to unleash their models in the wilderness of research for other scientists to test them, as they may be representative of more than the one biological context that they originally described. We believe this would be a precious shortcut to make the cross-context (mechanistic) traits of the model emerge, and would facilitate the identification and formalization of the relations among those contexts. At the cost of the supplementary work of providing a user-friendly GUI for their models, theorists would have a lot to gain and nothing to lose from this initiative.

For experimentalists: The experimentalist would approach the platform in search of inspiration in devising his/her experimental plan, gaining insight on which approach is more likely to be successful. After all, examining collected instances in one place offers synergy for interpreting the observations and uncovering patterns. The biologist will have the opportunity to get in touch with one or more publications of models that have dealt with the more (if applicable) similar conditions and premises of his/her own system, and receive a conceptual synthesis of which variables led to which results, gathering some guidance on how to proceed for his/her experiments.

We believe the literature harbors a plethora of models that can be useful to researchers in other fields, but those researchers may never become aware of the existence of such models. A lot is to be gained by facilitating and encouraging communication through the lens of mathematical representation. Actively pursuing the identification of common mechanisms across the different branches of biology holds great potential for life sciences in general, and microbiology in particular.

4. Conclusions and Final Remarks

To summarize, we believe that the great potential harbored by microbes can be unleashed through a close collaboration between theoreticians and experimentalists. Mathematical modeling is the vehicle towards this objective that requires investment and cooperation by both sides. Here, we have compiled suggestions to facilitate this cooperation between researchers from different backgrounds and disciplines. These suggestions come from experiencing interdisciplinary research within our own lab. In our opinion, there is a need to be aware of the differences among researchers from different disciplines, their outlooks, and their interests. To collaborate and cooperate, we need to make adjustments to accommodate these difference. A theoretician may have to balance the generality-realism trade-off and focus on functions and properties of practical applicability to experimentalists. An experimentalist, in turn, may have to adjust their experiments to collect and compile the data in a form that will be readily usable by theoreticians. Communications is key in this bilateral exchange and compromise. We advocate for practices that facilitate this communication: we encourage experimentalists to compile an easily accessible database of their data for theoreticians and encourage theoreticians to make their modeling frameworks welcoming to experimentalists. Finally, we propose a coordinated effort by the scientific community to lower the barriers between disciplines by focusing on processes and commonalities, built around the common language of mathematical modeling. The outcome will be a better understanding and an elevated intuition of microbial processes for both theoreticians and experimentalists, with a tremendous impact on applications from human health to industrial biotransformation to ecosystem sustainability.

Acknowledgments: The authors would like to thank Boston College for supporting this work. We thank Kathleen (Sayles) Day and Samantha Dyckman for their feedback on the manuscript.

Author Contributions: M.Z., S.D., and B.M. wrote the paper.

References

1. Bartlett, J.M.S.; Stirling, D. A Short History of thse Polymerase Chain Reaction. In *PCR Protocols*; Humana Press: New Jersey, NJ, USA, 2003; pp. 3–6.
2. Cong, L.; Ran, F.A.; Cox, D.; Lin, S.; Barretto, R.; Habib, N.; Hsu, P.D.; Wu, X.; Jiang, W.; Marraffini, L.A.; et al. Multiplex Genome Engineering Using CRISPR/Cas Systems. *Science* **2013**, *339*, 819–823. [CrossRef] [PubMed]
3. Baltimore, D.; Berg, P.; Botchan, M.; Carroll, D.; Charo, R.A.; Church, G.; Corn, J.E.; Daley, G.Q.; Doudna, J.A.; Fenner, M.; et al. A prudent path forward for genomic engineering and germline gene modification. *Science* **2015**, *348*, 36–38. [CrossRef] [PubMed]
4. Mali, P.; Yang, L.; Esvelt, K.M.; Aach, J.; Guell, M.; DiCarlo, J.E.; Norville, J.E.; Church, G.M. RNA-guided human genome engineering via Cas9. *Science* **2013**, *339*, 823–826. [CrossRef] [PubMed]
5. Jessup, C.M. Big questions, small worlds: Microbial model systems in ecology. *Trends Ecol. Evol.* **2004**, *19*, 189–197. [CrossRef] [PubMed]
6. Momeni, B.; Chen, C.-C.; Hillesland, K.L.; Waite, A.; Shou, W. Using artificial systems to explore the ecology and evolution of symbioses. *Cell. Mol. Life Sci.* **2011**, *68*, 1353–1368. [CrossRef] [PubMed]
7. Brenner, K.; You, L.; Arnold, F.H. Engineering microbial consortia: A new frontier in synthetic biology. *Trends Biotechnol.* **2008**, *26*, 483–489. [CrossRef] [PubMed]
8. Wintermute, E.H.; Silver, P.A. Dynamics in the mixed microbial concourse. *Genes Dev.* **2010**, *24*, 2603–2614. [CrossRef] [PubMed]
9. McGrady-Steed, J.; Harris, P.M.; Morin, P.J. Biodiversity regulates ecosystem predictability. *Nature* **1997**, *390*, 162–165.
10. Tanouchi, Y.; Smith, R.P.; You, L. Engineering microbial systems to explore ecological and evolutionary dynamics. *Curr. Opin. Biotechnol.* **2012**, *23*, 791–797. [CrossRef] [PubMed]
11. Sanchez, A.; Gore, J.; Frey, E.N.H., Jr.; Phillimore, A. Feedback between Population and Evolutionary Dynamics Determines the Fate of Social Microbial Populations. *PLoS Biol.* **2013**, *11*, e1001547. [CrossRef] [PubMed]
12. Dolinšek, J.; Goldschmidt, F.; Johnson, D.R. Synthetic microbial ecology and the dynamic interplay between microbial genotypes. *FEMS Microbiol. Rev.* **2016**, *40*, 961–979. [CrossRef] [PubMed]
13. Esty, J.R.; Meyer, K.F. The heat resistance of the spore of Bacillus botulinus and allied anaerobes, XI. *J. Infect. Dis.* **1922**, *31*, 650–663. [CrossRef]
14. Pérez-Rodríguez, F.; Valero, A. Predictive Microbiology in Foods. In *Predictive Microbiology in Foods*; Springer: New York, NY, USA, 2013; pp. 1–10.
15. Baldwin, R.L. *Modeling Ruminant Digestion and Metabolism*, 1st ed.; Chapman & Hall: London, UK, 1995.
16. Shou, W.; Bergstrom, C.T.; Chakraborty, A.K.; Skinner, F.K. Theory, models and biology. *eLife* **2015**, *4*, e07158. [CrossRef] [PubMed]

17. Whitesides, G.M. Whitesides' Group: Writing a Paper. *Adv. Mater.* **2004**, *16*, 1375–1377. [CrossRef]

18. Manuscript, A. NIH Public Access. *Changes* **2012**, *29*, 997–1003.

19. Kermack, W.O.; McKendrick, A.G. A Contribution to the Mathematical Theory of Epidemics. *Proc. R. Soc. A Math. Phys. Eng. Sci.* **1927**, *115*, 700–721. [CrossRef]

20. France, J.; Thornley, J.H.M. *Mathematical Models in Agriculture*; Butterworths: Oxford, UK, 1984.

21. Raman, K.; Chandra, N. Flux balance analysis of biological systems: Applications and challenges. *Brief. Bioinform.* **2009**, *10*, 435–449. [CrossRef] [PubMed]

22. Feist, A.M.; Herrgård, M.J.; Thiele, I.; Reed, J.L.; Palsson, B.Ø. Reconstruction of biochemical networks in microorganisms. *Nat. Rev. Microbiol.* **2009**, *7*, 129–143. [CrossRef] [PubMed]

23. Fuhrer, T.; Fischer, E.; Sauer, U. Experimental identification and quantification of glucose metabolism in seven bacterial species. *Society* **2005**, *187*, 1581–1590. [CrossRef] [PubMed]

24. Almaas, E.; Kovács, B.; Vicsek, T.; Oltvai, Z.N.; Barabási, A.-L. Global organization of metabolic fluxes in the bacterium Escherichia coli. *Nature* **2004**, *427*, 839–843. [CrossRef] [PubMed]

25. Hanemaaijer, M.; Röling, W.F.M.; Olivier, B.G.; Khandelwal, R.A.; Teusink, B.; Bruggeman, F.J. Systems modeling approaches for microbial community studies: From metagenomics to inference of the community structure. *Front. Microbiol.* **2015**, *6*, 213. [CrossRef] [PubMed]

26. Song, H.-S.; Cannon, W.; Beliaev, A.; Konopka, A. Mathematical Modeling of Microbial Community Dynamics: A Methodological Review. *Processes* **2014**, *2*, 711–752.

27. Zhang, T. Modeling Biofilms: From Genes to Communities. *Processes* **2017**, *5*, 5. [CrossRef]

28. Mahadevan, R.; Schilling, C.H. The effects of alternate optimal solutions in constraint-based genome-scale metabolic models. *Metab. Eng.* **2003**, *5*, 264–276. [CrossRef] [PubMed]

29. Schuetz, R.; Kuepfer, L.; Sauer, U. Systematic evaluation of objective functions for predicting intracellular fluxes in Escherichia coli. *Mol. Syst. Biol.* **2007**, *3*, 119. [CrossRef] [PubMed]

30. Widder, S.; Allen, R.J.; Pfeiffer, T.; Curtis, T.P.; Wiuf, C.; Sloan, W.T.; Cordero, O.X.; Brown, S.P.; Momeni, B.; Shou, W.; et al. Challenges in microbial ecology: Building predictive understanding of community function and dynamics. *ISME J.* **2016**, *10*, 2557–2568. [CrossRef] [PubMed]

31. Kessick, M. The kinetics of bacterial growth. *Biotechnol. Bioeng.* **1974**, *16*, 1545–1547. [CrossRef] [PubMed]

32. Oh, S.T.; Martin, A.D. Thermodynamic equilibrium model in anaerobic digestion process. *Biochem. Eng. J.* **2007**, *34*, 256–266. [CrossRef]

33. Hoh, C.Y.; Cord-Ruwisch, R. A practical kinetic model that considers endproduct inhibition in anaerobic digestion processes by including the equilibrium constant. *Biotechnol. Bioeng.* **1996**, *51*, 597–604. [CrossRef]

34. Hellweger, F.L.; Clegg, R.J.; Clark, J.R.; Plugge, C.M.; Kreft, J.-U. Advancing microbial sciences by individual-based modelling. *Nat. Rev. Microbiol.* **2016**, *14*, 461–471. [CrossRef] [PubMed]

35. Klapper, I.; Dockerty, J. Mathematical Description of Microbial Biofilms. *SIAM Rev.* **2010**, *50*, 221–265. [CrossRef]

36. Gujer, W.; Wanner, O. Modeling Mixed Population Biofilms. In *Biofilm*; Characklis, W.G., Marshall, K.C., Eds.; Wiley: New York, NY, USA, 1990; pp. 397–443.

37. Ferrer, J.; Prats, C.; López, D. Individual-based Modelling: An Essential Tool for Microbiology. *J. Biol. Phys.* **2008**, *34*, 19–37. [CrossRef] [PubMed]

38. Grimm, V.; Railsback, S.F. *Individual-Based Modeling and Ecology*; Princeton University Press: Princeton, NJ, USA, 2005.

39. Kreft, J.U.; Picioreanu, C.; Wimpenny, J.W.T.; van Loosdrecht, M.C.M. Individual-based modelling of biofilms. *Microbiology* **2001**, *147*, 2897–2912. [CrossRef] [PubMed]

40. Chopp, D.L.; Kirisits, M.J.; Moran, B.; Parsek, M.R. A mathematical model of quorum sensing in a growing bacterial biofilm. *J. Ind. Microbiol. Biotechnol.* **2002**, *29*, 339–346. [CrossRef] [PubMed]

41. Körstgens, V.; Flemming, H.-C.; Wingender, J.; Borchard, W. Uniaxial compression measurement device for investigation of the mechanical stability of biofilms. *J. Microbiol. Methods* **2001**, *46*, 9–17. [CrossRef]

42. Graf von der Schulenburg, D.A.; Pintelon, T.R.R.; Picioreanu, C.; van Loosdrecht, M.C.M.; Johns, M.L. Three-Dimensional Simulations of Biofilm Growth in Porous Media. *AIChE J.* **2009**, *55*, 494–504. [CrossRef]

43. Ebrahimi, A.; Or, D. Microbial community dynamics in soil aggregates shape biogeochemical gas fluxes from soil profiles—upscaling an aggregate biophysical model. *Glob. Chang. Biol.* **2016**, *22*, 3141–3156. [CrossRef] [PubMed]

44. Ebrahimi, A.N.; Or, D. Microbial dispersal in unsaturated porous media: Characteristics of motile bacterial cell motions in unsaturated angular pore networks. *Water Resour. Res.* **2014**, *50*, 7406–7429. [CrossRef]

45. Rosenzweig, R.; Furman, A.; Dosoretz, C.; Shavit, U. Modeling biofilm dynamics and hydraulic properties in variably saturated soils using a channel network model. *Water Resour. Res.* **2014**, *50*, 5678–5697. [CrossRef]

46. Picioreanu, C.; van Loosdrecht, M.C.M.; Heijnen, J.J. Effect of diffusive and convective substrate transport on biofilm structure formation: A two-dimensional modeling study. *Biotechnol. Bioeng.* **2000**, *69*, 504–515. [CrossRef]

47. Korolev, K.S.; Müller, M.J.I.; Karahan, N.; Murray, A.W.; Hallatschek, O.; Nelson, D.R. Selective sweeps in growing microbial colonies. *Phys. Biol.* **2012**, *9*, 26008. [CrossRef] [PubMed]

48. Datta, M.S.; Korolev, K.S.; Cvijovic, I.; Dudley, C.; Gore, J. Range expansion promotes cooperation in an experimental microbial metapopulation. *Proc. Natl. Acad. Sci. USA* **2013**, *110*, 7354–7359. [CrossRef] [PubMed]

49. Momeni, B.; Brileya, K.A.; Fields, M.W.; Shou, W. Strong inter-population cooperation leads to partner intermixing in microbial communities. *eLife* **2013**, *2*, e00230. [CrossRef] [PubMed]

50. Momeni, B.; Waite, A.J.; Shou, W. Spatial self-organization favors heterotypic cooperation over cheating. *eLife* **2013**, *2*, e00960. [CrossRef] [PubMed]

51. Xavier, J.B.; Martinez-Garcia, E.; Foster, K.R. Social Evolution of Spatial Patterns In Bacterial Biofilms: When Conflict Drives Disorder. *Am. Nat.* **2009**, *174*, 1–12. [CrossRef] [PubMed]

52. Mitri, S.; Xavier, J.B.; Foster, K.R. Social evolution in multispecies biofilms. *Proc. Natl. Acad. Sci. USA* **2011**, *108* (Suppl. 2), 10839–10846. [CrossRef] [PubMed]

53. Railsback, S. *Agent-Based and Individual-Based Modeling: A Practical Introduction*; Princeton Univerisy Press: Princeton, NJ, USA, 2011.

54. DeAngelis, D.L.; Mooij, W.M. Individual-Based Modeling of Ecological and Evolutionary Processes 1. *Annu. Rev. Ecol. Evol. Syst.* **2005**, *36*, 147–168. [CrossRef]

55. Edelstein-Keshet, L. *Mathematical Models in Biology*; Birkhauser-McGraw-Hill: New York, NY, USA, 1988.

56. Gurney, W.S.C.; Nisbet, R.M. *Ecological Dynamics*; Oxford University Press: New York, NY, USA, 1998.

57. Riggs, D. *The Mathematical Approach to Physiological Problems*; Elsevier: Baltimore, MD, USA, 1973; Volume 445.

58. Gunawardena, J. Models in biology: Accurate descriptions of our pathetic thinking. *BMC Biol.* **2014**, *12*, 29. [CrossRef] [PubMed]

59. Herrgard, M.J.; Swainston, N.; Dobson, P.; Dunn, W.B.; Arga, K.Y.; Arvas, M.; Borger, S.; Costenoble, R.; Heinemann, M.; Le Novere, N. A consensus yeast metabolic network reconstruction obtained from a community approach to systems biology. *Nat. Biotechnol.* **2008**, *26*, 1155–1160. [CrossRef] [PubMed]

60. McNamara, D.; France, J.P.; Beever, J. *Modelling Nutrient Utilization in Farm. Animals*; Elsevier Inc.; CABI Publishing: Oxon, UK, 2000.

61. Ingalls, B.P. Mathematical Modelling in Systems Biology: An Introduction. *J. Chem. Inf. Model.* **2014**, *53*, 1–396.

62. Horowitz, N.H.; Bonner, D.; Mitchell, H.K.; Tatum, E.L.; Beadle, G.W. Genic Control of Biochemical Reactions in Neurospora. *Am. Nat.* **1945**, *79*, 304–317. [CrossRef]

63. Beadle, E.; Tatum, G.W. Genetic control of biochemical reactions in neurospora. *Proc. Natl. Acad. Sci. USA* **1941**, *27*, 499–506. [CrossRef] [PubMed]

64. Watson, F.; Crick, J.D. The structure of DNA. *Cold Spring Harb. Symp. Quant. Biol.* **1953**, *18*, 123–131. [CrossRef] [PubMed]

65. Henry, C.S.; Bernstein, H.C.; Weisenhorn, P.; Taylor, R.C.; Lee, J.-Y.; Zucker, J.; Song, H.-S. Microbial Community Metabolic Modeling: A Community Data-Driven Network Reconstruction. *J. Cell. Physiol.* **2016**, *231*, 2339–2345. [CrossRef] [PubMed]

66. Morvay, Y.; Bannink, A.; France, J.; Kebreab, E.; Dijkstra, J. Evaluation of models to predict the stoichiometry of volatile fatty acid profiles in rumen fluid of lactating Holstein cows. *J. Dairy Sci.* **2011**, *94*, 3063–3080. [CrossRef] [PubMed]

67. Bergman, E. Energy contributions of volatile fatty acids from the gastrointestinal tract in various species. *Physiol. Rev.* **1990**, *70*, 567–590. [PubMed]

68. Murphy, M.R.; Baldwin, R.L.; Koong, L.J. Estimation of stoichiometric parameters for rumen fermentation of roughage and concentrate diets. *J. Anim. Sci.* **1982**, *55*, 411–421. [CrossRef] [PubMed]

Modeling of Microbial Communities: Theory and Practice

69. Koong, L.J.; Baldwin, R.L.; Ulyatt, M.J.; Charlesworth, T.J. Iterative computation of metabolic flux and stoichiometric parameters for alternate pathways in rumen fermentation. *Comput. Programs Biomed.* **1975**, *4*, 209–213. [CrossRef]

70. Kohn, R.A.; Boston, R.C. The Role of Thermodynamics in Controlling Rumen Metabolism. *Model. Nutr. Util. Farm. Anim.* **2000**, *1*, 11–24.

71. Argyle, J.L.; Baldwin, R.L. Argyle and Baldwin_1988_Modeling of rumen water kinetics and effects of rumen pH changes.pdf. *J. Dairy Sci.* **1988**, *71*, 1178–1188. [CrossRef]

72. Feist, A.M.; Henry, C.S.; Reed, J.L.; Krummenacker, M.; Joyce, A.R.; Karp, P.D.; Broadbelt, L.J.; Hatzimanikatis, V.; Palsson, B.Ø. A genome-scale metabolic reconstruction for *Escherichia coli* K-12 MG1655 that accounts for 1260 ORFs and thermodynamic information. *Mol. Syst. Biol.* **2007**, *3*, 1–18. [CrossRef] [PubMed]

73. Ellis, J.L.; Dijkstra, J.; Kebreab, E.; Bannink, A.; Odongo, N.E.; Mcbride, B.W.; France, J. Aspects of rumen microbiology central to mechanistic modelling of methane production in cattle. *J. Agric. Sci.* **2008**, *146*, 213–233. [CrossRef]

74. Krause, D.O.; Denman, S.E.; Mackie, R.I.; Morrison, M.; Rae, A.L.; Attwood, G.T.; McSweeney, C.S. Opportunities to improve fiber degradation in the rumen: Microbiology, ecology, and genomics. *FEMS Microbiol. Rev.* **2003**, *27*, 663–693. [CrossRef]

75. Nelson, M.; Aminov, K.; Forsberg, R.; Mackie, C.; Russell, R.I.; White, J.B.; Wilson, B.A.; Mulligan, D.B.; Tran, S.; Carty, K.; et al. The Fibrobacter succinogenes strain S85 genome sequencing project. In *Beyond Antimicrobials—The Future of Gut Microbiology, Proceedings of the 3rd RRI-INRA Symposium, Aberdeen, UK, 12–15 June 2002*; Rowett Research Institute: Aberdeen, UK, 2002; Volume 19.

76. Devillard, M.; Goodheart, E.; Morrison, D. Proteomics based analysis of Ruminococcus albus 8 adhesion-defective mutants. In *Beyond Antimicrobials—The Future of Gut Microbiology, Proceedings of the 3rd RRI-INRA Symposium, Aberdeen, UK, 12–15 June 2002*; Rowett Research Institute: Aberdeen, UK, 2002; Volume 37, pp. 777–788.

77. Morrison, D.; Devillard, M.; Goodheart, E. The effects of phenyl-substituted fatty acids and carbon source on the cellulose-binding sub-proteome of Ruminococcus albus strain 8. In Proceedings of the 102nd General Meeting of the American Society for Microbiology, Salt Lake City, UT, USA, 19–23 May 2002; pp. 3255–3266.

78. Rodríguez, J.; Kleerebezem, R.; Lema, J.M.; Van Loosdrecht, M.C.M. Modeling product formation in anaerobic mixed culture fermentations. *Biotechnol. Bioeng.* **2006**, *93*, 592–606. [CrossRef] [PubMed]

79. Larsen, P.; Hamada, Y.; Gilbert, J. Modeling microbial communities: Current, developing, and future technologies for predicting microbial community interaction. *J. Biotechnol.* **2012**, *160*, 17–24. [CrossRef] [PubMed]

80. Dubilier, N.; Mcfall-ngai, M.; Zhou, L. Create a global microbiome effort. *Nature* **2015**, *526*, 631–634. [CrossRef] [PubMed]

81. Murray, J.D. *Mathematical Biology: An. Introduction*; Springer: New York, NY, USA, 1989.

82. Hillis, W.D. Why physicists like models and why biologists should. *Curr. Biol.* **1993**, *3*, 79–81. [CrossRef]

83. Dobzhansky, T. Nothing in biology makes sense except in the light of evolution. *Am. Biol. Teach.* **1973**, *35*, 125–129. [CrossRef]

84. Jacob, F. Evolution and tinkering. *Science* **1977**, *196*, 1161–1166. [CrossRef] [PubMed]

85. Miyamoto, K. *Renewable Biological Systems for Alternative Sustainable Energy Production*; Food and Agriculture Organization of the United Nations: Rome, Italy, 1997.

86. Reverberi, M.; Ricelli, A.; Zjalic, S.; Fabbri, A.A.; Fanelli, C. Natural functions of mycotoxins and control of their biosynthesis in fungi. *Appl. Microbiol. Biotechnol.* **2010**, *87*, 899–911. [CrossRef] [PubMed]

87. Bennett, J.W.; Klich, M. Mycotoxins. *Clin. Microbiol. Rev.* **2003**, *16*, 497–516. [CrossRef] [PubMed]

88. Mitchell, N.J.; Bowers, E.; Hurburgh, C.; Wu, F. Potential economic losses to the USA corn industry from aflatoxin contamination. *Food Addit. Contam. Part A Chem. Anal. Control Expo. Risk Assess.* **2016**, *33*, 540–550. [CrossRef] [PubMed]

89. Vanhoutte, I.; Audenaert, K.; De Gelder, L. Biodegradation of Mycotoxins: Tales from Known and Unexplored Worlds. *Front. Microbiol.* **2016**, *7*, 1–20. [CrossRef] [PubMed]

90. Justman, Q. 1970s Nostalgia for the Modern Day. *Cell. Syst.* **2015**, *1*, 175. [CrossRef] [PubMed]

91. Lilja, E.E.; Johnson, D.R. Segregating metabolic processes into different microbial cells accelerates the consumption of inhibitory substrates. *ISME J.* **2016**, *10*, 1568–1578. [CrossRef] [PubMed]

92. Johnson, D.R.; Goldschmidt, F.; Lilja, E.E.; Ackermann, M. Metabolic specialization and the assembly of microbial communities. *ISME J.* **2012**, *6*, 1985–1991. [CrossRef] [PubMed]

93. HilleRisLambers, J.; Adler, P.B.; Harpole, W.S.; Levine, J.M.; Mayfield, M.M. Rethinking Community Assembly through the Lens of Coexistence Theory. *Annu. Rev. Ecol. Evol. Syst.* **2012**, *43*, 227–248. [CrossRef]

94. Minty, J.J.; Singer, M.E.; Scholz, S.A.; Bae, C.-H.; Ahn, J.-H.; Foster, C.E.; Liao, J.C.; Lin, X.N. Design and characterization of synthetic fungal-bacterial consortia for direct production of isobutanol from cellulosic biomass. *Proc. Natl. Acad. Sci. USA* **2013**, *110*, 14592–14597. [CrossRef] [PubMed]

95. Zuroff, T.R.; Xiques, S.B.; Curtis, W.R. Consortia-mediated bioprocessing of cellulose to ethanol with a symbiotic Clostridium phytofermentans/yeast co-culture. *Biotechnol. Biofuels* **2013**, *6*, 59. [CrossRef] [PubMed]

96. Mee, M.T.; Wang, H.H. Engineering ecosystems and synthetic ecologies. *Mol. Biosyst.* **2012**, *8*, 2470–2483. [CrossRef] [PubMed]

97. Chen, A.H.; Silver, P.A. Designing biological compartmentalization. *Trends Cell. Biol.* **2012**, *22*, 662–670. [CrossRef] [PubMed]

98. Mougi, A.; Kondoh, M. Diversity of Interaction Types and Ecological Community Stability. *Science* **2012**, *337*, 349–351. [CrossRef] [PubMed]

99. Thébault, E.; Fontaine, C. Stability of Ecological Communities and the Architecture of Mutualistic and Trophic Networks. *Science* **2010**, *329*, 853–856. [CrossRef] [PubMed]

100. Allesina, S.; Tang, S. Stability criteria for complex ecosystems. *Nature* **2012**, *483*, 205–208. [CrossRef] [PubMed]

101. Sole, R.V.; Bascompte, J. *Self-Organization in Complex Ecosystems*; Princeton University Press: Princeton, NJ, USA, 2006.

102. Pascual, M.; Dunne, J.A. *Ecological Networks: Linking Structure to Dynamics in Food Webs*; Oxford University Press: New York, NY, USA, 2005.

103. Paine, R.T. Food Webs: Linkage, Interaction Strength and Community Infrastructure. *J. Anim. Ecol.* **1980**, *49*, 666. [CrossRef]

104. Neutel, A.-M.; Heesterbeek, J.A.P.; De Ruiter, P.C. Stability in Real Food Webs: Weak Links in Long Loops. *Science* **2002**, *296*, 1120–1123. [CrossRef] [PubMed]

105. Bastolla, U.; Fortuna, M.A.; Pascual-Garcia, A.; Ferrera, A.; Luque, B.; Bascompte, J. The architecture of mutualistic networks minimizes competition and increases biodiversity. *Nature* **2009**, *458*, 1018–1020. [CrossRef] [PubMed]

106. Okuyama, T.; Holland, J.N. Network structural properties mediate the stability of mutualistic communities. *Ecol. Lett.* **2008**, *11*, 208–216. [CrossRef] [PubMed]

107. Rohr, R.P.; Saavedra, S.; Bascompte, J. On the structural stability of mutualistic systems. *Science* **2014**, *345*, 1253497. [CrossRef] [PubMed]

108. May, R.M. *Stability and Complexity in Model Ecosystems*; Princeton University Press: Princeton, NJ, USA, 1974.

109. Cohen, J.E.; Newman, C.M. The Stability of Large Random Matrices and Their Products. *Ann. Probab.* **1984**, *12*, 283–310. [CrossRef]

110. Coyte, K.Z.; Schluter, J.; Foster, K.R. The ecology of the microbiome: Networks, competition, and stability. *Science* **2015**, *350*, 663–666. [CrossRef] [PubMed]

111. Momeni, B.; Xie, L.; Shou, W. Lotka-Volterra pairwise modeling fails to capture diverse pairwise microbial interactions. *eLife* **2017**, *6*, e25051. [CrossRef] [PubMed]

112. Worthen, W.B.; Moore, J.L. Higher-Order Interactions and Indirect Effects: A Resolution Using Laboratory Drosophila Communities. *Am. Nat.* **1991**, *138*, 1092–1104. [CrossRef]

113. Wootton, J.T. Indirect effects in complex ecosystems: Recent progress and future challenges. *J. Sea Res.* **2002**, *48*, 157–172. [CrossRef]

114. Werner, E.E.; Peacor, S.D. A review of trait-mediated indirect interactions in ecological communities. *Ecology* **2003**, *84*, 1083–1100. [CrossRef]

115. Lendenmann, U.; Egli, T. Kinetic models for the growth of Escherichia coli with mixtures of sugars under carbon-limited conditions. *Biotechnol. Bioeng.* **1998**, *59*, 99–107. [CrossRef]

116. Hermsen, R.; Okano, H.; You, C.; Werner, N.; Hwa, T. A growth-rate composition formula for the growth of E. coli on co-utilized carbon substrates. *Mol. Syst. Biol.* **2015**, *11*, 801. [CrossRef] [PubMed]

117. Lipsitch, M.; Levin, B.R. The population dynamics of antimicrobial chemotherapy. *Antimicrob. Agents Chemother.* **1997**, *41*, 363–373. [PubMed]

118. Acar, J.F. Antibiotic synergy and antagonism. *Med. Clin. N. Am.* **2000**, *84*, 1391–1406. [CrossRef]

119. Chait, R.; Craney, A.; Kishony, R. Antibiotic interactions that select against resistance. *Nature* **2007**, *446*, 668–671. [CrossRef] [PubMed]

120. White, R.L.; Burgess, D.S.; Manduru, M.; Bosso, J.A. Comparison of three different in vitro methods of detecting synergy: Time-Kill, checkerboard, and E test. *Antimicrob. Agents Chemother.* **1996**, *40*, 1914–1918. [PubMed]

121. Ocampo, P.S.; Lázár, V.; Papp, B.; Arnoldini, M.; Abel zur Wiesch, P.; Busa-Fekete, R.; Fekete, G.; Pál, C.; Ackermann, M.; et al. Antagonism between bacteriostatic and bactericidal antibiotics is prevalent. *Antimicrob. Agents Chemother.* **2014**, *58*, 4573–4582. [CrossRef] [PubMed]

122. Sanders, C.C.; Sanders, W.E.; Goering, R.V. In vitro antagonism of beta-lactam antibiotics by cefoxitin. *Antimicrob. Agents Chemother.* **1982**, *21*, 968–975. [CrossRef] [PubMed]

123. Burgess, J.G.; Jordan, E.M.; Bregu, M.; Mearns-Spragg, A.; Boyd, K.G. Microbial antagonism: A neglected avenue of natural products research. *J. Biotechnol.* **1999**, *70*, 27–32. [CrossRef]

124. Yu, G.; Baeder, D.Y.; Regoes, R.R.; Rolff, J. Combination Effects of Antimicrobial Peptides. *Antimicrob. Agents Chemother.* **2016**, *60*, 1717–1724. [CrossRef] [PubMed]

125. Validation by Science Exchange—Identifying and Rewarding High-Quality Research. Available online: http://validation.scienceexchange.com/#/ (accessed on 23 June 2017).

126. Baker, M.; Dolgin, E. Cancer reproducibility project releases first results. *Nature* **2017**, *541*, 269–270. [CrossRef] [PubMed]

127. The challenges of replication. *eLife* **2017**, *6*, e23693.

128. Steinberg, G.; Perez-Martin, J. Ustilago maydis, a new fungal model system for cell biology. *Trends Cell. Biol.* **2008**, *18*, 61–67. [CrossRef] [PubMed]

129. Etxebeste, O.; Espeso, E.A. Neurons show the path: Tip-to-nucleus communication in filamentous fungal development and pathogenesisa. *FEMS Microbiol. Rev.* **2016**, *40*, 610–624. [CrossRef] [PubMed]

9

Individual-Based Modelling of Invasion in Bioaugmented Sand Filter Communities

Aisling J. Daly [1], Jan M. Baetens [1], Johanna Vandermaesen [2], Nico Boon [3], Dirk Springael [2] and Bernard De Baets [1],*

[1] KERMIT, Department of Data Analysis and Mathematical Modelling, Ghent University, Coupure links 653, B-9000 Ghent, Belgium; aisling.daly@ugent.be (A.J.D.); jan.baetens@ugent.be (J.M.B.)

[2] Division of Soil and Water Management, KU Leuven, Kasteelpark Arenberg 20 bus 2459, B-3001 Heverlee, Belgium; joke.vandermaesen@kuleuven.be (J.V.); dirk.springael@kuleuven.be (D.S.)

[3] Center for Microbial Ecology and Technology (CMET), Ghent University, Coupure links 653, B-9000 Ghent, Belgium; nico.boon@ugent.be

*   Correspondence: bernard.debaets@ugent.be

Abstract: Using experimental data obtained from in vitro bioaugmentation studies of a sand filter community of 13 bacterial species, we develop an individual-based model representing the in silico counterpart of this synthetic microbial community. We assess the inter-species interactions, first by identifying strain identity effects in the data then by synthesizing these effects into a competition structure for our model. Pairwise competition outcomes are determined based on interaction effects in terms of functionality. We also consider non-deterministic competition, where winning probabilities are assigned based on the relative intrinsic competitiveness of each strain. Our model is able to reproduce the key qualitative dynamics observed in in vitro experiments with similar synthetic sand filter communities. Simulation outcomes can be explained based on the underlying competition structures and the resulting spatial dynamics. Our results highlight the importance of community diversity and in particular evenness in stabilizing the community dynamics, allowing us to study the establishment and development of these communities, and thereby illustrate the potential of the individual-based modelling approach for addressing microbial ecological theories related to synthetic communities.

Keywords: individual-based model; invasion; bioaugmentation; engineered community

1. Introduction

1.1. Background

The composition, establishment and functional maintenance of any ecosystem are largely driven by the interactions between individuals [1], and microbial communities are no exception [2]. The fundamental basis of all studies of interactions between cell populations are synthetic co-culture systems: experimental set-ups where "two or more different populations of cells are grown with some degree of contact between them" [3]. Synthetic co-cultures have gained particular interest from microbiologists in recent years due to their reduced complexity and increased controllability, which favours them over more complex natural systems for examining ecological theories [4] and also for more specific industrial, medical and environmental applications such as industrial fermentation and the production of chemical compounds [5].

A more specific application of co-cultures is bioaugmentation, where the biomass in soil or water treatment plants is altered by the addition of certain microbial strains that have been selected for their ability to degrade specific chemical compounds [6]. From a microbial ecological perspective, bioaugmentation represents a kind of microbial invasion process, where the strains introduced to augment resident community functionality are the invaders. For example, during the treatment of drinking water, the common groundwater pollutant 2,6-dichlorobenzamide (2,6-BAM) must be removed below a threshold concentration of $0.1 \, \mu g \, L^{-1}$ to meet the EU Directive on Drinking Water [7]. However, the endogenous microbial communities in the sand filters (SFs) of such drinking water treatment plants are not capable of achieving sufficient BAM removal to respect this threshold [8]. Therefore, bioaugmentation of SFs has been proposed as an alternative strategy, by the addition of a specialized BAM mineralizer such as *Aminobacter* sp. MSH1 [9]. However, studies of this type of bioaugmentation of drinking water ecosystems rarely address how exactly the pesticide degrader interacts with the resident community, and other such fundamental ecological questions [10].

In Vandermaesen et al. [11], the authors hypothesize that the establishment of MSH1 and its subsequent BAM mineralization in SFs depend not only on exploitative competition effects, but also on other features such as interactions with resident community members. Therefore, the BAM mineralization activity of MSH1 was evaluated in sand microcosms in the presence of a selection of the 13 sand filter isolates (SFIs) described in Vandermaesen et al. [11]. Synthetic microbial communities of MSH1 combined with SFIs were subjected to an initial competition phase. Subsequently, BAM was added and the kinetics of BAM mineralization was evaluated as a measure of bioaugmentation success.

To characterize the interactions between resident community members, co-cultures of various combinations of SFIs with MSH1 were inoculated, and their mineralization kinetics was followed. However, given the total number of strains in the community, it is practically impossible to experimentally study all possible co-culture combinations. In such cases, mathematical modelling is becoming more and more appreciated as a tool for identifying possible co-cultures of interest [12–15].

1.2. Motivation and Scope

We use an individual-based modelling approach to construct the in silico counterpart of the in vitro synthetic community used in the experiments of Vandermaesen et al. [11], with the goal of qualitatively reproducing the observed dynamics. Due to their inherent flexibility and ability to reproduce complex system-level behaviour by capturing the interactions between individuals, individual-based models (IBMs) have proven useful for addressing fundamental microbial ecology questions, such as our questions related to fundamental interactions between invader and resident community members. Other examples include IBM studies of the evolution of cooperative behaviour in microbial communities [16] and the role of spatial aggregation in maintaining cooperation between cross-feeding microbial strains [17]. However, such models are typically restricted to only a few species, hence our model of 13 species would be an outlier in this respect [4].

Previous results with synthetic microbial communities with similar characteristics in terms of diversity and composition [18] and also for this particular synthetic community of MSH1 and 13 SFIs [11], highlighted the importance of the initial competition phase (before the addition of BAM) where all 13 SFIs are inoculated in the co-culture. Competition between the SFIs results in extinctions, leading to a stable subcommunity of reduced richness. It is this subcommunity that is present at the moment of the BAM spike and during the subsequent mineralization period that determines the bioaugmentation success. We aim to retrieve this behaviour with our modelling approach.

For this purpose, we make use of the data obtained from the Vandermaesen et al. experiments [11] (described in Section 2) to model the competitive interactions between individual microbes. We then present the results of inferring the strain interactions (Section 3.1), incorporating this information in an IBM (Section 3.2), as well as the results of the in silico experiments it is subsequently employed for (Section 3.3). In the final section, we summarize the conclusions of the modelling and simulation studies.

2. Materials and Methods

In this section, we summarize the experimental set-up and procedure used by Vandermaesen et al. [11] to obtain the dataset used for the modelling and simulation studies presented in this paper.

2.1. Experimental Set-Up

The hypothesis of this in vitro study was that the establishment of MSH1 and its subsequent BAM mineralization in SFs depend on interactions with and between resident community members. Therefore, the BAM mineralization activity of MSH1 was evaluated in sand microcosm co-cultures in the presence of different combinations of 13 SFIs. Synthetic microbial communities of MSH1 combined with SFIs were co-cultured, then BAM was added and the kinetics of BAM mineralization was evaluated as a measure of bioaugmentation success.

2.1.1. Bacterial Strains

The specific variant of the BAM mineralizing *Aminobacter* sp. MSH1 [9] used in this study, MSH1-GFP, was fluorescently tagged. The 13 SFIs used were isolated from SF material from two drinking water treatment plants [11]: *Acidovorax* sp. S9, *Undibacterium* sp. S22, *Brachybacterium* sp. S51, *Mesorhizobium* sp. S158, *Acidovorax* sp. S164, *Rhodococcus* sp. K27, *Acidovorax* sp. K52, *Aeromonas* sp. K62, *Paucibacter* sp. K67, *Pelomonas* sp. K89, *Rhodoferax* sp. K112, *Rhodoferax* sp. K129, and *Piscinibacter* sp. K169. None of the selected SFIs were capable of BAM mineralization, avoiding any confounding effects with the BAM mineralization performance of MSH1.

2.1.2. Microcosm Set-Up

Microcosms were created in deep 96-well plates, containing sterile sand in every well. MSH1 and SFIs were cultured and prepared as described in Vandermaesen et al. [11] and combined in synthetic communities in such a way that the number of cells of every strain was 10^7 cells/mL. Since each community included MSH1, the total richness of a community R_T is given by $R_T = R_{SFI} + 1$, where R_{SFI} is the number of SFIs present. In addition to all combinations of individual SFI with MSH1 (13 combinations at $R_{SFI} = 1$), all 78 different pair combinations of two SFIs with MSH1 ($R_{SFI} = 2$) were tested.

Sodium acetate was provided as the only carbon source at a concentration of 150 µg L^{-1} in MMO medium (MMO + Ac). Assuming that 50% of acetate-C is actually assimilated, this corresponds to an AOC (assimilable organic carbon) concentration of 22 µg C /L, which is within the range of AOC values in drinking water ecosystems (20–100 µg C /L) [19]. Of every synthetic community, 100 µL was inoculated in the sand microcosms. A reference microcosm inoculated with 100 µL MSH1 at 10^7 cells/mL ($R_{SFI} = 0$) was included in every deep well plate. In addition, to account for abiotic $^{14}CO_2$ production, one negative control ($R_T = 0$) was included, containing sand amended with 100 µL MMO + Ac. All synthetic communities and controls were replicated four times. No $^{14}CO_2$ production was observed in the abiotic control. The plates were sealed and incubated at 20 °C for 7 days.

After this initial competition phase, all wells were spiked with 5000 counts per minute ^{14}C-BAM, dissolved in 5 µL MMO, which corresponds to a final BAM concentration of 150 µg L^{-1}. BAM mineralization was then followed for approximately 130 h by trapping the produced $^{14}CO_2$ with Ca(OH)$_2$. Trapped $^{14}CO_2$ radioactivity was quantified by digital autoradiography. The cumulative percentage $^{14}CO_2$ was plotted relative to the total amount of ^{14}C added as a function of the incubation time, and hence cumulative mineralization curves were obtained.

2.2. Modelling of Mineralization Kinetics

To describe the kinetics of BAM mineralization, the modified Gompertz model [20] was used. This model is one of the most commonly used microbial growth models [21], and is given by

$$P = A \exp\left(-\exp\left(\frac{\mu e}{A}(\lambda - ct) + 1\right)\right) \tag{1}$$

where P (%) is the percentage mineralization at time t (h), A (%) is the total extent of mineralization after the exponential mineralization phase, λ (% h^{-1}) is the lag time, c (% h^{-1}) is the endogenous mineralization rate, and μ (% h^{-1}) is the maximum mineralization rate constant. The modified Gompertz model differs from the standard Gompertz model [20] in that its parameters each have a biological meaning.

The Gompertz parameters of the cumulative mineralization curves were determined by least squares curve fitting, using the Trust-Region-Reflective algorithm [22,23], at a termination tolerance of 10^{-14} and allowing at most 2×10^5 function evaluations and 3×10^5 iterations. Initial parameter estimates were set at 30, 5, 0.1, and 2 for A, μ, c, and λ, respectively [24]. This was implemented using Matlab R2012b (Mathworks, Natick, MA, USA). All values of c were zero or close to, and were hence excluded.

2.3. Description of the Dataset

From the experimental set-up described in Section 2.1, we obtained a dataset representing 13 monocultures (the individual strains) and 78 co-cultures (the pair combinations). For each of these 91 conditions, we have two types of mineralization data. First, a cumulative BAM mineralization time series consisting of achieved mineralization values at 13 time points, from $t = 0$ h to $t = 130$ h. There are four biological replicates of each time series, except where some outliers were removed as indicated in Vandermaesen et al. [11]. In total, 21 out of 364 time series were removed. After removal of these outliers, no condition had less than three replicates. The second data type consists of the fitted Gompertz parameters λ, μ and A describing the mineralization kinetics, namely one set of parameters per time series.

3. Results and Discussion

3.1. Assessing Strain Interactions

The experiments of Vandermaesen et al. [11] focused on bioaugmentation success and therefore collected data related to BAM mineralization and MSH1 survival. The data related to the SFIs themselves are their monoculture growth curves and their monoculture survival curves on acetate (see Appendix A). These data can give us an idea of how the SFIs grow and persist in isolation, and on this basis Vandermaesen et al. [11] classified the strains according to their "intrinsic competitiveness", a classification that we can use as an additional feature of the strains. However, these data do not give us any information about how the SFIs may interact, and in particular compete, when they are inoculated together in co-culture.

The information we do have regarding the interactions between SFIs is indirect. From the differences in mineralization parameters between the different co-culture combinations, we can infer when there are interaction effects occurring between strains, by comparing the mineralization performances of MSH1 alone, in co-culture with individual SFIs, and in co-culture with both strains. The mineralization performance was studied using the Gompertz model (see Section 2.2 for details). This model has four parameters: the lag time λ, the maximum mineralization rate μ, the total extent of mineralization A, and the endogenous mineralization rate c.

To study the strain interaction effects, we focus on two of these parameters: the lag time λ and the maximum mineralization rate μ. These two parameters have been highlighted as key to the success of bioaugmentation strategies and are more strongly linked with both positive and negative mineralization effects than the other mineralization parameters [25].

3.1.1. Identifying Strain Identity Effects

Since each synthetic community included MSH1, the total richness of a community R_T is given by $R_T = R_{SFI} + 1$, where R_{SFI} is the number of SFIs present. In addition to all combinations of individual SFIs with MSH1 (13 combinations at $R_{SFI} = 1$), all 78 different pair combinations of two SFIs with MSH1 ($R_{SFI} = 2$) were tested (further details given in Section 2). The 13 SFIs are assigned the following labels: S9, S22, S51, S158, S164, K27, K52, K62, K67, K89, K112, K129, and K169.

Previous studies have also used growth model parameters to identify different growth behaviours between microbial species, for example through the use of regression models [26]. We employ a statistical test known as the pairwise Tukey test [27] to compare values of the lag time λ and mineralization rate μ across different R_{SFI} levels. With this test it is possible to evaluate whether values of λ or μ observed for a specific synthetic community are significantly different from the respective parameter values observed for a different community.

The Tukey test statistic is $z = \frac{m_A - m_B}{S_E}$, where m_A and m_B are the respective means of the observations of two populations being compared, and S_E is the data's standard error [28]. The null hypothesis of the test is that the means are from the same population. The test statistic is then compared to a critical test statistic value z_{crit} which is obtained from the studentized range distribution [29]. If z is larger than z_{crit}, then the null hypothesis is rejected and it is concluded that the two populations are significantly different. Tests were performed at the 95% significance level, using Mathematica (version 11.0, Wolfram Research, Champaign, IL, USA).

Two types of tests were conducted. First, we compared values of λ or μ for $R_{SFI} = 1$ communities against $R_{SFI} = 0$ (i.e., MSH1 alone) as a benchmark population. To determine the sign of the change, we consider the biological interpretation of a positive or beneficial change in these parameters. For the lag time λ, a decrease in this parameter is considered a positive effect while an increase is considered a negative effect. For the mineralization rate μ, the opposite is true.

The second type of test required selecting one of the SFIs as the focal strain. The test then compared values of λ or μ for the $R_{SFI} = 2$ communities including this focal strain, against the values of λ or μ for the corresponding $R_{SFI} = 1$ community for the non-focal strain. For example, when S9 was the focal strain of the test and the parameter under consideration was λ, we selected all $R_{SFI} = 2$ communities containing S9. One such community contained S9, S22 and MSH1. We then compared the values of λ of this community against the values of λ of the community containing S22 and MSH1. This allowed us to conclude if in this case there were significant differences in lag time due to the inclusion of S9. This analysis was repeated for every strain other than the focal strain.

This test was done 13 times for each parameter, so that each of the strains was used once as the focal strain. The results of these tests are collected in the tables shown in Figures 1 and 2. In these tables, each row collects the results of Tukey tests with a particular focal strain, e.g., the first row shows the results of tests where S9 was the focal strain, and the columns indicate the other strains being tested for interaction effects with S9.

3.1.2. Building the Competition Structures

Using the information gathered in Figures 1 and 2, we represented the competition occurring between the SFIs using so-called tournament matrices. Such a matrix M for s species has dimensions $s \times s$. If the species represented by row i outcompetes the species represented by column j, then $M_{ij} = 1$. On the other hand, if the species represented by row i is outcompeted by the species represented by column j, we have $M_{ij} = -1$. If $i = j$, then $M_{ij} = 0$. Using the information in Figures 1 and 2, we can compile such a tournament or competition matrix. The question remains how precisely to do so.

We have two possibilities: to merge the information about the lag time λ and mineralization rate μ interaction effects, or to treat the parameters separately. The latter option is justified by considering that the parameters represent different biological attributes and different underlying processes [25]. This is most noticeable in their opposing effects on mineralization performance in particular; an increased parameter λ is considered a negative effect while an increased parameter μ is considered a positive effect.

	S9	S22	S51	S158	S164	K27	K52	K62	K67	K89	K112	K129	K169
S9		+	0	+	+	0	+	+	+	+	+	+	–
S22	+		–	+	+	0	+	0	+	0	+	+	–
S51	0	–		0	0	–	0	–	0	–	0	–	–
S158	+	+	0		+	0	+	+	+	+	+	+	–
S164	+	+	0	+		0	+	+	+	+	+	+	–
K27	0	0	–	0	0		0	–	0	–	0	–	–
K52	+	+	0	+	+	0		+	+	+	+	+	–
K62	+	0	–	+	+	–	+		+	0	+	0	–
K67	+	+	0	+	+	0	+	+		+	+	+	–
K89	+	0	–	+	+	–	+	0	+		+	0	–
K112	+	+	0	+	+	0	+	+	+	+		+	0
K129	+	+	–	+	+	–	+	0	+	0	+		–
K169	–	–	–	–	–	–	–	–	–	–	0	–	

Figure 1. Tukey test results for the lag time λ. Each row collects the results of Tukey tests with a particular focal strain, the columns then indicate the strains that were tested for interaction effects with it. The entry in cell (i,j) indicates the difference (if any) between the $R_{SFI} = 2$ community containing species i and species j, and the control $R_{SFI} = 0$ community: "+" indicates the $R_{SFI} = 2$ parameter values were significantly larger than the $R_{SFI} = 0$ values, "–" indicates they were significantly smaller, and "0" indicates no significant difference. The background colour of cell (i,j) indicates the difference (if any) between the $R_{SFI} = 2$ community containing species i and species j, and the $R_{SFI} = 1$ community containing species j: green indicates the $R_{SFI} = 2$ parameter values were significantly smaller than the $R_{SFI} = 1$ values, red indicates they were significantly larger, and no colour indicates no significant difference.

	S9	S22	S51	S158	S164	K27	K52	K62	K67	K89	K112	K129	K169
S9		–	0	–	–	0	–	–	–	–	–	–	0
S22	–		0	–	–	–	–	–	–	–	–	–	0
S51	0	0		–	–	0	–	–	–	–	–	0	0
S158	–	–	–		–	–	–	–	–	–	–	–	0
S164	–	–	–	–		–	–	–	–	–	–	–	–
K27	0	–	0	–	–		–	0	–	0	–	0	0
K52	–	–	–	–	–	–		–	–	–	–	–	0
K62	–	–	–	–	–	0	–		–	–	–	0	0
K67	–	–	–	–	–	–	–	–		–	–	–	0
K89	–	–	–	–	–	0	–	–	–		–	0	0
K112	–	–	–	–	–	–	–	–	–	–		–	0
K129	–	–	0	–	–	0	–	0	–	0	–		0
K169	0	0	0	0	–	0	0	0	0	0	0	0	

Figure 2. Tukey test results for the mineralization rate μ. Each row collects the results of Tukey tests with a particular focal strain, the columns then indicate the strains that were tested for interaction effects with it. The entry in cell (i,j) indicates the difference (if any) between the $R_{SFI} = 2$ community containing species i and species j, and the control $R_{SFI} = 0$ community: "+" indicates the $R_{SFI} = 2$ parameter values were significantly larger than the $R_{SFI} = 0$ values, "–" indicates they were significantly smaller, and "0" indicates no significant difference. The background colour of cell (i,j) indicates the difference (if any) between the $R_{SFI} = 2$ community containing species i and species j, and the $R_{SFI} = 1$ community containing species j: green indicates the $R_{SFI} = 2$ parameter values were significantly larger than the $R_{SFI} = 1$ values, red indicates they were significantly smaller, and no colour indicates no significant difference.

This approach results in two competition matrices, the first based on lag time λ interaction effects, and the second based on mineralization rate μ interaction effects. We look in Figure 1 (λ interaction effects) or Figure 2 (μ interaction effects) for pairs of SFIs that appear to interact with each other, and check what kind of interaction appears to be taking place: is it positive or negative with respect to each of the SFIs?

This corresponds in Figures 1 and 2 to both the cell entries and the cell background colours. The cell entries indicate which kind of difference (if any) exists between the control community and the $R_{SFI} = 2$ community containing the particular species corresponding to the cell row and column. These relationships can be positive, negative, or not significant. The cell background colours indicate the difference (if any) between the $R_{SFI} = 1$ community containing the species corresponding to the cell column, and the $R_{SFI} = 2$ community containing the particular species corresponding to the cell row and column. These relationships can also be positive, negative, or not significant.

We then obtain the following matrices representing competition between the SFIs. When considering interactions based on lag time λ effects, the matrix reads:

$$
M_\lambda =
\begin{pmatrix}
0 & 0 & 1 & 1 & 0 & 0 & 0 & 0 & 0 & 0 & 0 & 1 & 0 \\
0 & 0 & 0 & -1 & 1 & 0 & 0 & 0 & 0 & 0 & 1 & 1 & 0 \\
-1 & 0 & 0 & 0 & 0 & 0 & 0 & 0 & 0 & 0 & 0 & 0 & 0 \\
-1 & 1 & 0 & 0 & 1 & 0 & 0 & 1 & 0 & 0 & 1 & 1 & 0 \\
0 & -1 & 0 & -1 & 0 & 0 & 0 & 0 & 0 & 0 & 0 & 1 & 0 \\
0 & 0 & 0 & 0 & 0 & 0 & 1 & 0 & 0 & 0 & 0 & 0 & 0 \\
0 & 0 & 0 & -1 & 0 & 0 & 0 & 0 & 0 & 0 & 0 & 1 & 0 \\
0 & 0 & 0 & -1 & 0 & 0 & 0 & 0 & 0 & 0 & 0 & 0 & 0 \\
0 & 0 & 0 & 0 & 0 & 0 & 0 & 0 & 0 & 0 & 0 & 0 & 0 \\
0 & 0 & 0 & 0 & 0 & 0 & 0 & 0 & 0 & 0 & 0 & 0 & 0 \\
0 & -1 & 0 & -1 & 0 & 0 & 0 & 0 & 0 & 0 & 0 & -1 & 0 \\
-1 & -1 & 0 & -1 & -1 & 0 & -1 & 0 & 0 & 0 & 1 & 0 & 0 \\
0 & 0 & 0 & 0 & 0 & 0 & 0 & 0 & 0 & 0 & 0 & 0 & 0
\end{pmatrix}
\tag{2}
$$

When considering interactions based on mineralization rate μ effects, the matrix has the form:

$$
M_\mu =
\begin{pmatrix}
0 & 0 & 0 & 0 & 0 & 0 & 0 & 0 & 0 & 0 & 0 & 0 & 0 \\
0 & 0 & 0 & -1 & 1 & 0 & 0 & 0 & 0 & 0 & 1 & 0 & 0 \\
0 & 0 & 0 & 0 & 0 & 0 & -1 & 0 & 0 & 0 & 0 & 0 & 0 \\
0 & 1 & 0 & 0 & 0 & 0 & 0 & 1 & 0 & 1 & 0 & 1 & 0 \\
0 & -1 & 0 & 0 & 0 & 0 & 0 & -1 & 0 & -1 & 0 & 1 & -1 \\
0 & 0 & 0 & 0 & 0 & 0 & 0 & 0 & 0 & 0 & 0 & 0 & 0 \\
0 & 0 & 1 & 0 & 0 & 0 & 0 & 1 & 0 & 1 & 0 & 1 & 0 \\
0 & 0 & 0 & -1 & 1 & 0 & -1 & 0 & 1 & 0 & 1 & 0 & 0 \\
0 & 0 & 0 & 0 & 0 & 0 & 0 & -1 & 0 & 0 & 0 & 0 & -1 \\
0 & 0 & 0 & -1 & 1 & 0 & -1 & 0 & 0 & 0 & 0 & 0 & 0 \\
0 & -1 & 0 & 0 & 0 & 0 & 0 & -1 & 0 & 0 & 0 & -1 & 0 \\
0 & 0 & 0 & -1 & -1 & 0 & -1 & 0 & 0 & 0 & 1 & 0 & 0 \\
0 & 0 & 0 & 0 & 1 & 0 & 0 & 0 & 1 & 0 & 0 & 0 & 0
\end{pmatrix}
\tag{3}
$$

An additional extension of our modelling approach that will bring it closer to reality is to also consider non-deterministic competition. Deterministic competition assumes that, if the competition structure specifies that A beats B, this will always occur: it will never be possible for B to beat A. This is reflected in the competition matrices M_λ and M_μ, which contain only 1's (implying certain victory), -1's (certain defeat) and 0's (no competition). But this is not always realistic [30–32]. Variation between individuals can result in an individual of species A that is a particularly weak competitor, and an individual of species B that is a particularly strong competitor. If these two specific individuals meet, the outcome of the competition can be in doubt. It may be more realistic to specify a so-called winning probability [33,34]: the probability that A beats B. Including a winning probability allows for different competition outcomes to occur, and the value of the winning probability allows us to account for the relative strengths of the individuals.

Therefore we will also consider non-deterministic competition between the SFIs, not only in terms of its effects on the diversity and stability of the community (and possible subcommunity), but in comparison with the same effects due to deterministic competition. Our immediate question is then how to assign the winning probabilities to the different pairwise competitions.

Using data related to the SFIs' monoculture growth and survival curves, Vandermaesen et al. [11] classified the "intrinsic competitiveness" of the SFIs and on this basis grouped them into strong, intermediate and weak competitors. Using this information, we can assign winning probabilities to each pairwise competition based on the relative differences in intrinsic competitiveness between the two strains. For example, competition between a weak intrinsic competitor and a strong intrinsic competitor will most likely result in the success of the latter. It should also be clear that this winning probability should be higher than the winning probability assigned to an intermediate intrinsic competitor when faced with a weak intrinsic competitor. Using this approach, we replace the 1's and -1's populating our matrices M_λ and M_μ with rational numbers of absolute value less than 1, corresponding to the appropriate winning probability.

Using this approach, we obtain the following matrices representing non-deterministic competition. When considering interactions based on lag time λ effects, the matrix has the form:

$$M_\lambda^* = \begin{pmatrix}
0 & 0 & 0.9 & 0.9 & 0 & 0 & 0 & 0 & 0 & 0 & 0 & 0.9 & 0 \\
0 & 0 & 0 & -0.9 & 0.7 & 0 & 0 & 0 & 0 & 0 & 0.7 & 0.7 & 0 \\
-0.9 & 0 & 0 & 0 & 0 & 0 & 0 & 0 & 0 & 0 & 0 & 0 & 0 \\
-0.9 & 0.9 & 0 & 0 & 0.9 & 0 & 0 & 0.9 & 0 & 0 & 0.9 & 0.9 & 0 \\
0 & -0.7 & 0 & -0.9 & 0 & 0 & 0 & 0 & 0 & 0 & 0 & 0.9 & 0 \\
0 & 0 & 0 & 0 & 0 & 0 & 0 & 0 & 0 & 0 & 0 & 0 & 0 \\
0 & 0 & 0 & -0.9 & 0 & 0 & 0 & 0 & 0 & 0 & 0 & 0.9 & 0 \\
0 & 0 & 0 & -0.9 & 0 & 0 & 0 & 0 & 0 & 0 & 0 & 0 & 0 \\
0 & 0 & 0 & 0 & 0 & 0 & 0 & 0 & 0 & 0 & 0 & 0 & 0 \\
0 & -0.7 & 0 & -0.9 & 0 & 0 & 0 & 0 & 0 & 0 & 0 & -0.6 & 0 \\
-0.9 & -0.7 & 0 & -0.9 & -0.9 & 0 & -0.9 & 0 & 0 & 0 & 0.6 & 0 & 0 \\
0 & 0 & 0 & 0 & 0 & 0 & 0 & 0 & 0 & 0 & 0 & 0 & 0
\end{pmatrix} \tag{4}$$

When considering interactions based on mineralization rate μ effects, the matrix has the form:

$$M_\mu^* = \begin{pmatrix} 0 & 0 & 0 & 0 & 0 & 0 & 0 & 0 & 0 & 0 & 0 & 0 & 0 \\ 0 & 0 & 0 & -0.9 & 0.7 & 0 & 0 & 0 & 0 & 0 & 0.7 & 0 & 0 \\ 0 & 0 & 0 & 0 & 0 & 0 & -0.9 & 0 & 0 & 0 & 0 & 0 & 0 \\ 0 & 0.9 & 0 & 0 & 0 & 0 & 0 & 0.9 & 0 & 0.9 & 0 & 0.9 & 0 \\ 0 & -0.7 & 0 & 0 & 0 & 0 & 0 & -0.7 & 0 & -0.6 & 0 & 0.9 & -0.7 \\ 0 & 0 & 0 & 0 & 0 & 0 & 0 & 0 & 0 & 0 & 0 & 0 & 0 \\ 0 & 0 & 0.9 & 0 & 0 & 0 & 0 & 0.9 & 0 & 0.9 & 0 & 0.9 & 0 \\ 0 & 0 & 0 & -0.9 & 0.7 & 0 & -0.9 & 0 & 0.7 & 0 & 0.7 & 0 & 0 \\ 0 & 0 & 0 & 0 & 0 & 0 & 0 & -0.7 & 0 & 0 & 0 & 0 & -0.7 \\ 0 & 0 & 0 & -0.9 & 0.6 & 0 & -0.9 & 0 & 0 & 0 & 0 & 0 & 0 \\ 0 & -0.7 & 0 & 0 & 0 & 0 & 0 & -0.7 & 0 & 0 & 0 & -0.6 & 0 \\ 0 & 0 & 0 & -0.9 & -0.9 & 0 & -0.9 & 0 & 0 & 0 & 0.6 & 0 & 0 \\ 0 & 0 & 0 & 0 & 0.7 & 0 & 0 & 0 & 0.7 & 0 & 0 & 0 & 0 \end{pmatrix} \tag{5}$$

3.2. Constructing the Individual-Based Model

To understand how the different competition structures affect the dynamics of the system, we consider the in silico counterpart of the synthetic community of 13 SFIs. We model this community using an individual-based approach, which we describe using an established standard protocol known as the ODD protocol [35].

3.2.1. Overview

Purpose

The aim of the model is to study how more realistic competition structures affect the in silico dynamics, particularly in terms of community diversity and stability, and investigate whether this approach can qualitatively reproduce the dynamics observed in similar in vitro studies, namely a stable and persisting subcommunity.

State Variables and Scales

The model is a two-dimensional representation of an experimental domain divided into a regular grid of size $L \times L = N$, and populated by a community of 13 SFIs. We assign to each strain a numerical label between one and 13, in the order given in Section 3.1.1: S9, S22, S51, S158, S164, K27, K52, K62, K67, K89, K112, K129, K169. Each grid site is either occupied by a single individual, or is empty. Individuals are characterized by two state variables: grid position (x, y) and species identity $s \in \{1, \ldots, 13\}$.

Process Overview

We consider an in silico microbial community that is initially placed on the grid with a random spatial distribution. The community's initial species abundance distribution is completely even, to mimic the in vitro experimental set-up.

An individual can interact with its nearest neighbours, defined as those individuals in its von Neumann neighbourhood (the four grid cells with which it shares an edge). Three possible interactions can occur, representing the key demographic processes: reproduction, competition and mobility.

Reproduction can occur when an individual is located adjacent to an empty grid site, which is then filled with a new individual of the same species. In order to provide a form of mobility, all individuals can exchange their position with a nearest neighbour or move to a neighbouring empty site. Competition can occur between two neighbouring individuals that do not represent the same species. The outcome of the competition event is determined by the governing competition matrix; the defeated individual is removed from the grid and the grid site becomes empty.

Scheduling

The IBM proceeds using a modified version of the Gillespie algorithm [36], to determine which interaction occurs at each time step and calculate the interaction outcome. The algorithm iterates over the following steps:

(1)   Set time to $t = 0$ and set the event rate constants:

 (a) reproduction with rate constant μ
 (b) competition with rate constant σ
 (c) mobility with rate constant ϵ

(2)   Calculate the overall rate of events $r = \mu + \sigma + \epsilon$
(3)   Select an individual at random
(4)   Select one of the focal individual's nearest neighbours at random
(5)   Select an interaction event with the following probabilities, by drawing a random number from the interval $[0, r]$:

 (a) reproduction with probability $\frac{\mu}{r}$
 (b) competition with probability $\frac{\sigma}{r}$
 (c) mobility with probability $\frac{\epsilon}{r}$

(6)   Execute the selected interaction event on the selected individual (if permitted) and determine the outcome according to the governing rules:

 (a) reproduction occurs deterministically (it is always carried out if possible)
 (b) mobility occurs deterministically
 (c) competition can occur:

  i. deterministically: the winner is determined by the appropriate entry (being 1 or -1) in the competition matrix M_λ or M_μ
  ii. non-deterministically: a random number r_c is drawn from the unit interval and compared to the appropriate winning probability M_{ij} in the competition matrix M_λ^* or M_μ^*, where species i and species j are competing.
  If $M_{ij} > 0$:

   • species i wins the competitive event if $r_c < M_{ij}$
   • species j wins the competitive event if $r_c > M_{ij}$

  If $M_{ij} < 0$:

   • species i wins the competitive event if $r_c > |M_{ij}|$
   • species j wins the competitive event if $r_c < |M_{ij}|$

(7)   Update the grid according to the outcome of step 6
(8)   Update the time to $t = t + 1$
(9)   Return to step 3 and continue until $t = t_{end}$

This procedure is repeated for a specified number of generations, where a generation is defined as the number of steps required for each cell to be the subject of on average one interaction.

3.2.2. Design Concepts

- **Emergence:** the spatial patterns and population-level dynamics of the community emerge naturally from the interactions occurring between individuals.
- **Competition based on pairwise interaction effects:** the competition scheme is constructed based on pairwise interaction effects, encoded in a competition matrix.
- **Non-deterministic competition:** In addition to deterministic competition, we also investigate the effects of non-deterministic competition, where the victor of any competition event is not predetermined but is instead stochastic.
- **Interactions:** individuals interact with each other and their environment by reproducing if located next to an empty site, exchanging sites with their neighbours, or competing with their neighbours.
- **Stochasticity:** the stochasticity in the model arises from the initial spatial distribution of the grid; the interactions between individuals and the environment (reproduction); the interactions between individuals (mobility, competition); and from the non-deterministic competition.
- **Sensing:** if selected for reproduction, individuals can sense whether their selected neighbouring site is empty; if so, they will reproduce. If the site is occupied by an individual, no reproduction will occur.
- **Observation:** the data collected from the IBM includes the population count of each species, the community evenness and diversity, the spatial distribution of individuals, and their time to extinction. These are tracked for each time step.

3.2.3. Details

Initialization

The model is initialized with a random spatial distribution of individuals and empty sites. Initially, a certain proportion of grid sites is left empty; thus the system is initially below carrying capacity. The initial species abundance distribution is completely even, as is the typical approach in similar modelling studies [37–39]. Aside from the input variables, all other parameters used to initialize the model are fixed for all simulations, and are shown in Table 1. Note in particular that the mobility rate constant ϵ is set below the system's critical mobility rate, above which extinctions are certain due to the interactions between individuals being insufficiently localized. It has been shown for models of this type that coexistence of all species is only possible when mobility remains low and therefore individuals can only interact over small spatial scales (in our case, with their nearest neighbours) [40].

Table 1. Parameters of the individual-based model of 13 SFIs.

Parameter	Description	Value
L	Grid side length	200
$\varnothing$	Initial proportion of empty sites	0.1
μ	Reproduction rate constant	1
σ	Competition rate constant	1
ϵ	Mobility rate constant	4.25
T	Number of generations evolved	1000

3.2.4. Input

The model's input is the competition matrix. There are four different matrices:

(i) M_λ: deterministic competition based on λ interaction effects (Matrix (2))

(ii) M_μ: deterministic competition based on μ interaction effects (Matrix (3))

(iii) M_λ^*: non-deterministic competition based on λ interaction effects (Matrix (4))

(iv) M_μ^*: non-deterministic competition based on μ interaction effects (Matrix (5))

For each of these initial settings, we run 200 replicate simulations.

3.3. In Silico Community Dynamics

3.3.1. Richness

To study the effects of the different types of competition on the diversity of the in silico synthetic community, we first examine the richness effects, by determining the number of surviving species after 1000 generations to see what levels of richness are maintained under the different competition structures.

In Figure 3, we show the probability of observing a certain species richness after 1000 generations. for deterministic and non-deterministic competition based on lag time λ interaction effects. With this competition structure, we observe monocultures very rarely in the deterministic case, and never in the non-deterministic case. We find final richness levels as high as eight (deterministic case) or nine species (non-deterministic case). In the deterministic case, approximately 70% of simulations result in communities of five or six species, and the same for the non-deterministic case. The distribution of final richness is more skewed towards higher richness values for the non-deterministic case, indicating a stabilizing effect on the dynamics in terms of fewer extinctions and thus higher richness, an effect observed in other modelling studies comparing deterministic and non-deterministic effects [31]. This effect is not surprising, since non-deterministic competition results in fewer prey extinctions and more predator extinctions compared to deterministic competition, and thus decreasing extinction probabilities of the most vulnerable species.

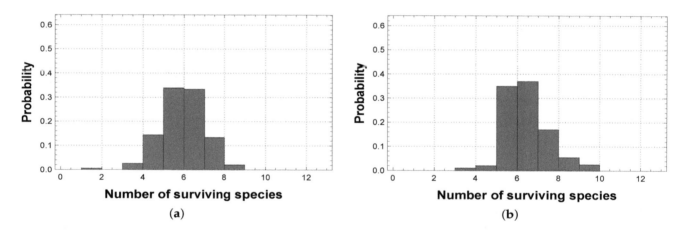

Figure 3. Probability of observing a particular species richness after 1000 generations for (**a**) deterministic and (**b**) non-deterministic competition based on λ effects. Probabilities calculated from 200 replicates.

The distribution of final richness for deterministic and non-deterministic competition based on mineralization rate μ interaction effects (Figure 4) is again more skewed towards higher richness values for the non-deterministic case, indicating a stabilizing effect on the dynamics in terms of fewer extinctions and thus higher richness. Additionally, higher richness levels are observed compared to the case of competition based on λ interaction effects. No monocultures are ever observed for competition based on μ interaction effects, and in fact community richness never drops below four (deterministic case) or five species (non-deterministic case). In the deterministic case, approximately 95% of simulations result in communities of five or six species, in the non-deterministic case approximately 95% of simulations result in communities of five, six or seven species.

Thus, in both cases (λ and μ interaction effects), we find similar behaviour in terms of community richness as was observed for the in vitro synthetic community of Vandermaesen et al. [11], namely the establishment of a stable community of reduced richness compared to the initial inoculation of 13 SFIs.

The increased in silico community diversity in the case of competition based on mineralization rate μ interaction effects, compared to lag time λ effects, can be ascribed to a more balanced competition structure in the former case, and more specifically its relatively higher intransitivity. A competition structure is transitive if the constituent species can be ranked in a strict competitive hierarchy, and hence intransitivity refers to the lack of such a strict hierarchy [41]. This characteristic can be quantified for example using a measure of relative intransitivity proposed by Laird and Schamp [42], denoted by R_I. This index takes values in the unit interval, with larger values corresponding to more intransitive competition structures. Using this index, we find that Matrix M_μ is more intransitive than Matrix M_λ, with a relative intransitivity of $R_I = 0.83$ compared to $R_I = 0.80$, respectively.

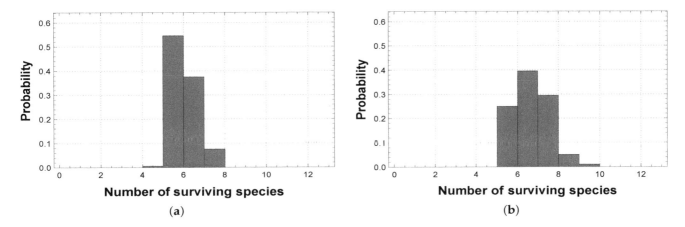

Figure 4. Probability of observing a particular species richness after 1000 generations for (a) deterministic and (b) non-deterministic competition based on μ effects. Probabilities calculated from 200 replicates.

3.3.2. Diversity

After observing the richness effects due to the different forms of competition, we now consider community diversity. We do so using the Leinster-Cobbold diversity index [43], an effective number index that also includes a sensitivity parameter q that determines how much weight is assigned to rare or common species. For $q < 1$, more weight is given to rare species ($q = 0$ corresponds to species richness), while for $q > 1$ more weight is given to common species. All species are weighed equally by their proportions for $q = 1$ [43].

For each of the four competition matrices, we calculate the Leinster-Cobbold diversity index over time, for different values of q, so that we may gather information about the composition and balance of the communities, as well as their changes in diversity as the different simulations evolve.

In Figure 5 we show the average Leinster-Cobbold diversity over time for deterministic and non-deterministic competition based on lag time λ interaction effects, for varying values of the sensitivity parameter q. With different values of q, we can infer changes in species richness (for low values of q), evenness (for high values of q) and diversity (for $q = 1$). Hence we calculate the diversity profiles for $q \in \{0, 1, 20\}$.

Initially, the community undergoes a sharp drop in evenness (seen in differences between the two curves for $q > 0$ relative to the $q = 0$ curve), while richness is maintained at its initial level. The time to the first species extinction is roughly similar for all replicates, namely around 250 generations. This period represents the time required for spiral spatial structures to begin to form (see Section 3.3.3), and the first species to be entirely surrounded by its predator(s) and killed off. Following the first extinction, others follow as they are enabled by the spatial structures as the species have aggregated sufficiently to begin to chase each other around the grid.

The $q = 1$ and $q = 20$ curves approach each other late in the simulation time, indicating that relatively high evenness is maintained for significant periods of time. However, the higher order diversities are significantly less than the zero order diversity (richness), indicating that in the later stages of the in silico experiments, multiple species continue to coexist but these communities are quite uneven, in agreement with the dynamics of the in vitro synthetic community [11]. Finally, we again observe a stabilizing effect when considering non-deterministic rather than deterministic in silico competition, in terms of time to first extinction and final community diversity.

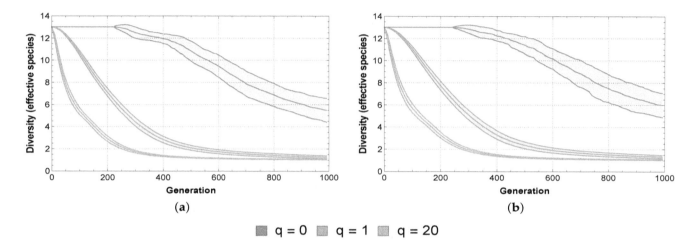

Figure 5. Mean diversity profiles (Leinster-Cobbold index) for $q \in \{0, 1, 20\}$, for (**a**) deterministic and (**b**) non-deterministic emergent competition based on λ effects. Mean and standard deviation calculated from 200 replicates.

In Figure 6, we compare the changes in diversity for communities subject to deterministic and non-deterministic competition based on mineralization rate μ interaction effects. Diversity is higher here than for the two previous competition matrices, for all values of q. Additionally, the communities are more even. Notably, in Figure 6 the $q = 1$ and $q = 20$ curves never overlap, indicating higher levels of evenness compared to the previous competition matrices which resulted in converging curves. This can also be inferred by the smaller distance between the $q = 0$ curve and the $q > 0$ curves in Figure 6, which indicates relatively more species coexisting in relatively more even communities. The minor stabilizing effect of non-deterministic competition compared to deterministic competition can also be observed in terms of diversity maintenance and time to first extinction.

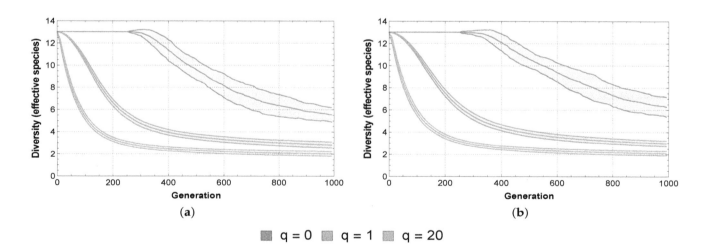

Figure 6. Mean diversity profiles (Leinster-Cobbold index) for $q \in \{0, 1, 20\}$, for (**a**) deterministic and (**b**) non-deterministic emergent competition based on μ effects. Mean and standard deviation calculated from 200 replicates.

3.3.3. Spatial Structures

These diversity effects, and the spatial dynamics underlying them, can also be observed in Figure 7, where we show two representative examples of the grid configuration at $T = 1000$ generations for non-deterministic competition based on lag time λ (Figure 7a) or mineralization rate μ (Figure 7b) interaction effects. As was observed in Figures 5 and 6, competition based on the former results in more uneven communities than competition based on the latter. In the former case, sufficient species are present in sufficient numbers to form the spiral patterns characteristic of this type of individual-based models, which have been shown to help maintain coexistence [40]. These patterns also qualitatively resemble those observed in in vitro experiments where a similar synthetic community of SFIs was co-cultured with MSH1 in the presence of BAM [18]. The spiral formations also enable spatial refuges, which have been observed to support species coexistence by allowing vulnerable species to persist at low but still significant levels [44]. Such refuges can be observed for example in Figure 7b for multiple species.

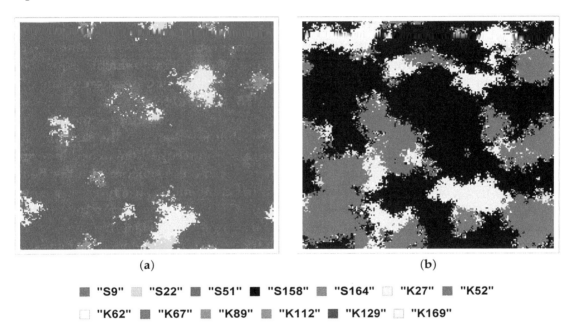

(a)   (b)

■ "S9"   ▩ "S22"   ▤ "S51"   ■ "S158"   ▩ "S164"   ▢ "K27"   ▦ "K52"
▢ "K62"   ▤ "K67"   ▦ "K89"   ▦ "K112"   ■ "K129"   ▢ "K169"

Figure 7. Examples of in silico communities at $T = 1000$ generations with emergent non-deterministic competition based on (**a**) λ effects, and (**b**) μ effects.

3.3.4. Community Composition

Having studied community diversity effects, we can now turn our attention to the composition of these persisting subcommunities. In Figure 8, we show the persistence probability for each SFI for deterministic and non-deterministic competition based on lag time λ interaction effects. The results reflect the dynamics illustrated in Figure 7a: S9 is the dominant strain, but it is a member of a subgroup of SFIs that are present in the majority of the simulations. This is unsurprising, since S9 was the strongest competitor in the two competition structures based on λ interaction effects (M_λ and M_λ^*) and thus it is the dominant SFI in the persisting subcommunity, which we recall is quite uneven (see e.g., Figures 5 and 7). In more than 80% of the simulations, we observe the same SFIs persisting together: S9, K67, K169, K27 and K89. This is true for both the deterministic and non-deterministic competition cases.

Thus our model is able to qualitatively reproduce the in vitro dynamics of a persisting smaller subcommunity [11]. These dynamics have also been observed for communities of microbial species [18] as well as for communities of higher organisms [45,46].

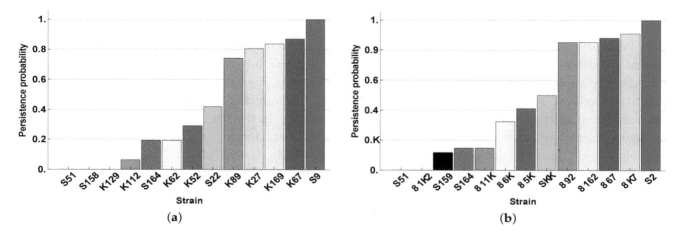

Figure 8. Probability of finding each strain in the community after 1000 generations) for (**a**) deterministic and (**b**) non-deterministic emergent competition based on λ effects.

Another subgroup of persisting SFIs is found for deterministic and non-deterministic competition based on mineralization rate μ interaction effects (Figure 9), once again matching qualitatively the dynamics observed in in vitro synthetic communities. The members of this subgroup are not entirely the same as for Figure 8. Instead we find K169, K52, S158, K27 and S9 coexisting in more than 80% of the simulations. The strains in the persisting subcommunity are also more equal in terms of their persistence probabilities (and hence their extinction probabilities) than was the case for competition based on λ interaction effects (Figure 8). These SFIs are also more equally matched in terms of their competitive strengths (see M_μ and M_μ^*). These factors result in these subcommunities being able to maintain significantly higher evenness levels than the other competition structures, as we noted when studying the diversity of these communities (Figure 6). The partial overlap in membership of the persisting subcommunities in the λ and μ cases may be ascribed to the fact that the dominance of a particularly adept competitor may be reflected in both the growth parameters under consideration.

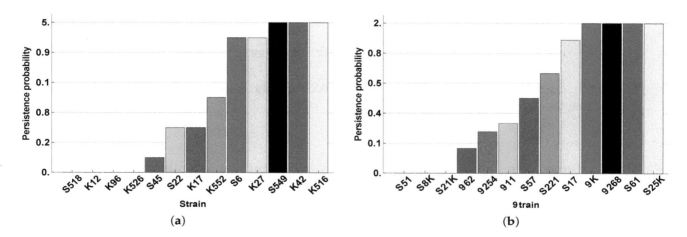

Figure 9. Probability of finding each strain in the community after 1000 generations) for (**a**) deterministic and (**b**) non-deterministic emergent competition based on μ effects.

Finally, we examine extinctions in our different communities. We have seen that extinctions are frequent, but generally limited to the same set of SFIs. In Figure 10, we show the average time to extinction for each SFI, for deterministic and non-deterministic competition based on lag time λ interaction effects. We note again that these are slightly longer for non-deterministic competition compared to deterministic competition, and always occur after an initial period of spiral formation (~300 generations). One strain, K129, collapses to extinction not long after spiral formation has commenced; this strain is the weakest in both competition structures. After it disappears, there is another lapse before extinctions recommence and thereafter proceed fairly regularly until the community is reduced to the persisting uneven subcommunity dominated by S9 (which never suffers any extinctions) and the other strains in small proportions.

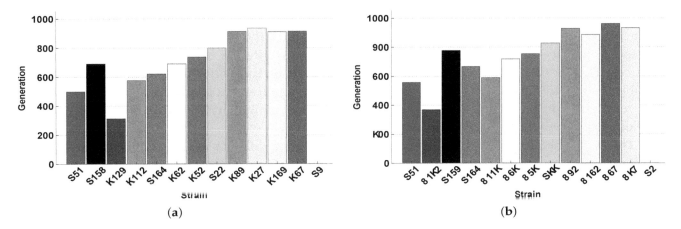

Figure 10. Mean time to extinction per strain for (**a**) deterministic and (**b**) non-deterministic emergent competition based on λ effects. Blue labels indicate strains for which no extinctions occurred. Means calculated from 200 replicates.

For deterministic and non-deterministic competition based on mineralization rate μ interaction effects (Figure 11), we notice a reduction in extinction times compared to competition based on lag time λ interaction effects. This may seem counterintuitive given that we have already observed these communities to be more stable, however the key point is that fewer species go extinct. Those that do collapse to extinction do so more quickly, but this does not affect the stability of the persisting subcommunity. Now S9 is not the only SFI to never suffer extinctions, but it is joined by the other members of the persisting subcommunity (S158, K52 and K169), again indicating that this subcommunity is more even and thus more stable than in the cases of competition based on λ interaction effects.

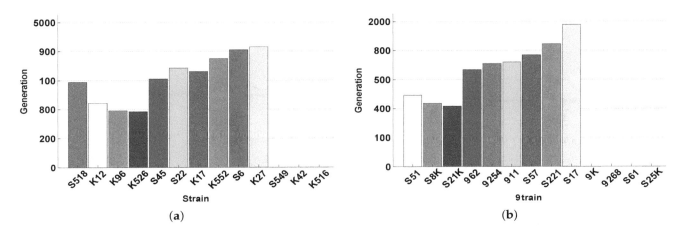

Figure 11. Mean time to extinction per strain for (**a**) deterministic and (**b**) non-deterministic emergent competition based on μ effects. Blue labels indicate strains for which no extinctions occurred. Means calculated from 200 replicates.

4. Conclusions

We have studied the in silico counterpart of an in vitro synthetic community of 13 SFIs in co-cultures of varying richness with MSH1. These SFIs had been selected based on their potential for improving the BAM mineralization performance of MSH1 for bioaugmentation applications. We developed an IBM representing the in silico counterpart of this synthetic community, where competition structures were constructed based on pairwise competition outcomes, using data related to lag time λ and mineralization rate μ interaction effects in terms of mineralization performance.

Our model was able to recover the qualitative dynamics observed in in vitro experiments with similar synthetic sand filter communities: the majority of the community collapsing to extinction and a subcommunity persisting [11,18]. The memberships of these subcommunities were consistent, and their presence could be explained based on their attributes as represented in the competition matrices. The simulation outcomes were explained based on the underlying competition structures (notably by their intransitivity) and the resulting spatial dynamics. Our results highlight the importance of diversity and in particular evenness in stabilizing the community dynamics, in agreement with previous experimental results [47,48].

This work therefore serves as a proof-of-concept for using IBMs as in silico counterparts of in vitro synthetic communities, as we were able to find a qualitative agreement between the in silico and in vitro dynamics, despite the in vitro experiments not being expressly designed for modelling purposes. For example, it would also be informative for this purpose to examine in more detail the interactions between the SFIs, not just in terms of effects on BAM mineralization. This could be done, for example, by tracking the growth and survival of SFIs in pairwise co-cultures. Despite this, our model was able to retrieve the qualitative in vitro dynamics, allowing us to interrogate their development, and thereby illustrating the potential of this modelling approach for addressing ecological theories relating to synthetic communities.

Acknowledgments: The authors gratefully acknowledge the financial support of the Belgian Science Policy Office (IUAP contract P7/25) and the FP7 project BIOTREAT (EU grant 266039).

Author Contributions: A.J.D., J.M.B. and B.D.B. conceived and designed the in silico experiments; A.J.D. performed the experiments and analyzed the data; J.V., N.B. and D.S. contributed data and analysis tools; A.J.D. wrote the paper.

Appendix A

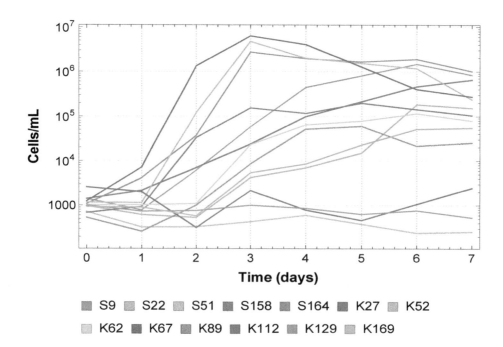

Figure A1. SFI monoculture growth curves on acetate [11].

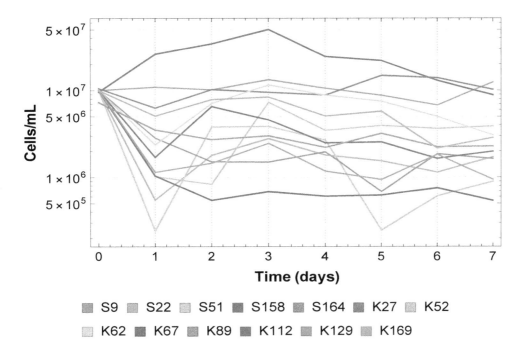

Figure A2. SFI monoculture survival curves on acetate [11].

References

1. Bairey, E.; Kelsic, E.; Kishony, R. High-order species interactions shape ecosystem diversity. *Nat. Commun.* **2016**, *7*, doi:10.1038/ncomms12285.

2. Faust, K.; Raes, J. Microbial interactions: From networks to models. *Nat. Rev. Microbiol.* **2012**, *10*, 538–550.

3. Goers, L.; Freemont, P.; Polizzi, K. Co-culture systems and technologies: Taking synthetic biology to the next level. *J. R. Soc. Interface* **2014**, *11*, doi:10.1098/rsif.2014.0065.

4. De Roy, K.; Marzorati, M.; Van den Abbeele, P.; Van de Wiele, T.; Boon, N. Synthetic microbial ecosystems: An exciting tool to understand and apply microbial communities. *Environ. Microbiol.* **2014**, *16*, 1472–1481.

5. Verstraete, W.; Wittebolle, L.; Heylen, K.; Vanparys, B.; De Vos, P.; Van de Wiele, T.; Boon, N. Microbial resource management: The road to go for environmental biotechnology. *Eng. Life Sci.* **2007**, *7*, 117–126.

6. Hairston, D.; Huban, C.; Plowman, R.; Chemicals, S. Bioaugmentation: Put microbes to work. *Chem. Eng.* **1997**, *104*, 74–85.

7. European Union. Directive 2006/118/EC of the European Parliament and of the council of 12 December on the protection of groundwater against pollution and deterioration. *Off. J. Eur. Union* **2006**, *372*, 19–31.

8. Björklund, E.; Anskær, G.; Hansen, M.; Styrishave, B.; Halling-Sørensen, B. Analysis and environmental concentrations of the herbicide dichlobenil and its main metabolite 2, 6-dichlorobenzamide (BAM): A review. *Sci. Total Environ.* **2011**, *409*, 2343–2356.

9. Sørensen, S.; Holtze, M.; Simonsen, A.; Aamand, J. Degradation and mineralization of nanomolar concentrations of the herbicide dichlobenil and its persistent metabolite 2, 6-dichlorobenzamide by *Aminobacter* spp. isolated from dichlobenil-treated soils. *Appl. Environ. Microbiol.* **2007**, *73*, 399–406.

10. Thompson, I.; van der Gast, C.; Ciric, L.; Singer, A. Bioaugmentation for bioremediation: The challenge of strain selection. *Environ. Microbiol.* **2005**, *7*, 909–915.

11. Vandermaesen, J. Pesticide mineralization and effect of endogeneous community diversity on bioaugmentation in sand filters used in drinking water treatment. Ph.D. Thesis, KU Leuven, Leuven, Belgium, 2016.

12. Esser, D.; Leveau, J.; Meyer, K. Modelling microbial growth and dynamics. *Appl. Microbiol. Biotechnol.* **2015**, *99*, 8831–8846.

13. Seshan, H.; Goyal, M.; Falk, M.; Wuertz, S. Support vector regression model of wastewater bioreactor performance using microbial community diversity indices: Effect of stress and bioaugmentation. *Water Res.* **2014**, *53*, 282–296.

14. Poschet, F.; Vereecken, K.; Geeraerd, A.; Nicolaï, B.; Van Impe, J. Analysis of a novel class of predictive microbial growth models and application to coculture growth. *Int. J. Food Microbiol.* **2005**, *100*, 107–124.

15. Widder, S.; Allen, R.; Pfeiffer, T.; Curtis, T.; Wiuf, C.; Sloan, W.; Cordero, O.; Brown, S.; Momeni, B.; Shou, W.; et al. Challenges in microbial ecology: Building predictive understanding of community function and dynamics. *ISME J.* **2016**, *10*, 1–12.

16. Nadell, C.; Foster, K.; Xavier, J. Emergence of spatial structure in cell groups and the evolution of cooperation. *PLoS Comput. Biol.* **2010**, *6*, doi:10.1371/journal.pcbi.1000716.

17. Momeni, B.; Waite, A.; Shou, W. Spatial self-organization favors heterotypic cooperation over cheating. *eLIFE* **2013**, *2*, doi:10.7554/elife.00960.

18. Horemans, B.; Vandermaesen, J.; Sekhar, A.; Springael, D. *Aminobacter* sp. MSH1 invades sand filter community biofilms while retaining 2,6-dichlorobenzamide degradation functionality under C and N limiting conditions. *FEMS Microbiol. Ecol.* **2017**, *93*, doi:10.1093/femsec/fix064.

19. Lehtola, M.; Miettinen, I.; Vartiainen, T.; Martikainen, P. Changes in content of microbially available phosphorus, assimilable organic carbon and microbial growth potential during drinking water treatment processes. *Water Res.* **2002**, *36*, 3681–3690.

20. Zwietering, M.; Jongenburger, I.; Rombouts, F.; Van't Riet, K. Modelling of the bacterial growth curve. *Appl. Environ. Microbiol.* **1990**, *56*, 1875–1881.

21. Buchanan, R.; Whiting, R.; Damert, W. When is simple good enough: A comparison of the Gompertz, Baranyi, and three-phase linear models for fitting bacterial growth curves. *Food Microbiol.* **1997**, *14*, 313–326.

22. Coleman, T.; Li, Y. On the convergence of interior-reflective Newton methods for nonlinear minimization subject to bounds. *Math. Program.* **1994**, *67*, 189–224.

23. Coleman, T.; Li, Y. An interior trust region approach for nonlinear minimization subject to bounds. *SIAM J. Optim.* **1996**, *6*, 418–445.

24. Vandermeeren, P.; Baken, S.; Vanderstukken, R.; Diels, J.; Springael, D. Impact of dry-wet and freeze-thaw events on pesticide mineralizing populations and their activity in wetland ecosystems: A microcosm study. *Chemosphere* **2016**, *146*, 85–93.

25. Ekelund, F.; Harder, C.; Knudsen, B.; Aamand, J. Aminobacter MSH1-mineralisation of BAM in sand-filters depends on biological diversity. *PLoS ONE* **2015**, *10*, e0128838.

26. Tonner, P.; Darnell, C.; Engelhardt, B.; Schmid, A. Detecting differential growth of microbial populations with Gaussian process regression. *Genome Res.* **2017**, *27*, 320–333.

27. Tukey, J. Comparing individual means in the analysis of variance. *Biometrics* **1949**, *5*, 99–114.

28. Haynes, W. Tukey's Test. In *Encyclopedia of Systems Biology*; Dubitzky, W., Wolkenhauer, O., Cho, K.H., Yokota, H., Eds.; Springer: New York, NY, USA, 2013; pp. 2303–2304.

29. Keuls, M. The use of the "studentized range" in connection with an analysis of variance. *Euphytica* **1952**, *1*, 112–122.

30. Mullon, C.; Fréon, P.; Cury, P.; Shannon, L.; Roy, C. A minimal model of the variability of marine ecosystems. *Fish Fish.* **2009**, *10*, 115–131.

31. Planque, B.; Lindstrom, U.; Subbey, S. Non-deterministic modelling of food-web dynamics. *PLoS ONE* **2014**, *9*, e108243.

32. Lindstrom, U.; Planque, B.; Subbey, S. Multiple Patterns of Food Web Dynamics Revealed by a Minimal Non-deterministic Model. *Ecosystems* **2017**, *20*, 163–182.

33. Ulrich, W.; Soliveres, S.; Kryszewski, W.; Maestre, F.; Gotelli, N. Matrix models for quantifying competitive intransitivity from species abundance data. *Oikos* **2014**, *123*, 1057–1070.

34. Moon, J. *Topics on Tournaments in Graph Theory*; Courier Dover Publications: Mineola, NY, USA, 2015.

35. Grimm, V.; Berger, U.; Bastiansen, F.; Eliassen, S.; Ginot, V.; Giske, J.; Goss-Custard, J.; Grand, T.; Heinz, S.; Huse, G.; et al. A standard protocol for describing individual-based and agent-based models. *Ecol. Model.* **2006**, *198*, 115–126.

36. Gillespie, D. A general method for numerically simulating the stochastic time evolution of coupled chemical reactions. *J. Comput. Phys.* **1976**, *22*, 403–434.

37. Case, S.O.; Durney, C.H.; Pleimling, M.; Zia, R. Cyclic competition of four species: Mean-field theory and stochastic evolution. *EPL Eur. Lett.* **2010**, *92*, doi:10.1209/0295-5075/92/58003.

38. Cheng, H.; Yao, N.; Huang, Z.G.; Park, J.; Do, Y.; Lai, Y.C. Mesoscopic interactions and species coexistence in evolutionary game dynamics of cyclic competitions. *Sci. Rep.* **2014**, *4*, 1–7.

39. Frachebourg, L.; Krapivsky, P.L.; Ben-Naim, E. Spatial organization in cyclic Lotka-Volterra systems. *Phys. Rev.* **1996**, *54*, 6186–6200.

40. Reichenbach, T.; Mobilia, M.; Frey, E. Mobility promotes and jeopardizes biodiversity in rock-paper-scissors games. *Nature* **2007**, *448*, 1046–1049.

41. Laird, R.; Schamp, B. Species coexistence, intransitivity, and topological variation in competitive tournaments. *J. Theor. Biol.* **2009**, *256*, 90–95.

42. Laird, R.; Schamp, B. Competitive intransitivity, population interaction structure, and strategy coexistence. *J. Theor. Biol.* **2015**, *365*, 149–158.

43. Leinster, T.; Cobbold, C. Measuring diversity: The importance of species similarity. *Ecology* **2012**, *93*, 477–489.

44. Daly, A.; Baetens, J.; De Baets, B. The impact of resource dependence of the mechanisms of life on the spatial population dynamics of an in silico microbial community. *Chaos* **2016**, *26*, 123121.

45. Dunne, J.; Williams, R. Cascading extinctions and community collapse in model food webs. *Philos. Trans. R. Soc. Lon. B Biol. Sci.* **2009**, *364*, 1711–1723.

46. Ebenman, B.; Jonsson, T. Using community viability analysis to identify fragile systems and keystone species. *Trends Ecol. Evol.* **2005**, *20*, 568–575.

47. De Roy, K.; Marzorati, M.; Negroni, A.; Thas, O.; Balloi, A.; Fava, F.; Verstraete, W.; Daffonchio, D.; Boon, N. Environmental conditions and community evenness determine the outcome of biological invasion. *Nat. Commun.* **2013**, *4*, doi:10.1038/ncomms2392.

48. Daly, A.J.; Baetens, J.M.; De Baets, B. The impact of initial evenness on biodiversity maintenance for a four-species in silico bacterial community. *J. Theor. Biol.* **2015**, *387*, 189–205.

10

Modeling Biofilms: From Genes to Communities

Tianyu Zhang

Department of Mathematical Sciences, Montana State University, Bozeman, MT 59717, USA;
tianyu.zhang@montana.edu

Academic Editor: Hyun-Seob Song

Abstract: Biofilms are spatially-structured communities of different microbes, which have a huge impact on both ecosystems and human life. Mathematical models are powerful tools for understanding the function and evolution of biofilms as diverse communities. In this article, we give a review of some recently-developed models focusing on the interactions of different species within a biofilm, the evolution of biofilm due to genetic and environmental causes and factors that affect the structure of a biofilm.

Keywords: biofilm; mathematical modeling; gene; community

1. Introduction

Despite the common view of microbes in their free state, pure culture planktonic growth is rarely how microbes exist in nature. Instead, most microbial species in nature live in the form of biofilms, which are described as a multicellular consortium of microbial cells that are attached to a surface and encased in a self-secreted, extracellular polymeric matrix [1,2]. Biofilms are usually heterogeneous in both their spatial structures and component species and interact with the surrounding environment in a complicated way.

Microbial biofilms are ubiquitous in both natural and industrial settings, and bacteria living in a biofilm often behave very differently from their planktonic counterpart. Bacteria inside a biofilm are usually more resistant to antimicrobial agents [3] and usually possess big competitive advantage over bacteria growing in suspension. This means that it is often difficult to remove biofilms efficiently.

Biofilms can cause many severe problems, such as chronic infections, food contamination and equipment damage due to bio-fouling. Biofilms can also be used for good and constructive purposes, such as waste water treatment, heavy metal removal from hazardous waste sites, biofuel production and microbial fuel cells. From a neutral point of view, since much of the microbial biomass appears in the form of biofilm and due to their ability to produce and consume organic materials, the biofilm communities also have a big impact on the global ecosystem and geochemical system.

In order to promote good biofilms and prevent bad biofilms, it is important to understand the mechanisms for biofilm formation, growth and its removal. The development of a biofilm is a complicated process affected by many biological, physical and chemical factors, and understanding it requires both experimental and modeling efforts. Experiments provide directly-measured qualitative or quantitative data of biofilm properties that are of interest, such as cell counts, cell viability, biofilm morphology and EPS structure, nutrient profile, as well as genetic information. A mathematical model translates the conceptual understanding of the biofilm system into mathematical terms, usually by combining the important processes involved, but omitting the less important ones, and the solutions (either analytical or numerical) are obtained by using available mathematical or statistical tools. Since a model can connect different processes and assess their relative importance, modeling results can help us to understand the biofilm system, facilitate experimental design and make predictions that can be tested by experiments. In this sense, progresses made in experimental and modeling research always promote the development of each other.

Mathematical modeling of biofilm started in the 1970s with models studying substrate utilization and mass transport in a homogeneous slab of biofilm [4,5]. In the 1980s, models including multispecies and the non-uniform distribution of different biomass types started to emerge [6,7], but they were still primarily for one space dimension and steady-state growth dynamics. Starting in the 1990s and up to today, aided by the fast advancement in computing power and better understanding of biofilms through experimental data, multidimensional, multispecies, multisubstrate models are being developed to incorporate realistic biofilm morphology, biofilm mechanics, interactions between biofilm and the environment and interactions between different species within a biofilm, as well as various time scales involved in biofilm-related processes [8–12].

There is a rich literature on the review of mathematical models for biofilms. A few representative ones are listed below.

The IWAtask group gives an excellent review of mathematical modeling of biofilms [13]. The book explains the basic steps in creating a mathematical model, emphasizes that the "golden rule" of modeling is that "a model should be as simple as possible, and only as complex as needed" and presents the model derivation based on mass conservation in detail. Models are classified into analytical models (A), Pseudo-Analytic models (PA), Numerical one-dimensional (N1) and multi-dimensional Numerical models (N2 and N3). The features, definitions and equations, as well as the application of each type of model are discussed. The performances of all models for solving three characteristic benchmark problems are compared, which help identify the trade-offs inherent to using different types of models. The book also points out the significance of the definitions and units of model parameters.

Klapper and Dockery [14] discuss how macroscale physical factors might influence the composition, structure and function of ecosystems within microbial communities from the modeling perspective and emphasizes that despite its difficulty and complexity, it is important to include the physical, chemical and biological processes at a variety of time and length scales in the model to fully understand the physiology and ecology of the microbial communities. Specific modeling aspects discussed include quorum sensing, growth, mechanics and antimicrobial tolerance mechanisms.

Wang and Zhang [15] give a chronological review of some biofilm models developed from the 1980s to the early 2000s. Based on their dimensionality, the way in which diffusion is treated and the complexity in terms of the incorporation of the physics, chemistry and biological effects, models are classified into four main categories: one-dimensional continuum models, diffusion-limited aggregate models, continuum-discrete diffusion models and biofilm-fluid coupled models.

In recent years, fast advancement in experimental technologies, such as microscopy and high-throughput sequencing, has provided an abundance of data at both the genetic and community level and helped researchers to understand biofilms much better. In particular, it is a common belief that biofilms should be viewed as spatially-structured communities of microbes, and the structure and function of the communities are determined by both the surrounding environment and the local interactions between different species within biofilms via complex metabolic networks [16–20]. To understand biofilms as a diverse community and the evolution of different species and its genetic causes, mathematical modeling again serves as a powerful tool. Experimental data with a high level of detail provide both opportunities and challenges for researchers working on biofilm modeling. On the one hand, there are more data available to improve conceptual understanding of biofilm and to compare with model predictions; on the other hand, more sophisticated models are demanded in order to accommodate the data. Song et al. [21] give a methodological review on mathematical modeling of microbial community dynamics. Widder et al. [22] address the challenges in building predictive models for understanding the function and dynamics of Microbial Communities (MCs). Several specific examples where model-experiment integration has already resulted in important insights into MC function and structure are discussed. These include inferring species interactions from proximal data, predicting species interactions using stoichiometric models and kinetic models for community dynamics. The conclusion is that addressing this challenge requires close coordination of experimental data collection and method development with mathematical model building.

In this article, we review some recent mathematical models that focus on studying biofilms as a diverse community. The rest of the article is organized as follows. Section 2 discusses the genetic basis for biofilm development based on experimental results and its mathematical modeling with emphasis on the models based on the idea of Quorum Sensing (QS); Section 3 discusses models based on Flux Balance Analysis (FBA) and stoichiometry; Section 4 discusses models based on statistical inference; Section 5 discusses models with novel growth kinetics and the ability to resolve the complex spatial structure of biofilm. Table 1 gives a brief overview of the models discussed in the article.

Table 1. Summary of biofilm models discussed in the article. QS, Quorum Sensing; FBA, Flux Balance Analysis; IbM, Individual-based Model.

Model Category	Specific Models/Mathematical Tools	Biofilm Aspects Modeled
Genetic modeling	gene-centric model trait-based model QS model	biofilm structure, genetic composition species interaction, ecosystem diversity species interaction, biofilm structure
FBA model	constraint optimization problem	species interaction, biofilm structure
Statistical inference model	similarity-based method, regression-based method	species interaction, community stability
Kinetic growth model	IbM continuum model	biofilm structure, interaction with environment, mechanical property

2. Modeling the Genetic Basis of Biofilm Development

Costerton et al. [3] pointed out that biofilms consist of microcolonies on a surface and that within these microcolonies, the bacteria have developed into organized communities with functional heterogeneity. Clinical characteristics of biofilm infections are discussed, and multiple mechanisms of biofilm resistance to antimicrobial agents are proposed. Furthermore, *P. aeruginosa* and the chronic lung infections it causes in most patients afflicted with the recessive genetic disease Cystic Fibrosis (CF) are used as a model to reveal information about the molecular and genetic basis of biofilm development. There is evidence [23,24] showing that during the attachment phase of biofilm development, the transcription of specific genes (such as the genes required for the synthesis of the extracellular polysaccharide) is activated. Research on quorum sensing in Gram-negative bacteria [25,26] has shown that acyl homoserine lactone signals are produced by individual bacterial cells. At a critical cell density, these signals can accumulate and trigger the expression of specific sets of genes. Detachment and dispersal of planktonic cells from biofilms could also have a genetic basis. It has been suggested that increased expression of the alginate lyase in the mucoid strain of *P. aeruginosa* led to alginate degradation and increased cell detachment [27,28]. Antibiotic therapy in patients colonized with *P. aeruginosa* often gives a measure of relief from symptoms, but fails to cure the basic ongoing infection [29,30]. One interpretation of this is that the antibiotics act on the planktonic cells that are shed by the biofilms, but cannot eliminate the antibiotic-resistant sessile biofilm communities, and the microcolonies of sessile bacteria in the lung act as niduses for the spread of the infection [31,32]. The conclusion from [3] is that the effective control of biofilm infections will require a concerted effort to develop therapeutic agents that target the biofilm phenotype and community signaling-based agents that prevent the formation, or promote the detachment, of biofilms.

Monds and O'Toole [33] give a critical review of the causal basis of biofilm formation and its molecular underpinnings. It discusses the concept of biofilm formation as a developmental process by evaluating experimental data and concludes that the developmental model of biofilm formation must be approached as a model in need of further validation, rather than coveted as a robust platform on which to base scientific inference. Here, the definition of developmental process is a "series of stable and meta-stable changes in the form and function of a cell, where those changes are part of the normal life cycle of the cell" according to [34]. Furthermore, the molecular requirement implicit in a developmental process is that a series of hierarchically-ordered genetic elements control temporal transition through the developmental pathway in response to specific cues. The review starts with the origin of the developmental model of biofilm formation as an analogy with *Myxococcus xanthus* fruiting-body formation [35], then provides some unequivocal support for the developmental model, including both structural transitions occurring during biofilm formation [36] and various phenotypes with biofilm-specific properties, such as increased antibiotic tolerance [37]. After that, two specific case studies are presented as evidence that bacteria have evolved genetic pathways that serve to directly link environmental cues to the regulation of stage-specific transitions in biofilm formation. One case is that low extracellular phosphate blocks microcolony formation by *Pseudomonas fluorescens* [38,39]; the other case is intracellular iron as a signal for biofilm maturation [40,41]. However, there is still lack of success in uncovering comprehensive genetic programs specific for the regulation of biofilm development [42,43]. Moreover, the development model also requires that groups of biofilm pathways are connected in a hierarchically-ordered genetic network, and there are causal links between form and function. Folkesson et al. [44] provide valuable evidence in this direction by showing that *Escherichia coli* biofilms formed by F-plasmid-containing cells were more structured and had increased tolerance to colistin relative to cells without the F-plasmids, but it is still not possible to say that pathways controlling the structural development of an *E. coli* biofilm are directly coupled with pathways for the formation of a subpopulation of cells with increased tolerance to colistin. Next, two examples are used to demonstrate biofilms as multicellular organisms with functional differentiation and alternative cell fates. The first one by Klausen et al. [45] investigated differential roles for motile and non-motile subpopulations in determining the topology (mushroom-like or flat) of *P. aeruginosa* biofilms. Burrows [46] gave a thorough review on the twitching motility of *P. aeruginosa* through Type IV pili (T4P) structure and function. The second one [47] examined the spatiotemporal patterns of cell-specific gene expression in *B. subtilis* biofilms and indicated that different cell types vary in abundance and location in the biofilm over time. Finally, an alternative experimental model for biofilm formation based on the ecological adaptation of individuals was proposed by Klausen et al. [48]. In this experiment, deterministic responses are integrated with stochastic interactions with the environment to shape biofilm form and function, where biofilm evolution has been driven by the selection for individual competitiveness in complex and dynamic environments.

Many models have been proposed to describe the genetic processes that regulate biofilm development, and a few representative ones are discussed below.

Reed et al. [49] proposed the gene-centric approach for integrating environmental genomics and biogeochemical models. In this model, the production rate or *j*-th gene is given by:

$$R_j = \Gamma_j \cdot F_T \cdot \mu_j \cdot \Pi_s \left(\frac{C_s}{K_s + C_s} \right) \cdot \Pi_x \left(\frac{K_x}{K_x + C_x} \right), \tag{1}$$

where Γ_j is gene abundance (genes per unit volume), F_T is the thermodynamic potential factor accounting for the chemical energy available to drive the metabolism, μ_j is the specific growth rate, C_s is the concentration of a reactant or nutrient s, K_s is the half-saturation constant of the reactant or nutrient s, C_x is the concentration of inhibitor x and K_x is the half-saturation constant of inhibitor x. Furthermore, metabolic plasticity, whereby growth via one metabolism can lead to the propagation of functional genes associated with other metabolisms, is incorporated into the model by introducing the following governing equation for the gene abundance:

$$\frac{d\Gamma_i}{dt} = \Sigma_j \left(\frac{n_i}{n_j} \cdot \sigma_{ij} \cdot R_j \right) - \lambda \cdot \Gamma_i, \qquad (2)$$

where n_i is the number of the i-th gene per unit mass of cells that contains this gene, σ_{ij} is a probabilistic measure of the co-occurrence of genes i and j within a genome and λ is the mortality rate constant of a gene. Equations (1)–(2) describing the microbial community are coupled to chemical dynamics by usual reaction equations. There are several advantages of this model: most of its parameters are either directly measurable (μ_j, K_s) or easily obtained by calculation (F_T); numerical solutions from the model give gene abundances and chemical concentrations that allow direct comparisons between model predictions and experimental results; and the metabolic plasticity could be important for understanding the complex microbial community dynamics. Zhang et al. [50] developed a theory for the analysis and prediction of the spatial and temporal patterns of gene and protein expression within microbial biofilms based on similar ideas. The theory integrates the phenomena of solute reaction and diffusion, microbial growth, mRNA or protein synthesis, biomass advection and gene transcript or protein turnover. Case studies illustrate the capacity of the theory to simulate heterogeneous spatial patterns and predict microbial activities in biofilms that are qualitatively different from those of planktonic cells.

Genetic changes via horizontal gene transfer often make taxonomic distinction among species obscure; thus, sometimes, it is convenient to characterize the dynamics of microbial communities by different traits. Trait-based models were developed based on this idea and applied for analyzing the diversity of ecosystems. Shipley et al. [51] developed the maximum entropy (MaxEnt) model, and Laughlin et al. [52] developed the the Traitspace model; both have been applied to predict the relative abundance of species for plant communities. Both approaches are based on statistical methods and are composed of three key elements: an underlying trait distribution, a performance filter defining the fitness of traits in different environments and a projection of the performance filter along some environmental gradient. The objective of the modeling is to estimate the relative abundance of the some species in a given environment by incorporating information about individual-level functional traits. The MaxEnt model tends to overestimate the relative abundance of species since it maximizes the evenness of their distribution. On the other hand, the Traitspace model tends to underestimate the relative abundance of species since it is based on Bayesian theory and predicts a low probability of abundances for functional groups that do not pass through environmental filters. Though the trait-based models have not be widely adopted for modeling of the biofilm yet, it is certainly promising to apply the methodology to study biofilm as a microbial community. For example, Lennon et al. [53] have found that certain traits are related to the biofilm-producing capability of strains and identified functional groups of microorganisms that will help predict the structure and functioning of microbial communities under contrasting soil moisture regimes. Furthermore, current biofilm models can provide accurate predictions of the environmental gradients inside the biofilm, which can be used as the input of the trait-based models.

Quorum sensing is the regulation of gene expression in response to fluctuations in cell-population density [54–56]. Quorum sensing bacteria produce and release chemical signal molecules called autoinducers, and the concentrations of autoinducers increase as the cell density increases. Once the concentration of an autoinducer reaches a threshold, an alteration in gene expression is triggered. Recent research on many different bacterial species has shown that quorum sensing systems play an important role in regulating the expression of genes involved in biofilm formation, biofilm maturation, biofilm dispersal and detachment [57–59]. Naturally, modeling of QS is an important part of the general effort in modeling the genetic processes involved in biofilm development. Ward [60] gives a good review of early mathematical modeling of QS.

The modeling of QS starts with a circuit describing the gene regulation involved in the QS system, which is usually given by a schematic diagram showing all of the genes, autoinducers and the corresponding positive and negative interactions. For example, the *las* and *rhl* systems in

P. aeruginosa [61] are extensively studied. James et al. [62] and Dockery and Keener [63] pioneered the work on modeling QS at the molecular level. These early QS models use a system of coupled Ordinary Differential Equations (ODEs) to describe the dynamics of the intracellular concentrations of genes (or proteins), autoinducers and substrates, where the reaction kinetics are carefully designed to reflect the interactions within the QS system. Due to their simple forms, these models can be investigated by both numerical and analytical tools, and results suggest that QS works as a biochemical switch between two stable steady states of the system, one with low levels of autoinducer and one with high levels of autoinducer.

The signal production in QS in biofilm can be affected by many physical, chemical and biological factors [64]. Examples include diffusion of nutrients and QS molecules inside the biofilm and mass transfer affected by the hydrodynamics of the bulk fluid and biofilm structure. For example, Kirisits et al. [65] studied the influence of the hydrodynamic environment on QS in a *P. aeruginosa* biofilm and concluded that the amount of biofilm biomass required for full QS induction of the population increases as the flow rate increases.

These more advanced QS models use either the continuum approach with Partial Differential Equations (PDEs) or the individual-based approach, both capable of capturing the spatial structure of the biofilm and its interaction with the surrounding environment, to study the effect of QS on either the biofilm structure or interactions among species within a biofilm. A continuum model involving QS will be discussed in Section 5, and here, we describe the work of Nadell et al. [66], which implemented detailed simulations using individual-based modeling methods [67–69] to investigate evolutionary competitions between strains that differ in their polymer production and quorum-sensing phenotypes. It is known that EPS secretion in the process of biofilm formation is under quorum-sensing control in a number of bacterial model systems in very different ways. For example, *P. aeruginosa* activates EPS production at high cell density [70]. In contrast, *V. cholerae* initiates EPS secretion after attaching to a surface and losing flagellar activity, but halts EPS secretion once it reaches its high cell density quorum-sensing threshold [71]. The model presented in [66] focuses on three strains with the following behavior: (1) no polymer secretion and no quorum sensing (EPS$^-$); (2) constitutive polymer secretion and no quorum sensing (EPS$^+$); and (3) polymer secretion under negative quorum-sensing control, such that EPS secretion stops at high cell density (QS$^+$). Cells consume substrate according to their strain-specific metabolism kinetics and produce additional biomass; all cells secrete an autoinducer without cost and at a constant rate, and QS$^+$ cells synthesize EPS only when the local autoinducer concentration is below the quorum-sensing threshold concentration, which is represented by a single dimensionless parameter. Results from Nadell et al. [66] suggest that QS$^+$ cells have a competitive advantage over EPS$^+$, but only for a limited time window. In contrast, QS$^+$ cells suffer an initial disadvantage due to a lower growth rate when competing with EPS$^-$ cells, then rapidly ascend to a majority in the biofilm and remain there indefinitely. In addition, the QS$^+$ strain can invade populations composed mostly of either EPS$^+$ or EPS$^-$ cells, but not vice versa. The importance of the work in [66] is that it provides an evolutionary model that can be used to make predictions on the evolution of specific biological outcomes based on the biological constraints, and these predictions can be tested by experiments. In particular, it predicts that pathogenic strains, such as *V. cholerae*, selected for rapid colonization of, and efficient dispersal from, human hosts or other temporary environments, will exhibit negative quorum-sensing-regulated EPS production. In contrast, upregulation of EPS secretion at high cell density, which focuses resource investment into sustained local competitive ability, is more likely to be favored for organisms occupying specific niches long term, such as *P. aeruginosa* in chronic infections.

Mathematical tools used for modeling genetic processes related to biofilm include differential equations, statistical methods and individual-based approaches. These models usually enjoy success, but face challenges at the same time. For example, it is challenging to apply the gene-centric model to complex ecological systems since it is not easy to obtain associations between functional genes and reactions. Among many models for studying QS in biofilm systems, most of them focus on upregulation and downregulation of certain genes, and only a few emphasize the effect of QS on biofilm structure, function and its interaction with the environment, which leaves much room for model improvement.

3. Models Based on Flux Balance Analysis and Stoichiometry

Interactions of different species within a biofilm are closely related to the substrate consumption and metabolite exchange, and mathematical models based on FBA are excellent tools for predicting these interactions.

Early work in this direction involves the synthesis of metabolic pathways. Seressiotis and Bailey [72] developed a computer software system for metabolic pathway synthesis, which can be used to identify biochemical pathways, to predict on a qualitative basis the effects of adding or deleting enzymatic activities to or from the cellular environment, to classify pathways with respect to cellular objectives and to extract information about metabolic regulation. Mavrovouniotis et al. [73] extended the work in [72] by including stoichiometric constraints. Schilling et al. [74] gave a review on the development of computer-aided algorithms for the synthesis of metabolic pathways and explained the important algebraic concepts used in pathway analysis, such as null space and convex cone.

Orth et al. [75] covered the theoretical basis of FBA and provide several practical examples and a software toolbox for performing the calculations. Figure 1 from [75] explains the conceptual basis of FBA as a constraint optimization problem. In FBA, metabolic reactions are represented as a stoichiometric matrix ($\mathbf{S}$) of size $m \times n$. Each row of $\mathbf{S}$ represents one unique compound (for a system with m compounds), and each column represents one reaction (n reactions). The entry S_{ij} of $\mathbf{S}$ is the stoichiometric coefficient denoting the number of moles of the i-th compound formed in the j-th reaction. The coefficient is positive if the metabolite is produced, negative if the metabolite is consumed and zero if the metabolite does not participate in a particular reaction. The flux through all of the reactions in a network is represented by the vector $\mathbf{v}$ of length n, and the concentrations of all metabolites are represented by the vector $\mathbf{x}$ of length m. The system of mass balance equations at steady state ($d\mathbf{x}/dt = 0$) gives a set of equality constraints $\mathbf{Sv} = 0$. Each reaction also has upper and lower bounds, which gives a set of inequality constraints on the flux components, namely $a_i < v_i < b_i, 1 \leq i \leq n$. FBA seeks to maximize or minimize an objective function $Z = \mathbf{c}^T\mathbf{v}$, which can be any linear combination of fluxes, where $\mathbf{c}$ is a vector of weights indicating how much each reaction (such as the biomass reaction when simulating maximum growth) contributes to the objective function. The constraint optimization problem is usually solved by linear programming. It is important to note that the stoichiometric matrix $\mathbf{S}$ can be directly constructed from knowledge of an organism's metabolic genotype, which in turn can be efficiently determined from the results of genome annotation.

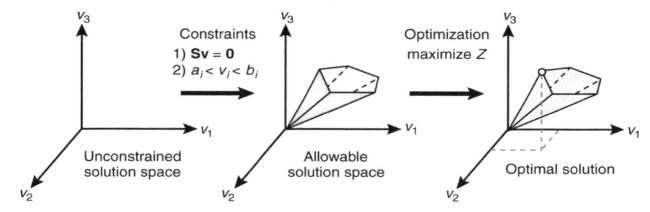

Figure 1. The conceptual basis of FBA as constraint-based modeling. Reprinted from [75] with permission from Nature Publishing Group.

Models based on FBA are used to study the role of interspecies exchange of metabolites in determining the spatiotemporal dynamics of microbial communities. Harcombe et al. [76] developed a model that integrates dynamic Flux Balance Analysis (dFBA) [77] with diffusion on a lattice and applied it to engineered communities. Simulations from the model predict the species ratio to which a two-species (*E. coli/S. enterica*) mutualistic consortium converges, the equilibrium composition of an engineered three-member (*E. coli/S. enterica/M. extorquens AM1*) community and the beneficial effect of a competitor in spatially-structured mutualism. All predictions are confirmed by experimental results. The strength of the model is highlighted by the fact that it requires very few free parameters and no a priori assumptions on whether or how species would interact.

Phalak et al. [78] developed a model to investigate the multispecies metabolism of a biofilm consortium comprised of two common chronic wound isolates: the aerobe *P. aeruginosa* and the facultative anaerobe *S. aureus*. The model combines genome-scale metabolic reconstructions for growth rates via FBA and partial differential equations for metabolite diffusion and provides both temporal and spatial predictions with genome-scale resolution. In particular, the two-species system was predicted to support a maximum biofilm thickness much greater than *P. aeruginosa* alone, but slightly less than *S. aureus* alone, suggesting an antagonistic metabolic effect of *P. aeruginosa* on *S. aureus*.

Sigurdsson et al. [79] used a systems biology approach to identify candidate drug targets for biofilm-associated *P. aeruginosa*. This study employed the published reconstruction of *P. aeruginosa* iMO1056 [80] and used FBA to simulate different medium and oxygen conditions. The effect of single and double gene deletion on bacterial growth in planktonic and biofilm-like environmental conditions was investigated. Condition-dependent genes were found that could be used to slow growth specifically in biofilm-associated *P. aeruginosa*. In particular, eight gene pairs were found to be synthetically lethal in oxygen-limited environments, and these gene sets may serve as metabolic drug targets to combat biofilm-associated *P. aeruginosa*. Results from [79] show that FBA can be used to determine key metabolic differences between planktonic and biofilm colonies and shed light on searching for novel drug targets.

The application of models based on FBA in studying biofilm as a microbial community is very promising, but also challenging. In particular, efficient and standardized methods are necessary for generating reliable stoichiometric models when a large number of species is involved. Furthermore, it is important to develop mathematical tools that can effectively incorporate omics-based metabolic pathway information into kinetic functions, which can be used directly in kinetic growth models. The cybernetic approach developed by Song et al. sheds some light on future research in this direction, and a review of this approach is given by [81].

4. Models Based on Statistical Inference

Species within a biofilm rarely live in isolation; instead, they often coexist and have complex interactions that affect the community structure and function [82–84]. The types of interactions include win-win (mutualism), win-zero (commensalism), win-lose (predation , parasitism), zero-lose (amensalism) and lose-lose (competition). Community-wide information on microbial interactions can be obtained using statistical inference based on correlations between taxon abundances from high-throughput sequence data [85,86].

Faust and Raes [87] reviewed strategies to construct community models from abundance data and use the models to predict the outcome of community alterations and the effects of perturbations. The prediction of microbial association networks from abundance data is known as the network inference problem [88]. The network inference methods can be classified into two categories: the similarity-based methods, which predict pairwise relationships, and the regression- and rule-based methods, which predict complex relationships. Figure 2 from [87] explains the principle of similarity- and regression-based network inference. Network inference starts from an incidence or an abundance matrix. Pairwise scores between two taxa are then computed using a suitable similarity or distance measure, and relationships involving more than two taxa are detected by either multiple regression or association rule mining [89]. Then, a random score distribution is generated by repeating the scoring step, and the P-value is computed to measure the significance of the predicted relationship. Finally, taxon pairs with P-values below a given threshold are visualized as a network. Inferred networks can be considered as static models of microbial communities, which describes the community status at a particular time. However, time series data obtained by network inference methods can provide important input (such as growth rates or interaction strengths) for dynamic models of microbial interactions; see [90] for modeling of cheese fermentation community interactions with generalized Lotka–Volterra equations. Network inference has several strengths. It is generic; it can integrate different data types; and it can identify community properties that are encoded in the network structure. However, network inference also suffers from several pitfalls, such as normalization, similarity measure biases, the choice of appropriate null models and multiple testing issues. Despite these pitfalls, network inference is a versatile tool for studying microbial interactions, can be used to build dynamic models that can predict community stability, alternative stable states and microbial succession and ultimately shed light on the manipulation of microbial communities to enhance the abundances of beneficial species and to suppress harmful ones.

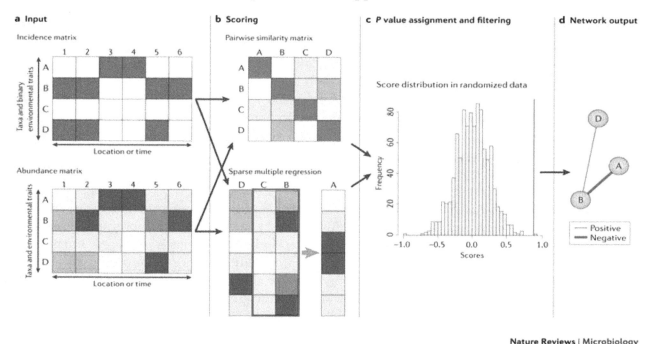

Nature Reviews | Microbiology

Figure 2. Principle of similarity- and regression-based network inference. Reprinted from [87] with permission from Nature Publishing Group.

Faust et al. [91] applied an ensemble method based on multiple similarity measures [92] in combination with Generalized Boosted Linear Models (GBLMs) [93] to taxonomic marker (16S rRNA gene) profiles of the Human Microbiome Project (HMP) cohort [94], resulting in a global network of 3005 significant co-occurrence and co-exclusion relationships between 197 clades occurring throughout the human microbiome. Analysis of the network revealed strong organization of the human microbiota into body area niches, mostly among closely-related individual body sites representing microbial habitats. For example, *Fusobacterium* species can bridge organisms in the development and maturation of oral biofilms by co-aggregation through physical contact, allowing a more complex use of resources, such as sugars and proteins. The approach in [91] provides a starting point for future mechanistic studies of the microbial ecology of the human microbiome.

Widder et al. [95] investigated the effect of features inherent to fluvial networks on the structure and function of biofilm communities in these ecosystems by combining co-occurrence analyses of biofilms based on pyrosequencing profiling and a probabilistic hydrological model. Co-occurrence networks were constructed using 454 pyrosequencing data of the 16S rRNA gene from benthic biofilms from 114 streams of the pre-alpine Ybbs catchment. Results suggested that hydrological disturbance and metacommunity dynamics affect the co-occurrence patterns of benthic biofilm communities in fluvial networks. In particular, the removal of gatekeepers disproportionately contributed to network fragmentation. This study provides a linkage between the biofilm communities and flow dynamics across fluvial networks, which are important for understanding the whole ecosystem processes.

Although network inference methods can uncover previously-unknown interactions, they still require validation by experimental data, which could be challenging. Some of these interactions could be indirect, such as bacteria modifying their environments via the secretion of metabolically-costly proteins and metabolites [96]. Since these indirect interactions usually happen on larger spatial scales than direct interactions, it is very important to develop non-invasive spatially-resolved experimental techniques to collect structure and population data.

5. Kinetic Growth Models and Spatial Heterogeneity

Kinetic models predict growth rates of different species within a biofilm based on the concentrations of growth-limiting nutrients and species-dependent parameters, such as maximum growth rate. Probably the most widely-used kinetic model is the Michaelis–Menten (or Monod) kinetic model [97] with the growth rate μ given by:

$$\mu = \mu_{max} \frac{S}{K_S + S} \qquad (3)$$

where μ_{max} is the maximum growth rate, S is the concentration of the growth-limiting nutrient and K_S is the half-saturation constant. The formula given by (3) can be readily generalized to cases with more than one limiting nutrient [13] by multiplying the contribution from each nutrient together ($S_i / (K_{S_i} + S_i)$ for the i-th nutrient), but usually, μ_{max} and K_{S_i} are considered constants. Under the nutrient-saturation condition, (3) can be approximated by the zero order kinetics, where μ is independent of S ($\mu = \mu_{max}$). Under the very low nutrient condition, (3) can be approximated by the first order kinetics, where μ is proportional to S ($\mu = \mu_{max} \cdot S$). These two approximations provide convenient mathematical bounds on the Monod kinetic forms and have the advantages of allowing analytic solutions [50] to the model equations for simple scenarios (ODE model or 1D in space).

The success of the kinetic models depends crucially both on their particular formula and parameter values. Since the formula is often empirical and the parameter values are usually measured from pure and mixed cultures growing in laboratory reactors, they may miss important factors of the growth kinetics in the biofilm community developed in the natural environment, and their application may require validation. Recently, new kinetic models have been developed to address this problem. Quéméner and Bouchez [98] and Jin et al. [99] developed kinetic models with thermodynamics included. The model proposed in [98] is based on the theory of Boltzmann statistics and builds a relationship between microbial growth rate and available energy, thus connecting microbial population dynamics to the thermodynamic driving forces of the surrounding ecosystem. The work in [99] modified the Monod kinetics by a thermodynamic potential factor, which accounts for the chemical energy available from the reaction (acetate oxidation and sulfate reduction in this case) and evaluated the feasibility of applying experimentally-obtained parameters to the natural environment. The results suggest that some parameters, such as maximum growth rate, can be applied directly to the environment; but others, such as half-saturation constants, should be determined using data from the environment of interest. Bonachela et al. [100] relaxed the requirement that the maximum nutrient uptake rate be a constant; instead, the maximum uptake rate was assumed to increase monotonically as the external nutrient concentration decreases. The model predicts larger uptake and growth rates than the standard Monod kinetic, which explains the ability of marine microbes to persist under extreme nutrient limitation.

Biofilms usually have a complex spatial structure, which contributes to their distinctive properties, such as strong antibiotic resistance and diverse population (niches for different species). Mathematical models designated for characterizing the spatial structure of biofilms include Individual-based Models (IbMs) [8,67], continuum models [101,102] and hybrid models [9], which combine both. For IbMs, biofilms are represented as a collection of individual microbes (usually hard spheres) whose growth and movement are determined by a set of rules depending on the local environment, such as the availability of nutrients and space. Results from IbMs usually provide more detailed information and are generally considered superior for studying interactions at the microbe level. Continuum models represent the biomass by functions depending continuously on time and space, and these functions are governed by differential equations derived by using physical, chemical and biological principles. Continuum models allow both numerical solutions obtained using readily-available numerical methods and theoretical analysis of the qualitative behavior of the solutions, such as stability [103], and they are often considered more applicable at larger spatial scales.

Recently, models incorporating novel features have been successfully applied to study biofilms as a microbial community. Below, we discuss an individual-based model and a continuum model, respectively.

Storck et al. [104] developed an individual-based, mass-spring modeling framework to study the effect of cell properties on the structure of biofilms. In this model, cells are represented by a collection of particles connected by springs, which allows variable morphology (e.g., cocci, bacilli and filaments). Three types of structures are considered: the primary structure, which defines the shape of individual cells; the secondary structure, which defines microbial assemblies related by filial links between immediate siblings; and the tertiary structure, which defines non-filial cell-cell and cell-substratum links, such as sticking and anchoring connections. Forces acted on the cells include both elastic force from the springs they are attached to and DLVOforce (combination of the van der Waals and electrostatic force). Simulation results of the growth of rod-shaped cells on a planar surface suggest that the biofilm may grow as a monolayer if there are no anchoring and filial links; the biofilm can be much thicker and much less spread if there are cell-substratum anchoring; and with filial links, but no anchoring to the substratum, a biofilm with an irregular shape (less circularity) is more likely to develop. Simulation results of the activated sludge floc structure suggest that in the floc with filament branching, the filaments are shorter than that of a floc made of straight filaments growing at a similar rate, therefore resulting in greater floc density and attenuating the bulking tendency of filamentous sludge. Furthermore, simulated flocs with spherical floc formers (in contrast to rod-shaped floc formers) were less dense, since the denser packing of spherical cells in a colony leads to a smaller cluster volume, which lowers the chance to encounter a filament former. These simulations demonstrate the close relationship between the fundamental controlling mechanisms, such as the intracellular, intercellular and cell-substratum links, and the diverse biofilm structures.

Emerenini et al. [105] developed a continuum model that includes biofilm growth, production of quorum sensing molecules, cell dispersal triggered by quorum sensing molecules and reattachment of cells. In this model, two distinct cell types are considered: the sessile cells in the biofilm and the motile cells, which can move into and in the liquid phase. The volume fraction of sessile cells and EPS is denoted by M; the motile cell density is denoted by N; and the concentrations of growth-controlling nutrients and autoinducers are denoted by C and A, respectively. Dispersal of cells from the biofilm is controlled by the local autoinducer concentration through Hill kinetics with switching threshold τ and maximum dispersal rate η_1. Re-attachment of cells in the biofilm is controlled by the local biofilm density M through Monod kinetics with maximum rate η_2. The autoinducer production rate is controlled by the local autoinducer concentration, which implicitly represents the switch between down- and up-regulated cells. The governing equations for M, N, C, A are reaction-diffusion-type equations with density-dependent (M) diffusion coefficients. Simulation results suggest that single quorum sensing-based mechanism can explain both periodic dispersal in discrete events and continuous dispersal, depending on the value of switching threshold parameter

τ. For smaller values of τ, the switching threshold is reached quickly, leading to a rapid dispersal of the biomass before the biofilm can grow into a large size. After the first dispersal event, the biofilm population starts growing again, and the autoinducer concentration increases again, resulting in an almost periodic pattern of discrete dispersal events. For bigger values of τ, it takes a much longer time to reach the switching threshold, and the biofilm develops into a stronger colony before the onset of dispersal. Release of cells from the biofilm into the liquid phase appears continuous, and the biofilm population reaches a plateau. Simulation also suggests that re-attachment of dispersed cells is negligible. The study in [105] indicated that important properties, such as biofilm mass and thickness, can be modified by changing the QS threshold and dispersal rates, and the systems can change between continuous and oscillating behavior. The findings can also help to optimize treatment strategies. For example, promoting quorum sensing can enhance cell dispersal and limit biofilm thickness, which could increase the efficacy of antibiotic treatment, since planktonic cells are generally assumed to be more vulnerable to antibiotic treatment.

IbMs have the obvious strength of describing the detailed biofilm structure and interactions at the cell level, and the forces between individual particles can be derived from first principles. On the other hand, IbMs also suffer the drawback of high computational cost, especially when the goal is to model a biofilm containing a very large number of cells. Therefore, it is important to use efficient numerical methods in the implementation of IbMs, and parallel computing is often the choice [106]. The continuum models can often be analyzed using well-established differential equation theory, and there are many numerical packages available for solving the corresponding discretized system of algebraic equations. However, continuum models often depend on some empirical formula; examples include the effective diffusion coefficient of nutrients inside the biofilm and the constitutive equation for the stress-strain relation when modeling the biofilm as a viscoelastic fluid, and derivation of such a formula is often nontrivial.

6. Conclusions

We present a review of some recently-developed mathematical models that focus on studying biofilms as diverse communities. Despite obvious overlapping, these models are categorized based on their principal methodologies, such as trait-based models, QS, FBA, statistical inference and spatially-resolved models with specific growth kinetics. There are some important topics that have been left out, such as models incorporating stochasticity and evolutionary game theory.

Even though currently-available models can describe many aspects of biofilms accurately, it still remains a challenge to build models that can predict the overall behavior of biofilms as complex and evolving communities. First, time scales involved in biofilm-related processes can vary in as many as ten orders of magnitude, ranging from the fast scale for fluid dynamics to the slow scale for biofilm growth, which is an obvious challenge for modeling. To address this, it is often necessary to assume equilibrium in the fast processes or a quasi-static biofilm profile, depending on the problem of interest. Second, fast advancement in the experimental technologies has provided an abundance of omics data at both the genetic and community level, and many community-scale models have been proposed to describe the interaction between biology, chemistry and physics inside biofilms. However, there is still the lack of a systematic approach to link the observational data to the community-level understanding, namely to tie system kinetics to omics data in a tractable and general way by translating omics to rate functions at the cellular level. The method based on the FBA approach is very promising in this direction. Third, long-term challenges for modeling biofilm as an MC include the necessity to incorporated evolutionary processes, social evolution and bacterial strategies, community assembly and historical contingency, as well as the importance of spatial structure. Addressing these challenges would inevitably require an integrated approach that not only selectively combines multiple relevant models by adding their strengths, but also combines modeling effort and experimental findings.

Acknowledgments: This work was supported by NSF Award DMS-1516951.

References

1.   Costerton, J.W. *The Biofilm Primer*; Springer Science & Business Media: Berlin, Germany, 2007; Volume 1.
2.   Hall-Stoodley, L.; Costerton, J.W.; Stoodley, P. Bacterial biofilms: From the natural environment to infectious diseases. *Nat. Rev. Microbiol.* **2004**, *2*, 95–108.
3.   Costerton, J.W.; Stewart, P.S.; Greenberg, E.P. Bacterial biofilms: A common cause of persistent infections. *Science* **1999**, *284*, 1318–1322.
4.   Harremoës, P. Biofilm kinetics. In *Water Pollution Microbiology*; John Wiley & Sons Ltd.: Hoboken, NJ, USA, 1978.
5.   Harris, N.P.; Hansford, G. A study of substrate removal in a microbial film reactor. *Water Res.* **1976**, *10*, 935–943.
6.   Rittmann, B.E.; McCarty, P.L. Evaluation of steady-state biofilm kinetics. *Biotechnol. Bioeng.* **1980**, *22*, 2359–2373.
7.   Wanner, O.; Gujer, W. A multispecies biofilm model. *Biotechnol. Bioeng.* **1986**, *28*, 314–328.
8.   Kreft, J.U.; Booth, G.; Wimpenny, J.W.T. BacSim, a simulator for individual-based modelling of bacterial colony growth. *Microbiology* **1998**, *144*, 3275–3287.
9.   Picioreanu, C.; Loosdrecht, M.C.M.; Heijnen, J.J. Mathematical modeling of biofilm structure with a hybrid differential-discrete cellular automaton approach. *Biotechnol. Bioeng.* **1998**, *58*, 101–116.
10.   Picioreanu, C.; Kreft, J.U.; Loosdrecht, M.C. Particle-Based Multidimensional Multispecies Biofilm Model. *Appl. Environ. Microbiol.* **2004**, *70*, 3024–3040.
11.   Alpkvist, E.; Picioreanu, C.; Loosdrecht, M.C.M.; Heyden, A. Three-dimensional biofilm model with individual cells and continuum EPS matrix. *Biotechnol. Bioeng.* **2006**, *94*, 961–979.
12.   Alpkvist, E.; Klapper, I. Description of Mechanical Response Including Detachment Using a Novel Particle Method of Biofilm/Flow Interaction. *Water Sci. Technol.* **2007**, *55*, 265–273.
13.   Wanner, O.; Eberl, H.; Morgenroth, E.N.D.; Picioreanu, C.; Rittmann, B.; Van Loosdrecht, M. *Mathematical Modeling of Biofilms*; IWA Publishing: London, UK, 2006; Volume 18.
14.   Klapper, I.; Dockery, J. Mathematical description of microbial biofilms. *SIAM Rev.* **2010**, *52*, 221–265.
15.   Wang, Q.; Zhang, T. Review of mathematical models for biofilms. *Solid State Commun.* **2010**, *150*, 1009–1022.
16.   May, R.M. Species coexistence and self-organizing spatial dynamics. *Nature* **1994**, *37*, 28.
17.   Davey, M.E.; O'toole, G.A. Microbial biofilms: From ecology to molecular genetics. *Microbiol. Mol. Biol. Rev.* **2000**, *64*, 847–867.
18.   Tilman, D. Niche tradeoffs, neutrality, and community structure: A stochastic theory of resource competition, invasion, and community assembly. *Proc. Natl. Acad. Sci. USA* **2004**, *101*, 10854–10861.
19.   De Meester, N.; Derycke, S.; Rigaux, A.; Moens, T. Active dispersal is differentially affected by inter-and intraspecific competition in closely related nematode species. *Oikos* **2015**, *124*, 561–570.
20.   Ellis, C.N.; Traverse, C.C.; Mayo-Smith, L.; Buskirk, S.W.; Cooper, V.S. Character displacement and the evolution of niche complementarity in a model biofilm community. *Evolution* **2015**, *69*, 283–293.
21.   Song, H.S.; Cannon, W.R.; Beliaev, A.S.; Konopka, A. Mathematical modeling of microbial community dynamics: A methodological review. *Processes* **2014**, *2*, 711–752.
22.   Widder, S.; Allen, R.J.; Pfeiffer, T.; Curtis, T.P.; Wiuf, C.; Sloan, W.T.; Cordero, O.X.; Brown, S.P.; Momeni, B.; Shou, W.; et al. Challenges in microbial ecology: Building predictive understanding of community function and dynamics. *ISME J.* **2016**, *10*, 2557–2568.
23.   Davies, D.; Geesey, G. Regulation of the alginate biosynthesis gene algC in *Pseudomonas aeruginosa* during biofilm development in continuous culture. *Appl. Environ. Microbiol.* **1995**, *61*, 860–867.
24.   Petrova, O.E.; Sauer, K. Sticky situations: Key components that control bacterial surface attachment. *J. Bacteriol.* **2012**, *194*, 2413–2425.
25.   Fuqua, W.C.; Winans, S.C.; Greenberg, E.P. Quorum sensing in bacteria: The LuxR-LuxI family of cell density-responsive transcriptional regulators. *J. Bacteriol.* **1994**, *176*, 269.
26.   Papenfort, K.; Bassler, B.L. Quorum sensing signal-response systems in Gram-negative bacteria. *Nat. Rev. Microbiol.* **2016**, *14*, 576–588.
27.   Boyd, A.; Chakrabarty, A.M. Role of alginate lyase in cell detachment of *Pseudomonas aeruginosa*. *Appl. Environ. Microbiol.* **1994**, *60*, 2355–2359.

28. Lamppa, J.W.; Griswold, K.E. Alginate lyase exhibits catalysis-independent biofilm dispersion and antibiotic synergy. *Antimicrob. Agents Chemother.* **2013**, *57*, 137–145.

29. Chen, X.; Stewart, P.S. Chlorine penetration into artificial biofilm is limited by a reaction-diffusion interaction. *Environ. Sci. Technol.* **1996**, *30*, 2078–2083.

30. Pabst, B.; Pitts, B.; Lauchnor, E.; Stewart, P.S. Gel-Entrapped *Staphylococcus aureus* Bacteria as Models of Biofilm Infection Exhibit Growth in Dense Aggregates, Oxygen Limitation, Antibiotic Tolerance, and Heterogeneous Gene Expression. *Antimicrob. Agents Chemother.* **2016**, *60*, 6294–6301.

31. Smith, J.J.; Travis, S.M.; Greenberg, E.P.; Welsh, M.J. Cystic fibrosis airway epithelia fail to kill bacteria because of abnormal airway surface fluid. *Cell* **1996**, *85*, 229–236.

32. Pezzulo, A.A.; Tang, X.X.; Hoegger, M.J.; Alaiwa, M.H.A.; Ramachandran, S.; Moninger, T.O.; Karp, P.H.; Wohlford-Lenane, C.L.; Haagsman, H.P.; van Eijk, M.; et al. Reduced airway surface pH impairs bacterial killing in the porcine cystic fibrosis lung. *Nature* **2012**, *487*, 109–113.

33. Monds, R.D.; O'Toole, G.A. The developmental model of microbial biofilms: Ten years of a paradigm up for review. *Trends Microbiol.* **2009**, *17*, 73–87.

34. Dworkin, M.; Dworkin, M. *Developmental Biology of the Bacteria*; Benjamin/Cummings Publishing Company; Reading, MA, USA, 1985.

35. O'Toole, G.; Kaplan, H.B.; Kolter, R. Biofilm formation as microbial development. *Annu. Rev. Microbiol.* **2000**, *54*, 49–79.

36. Stoodley, P.; Sauer, K.; Davies, D.; Costerton, J.W. Biofilms as complex differentiated communities. *Annu. Rev. Microbiol.* **2002**, *56*, 187–209.

37. Costerton, J.W.; Lewandowski, Z.; Caldwell, D.E.; Korber, D.R.; Lappin-Scott, H.M. Microbial Biofilms. *Annu. Rev. Microbiol.* **1995**, *49*, 711–745.

38. Jenal, U.; Malone, J. Mechanisms of cyclic-di-GMP signaling in bacteria. *Annu. Rev. Genet.* **2006**, *40*, 385–407.

39. Römling, U.; Galperin, M.Y.; Gomelsky, M. Cyclic di-GMP: The first 25 years of a universal bacterial second messenger. *Microbiol. Mol. Biol. Rev.* **2013**, *77*, 1–52.

40. Banin, E.; Vasil, M.L.; Greenberg, E.P. Iron and *Pseudomonas aeruginosa* biofilm formation. *Proc. Natl. Acad. Sci. USA* **2005**, *102*, 11076–11081.

41. Wiens, J.R.; Vasil, A.I.; Schurr, M.J.; Vasil, M.L. Iron-regulated expression of alginate production, mucoid phenotype, and biofilm formation by *Pseudomonas aeruginosa*. *MBio* **2014**, *5*, e01010-13.

42. Lazazzera, B.A. Lessons from DNA microarray analysis: The gene expression profile of biofilms. *Curr. Opin. Microbiol.* **2005**, *8*, 222–227.

43. Stewart, P.S.; Franklin, M.J.; Williamson, K.S.; Folsom, J.P.; Boegli, L.; James, G.A. Contribution of stress responses to antibiotic tolerance in *Pseudomonas aeruginosa* biofilms. *Antimicrob. Agents Chemother.* **2015**, *59*, 3838–3847.

44. Folkesson, A.; Haagensen, J.A.; Zampaloni, C.; Sternberg, C.; Molin, S. Biofilm induced tolerance towards antimicrobial peptides. *PLoS ONE* **2008**, *3*, e1891.

45. Klausen, M.; Heydorn, A.; Ragas, P.; Lambertsen, L.; Aaes-Jørgensen, A.; Molin, S.; Tolker-Nielsen, T. Biofilm formation by *Pseudomonas aeruginosa* wild type, flagella and type IV pili mutants. *Mol. Microbiol.* **2003**, *48*, 1511–1524.

46. Burrows, L.L. *Pseudomonas aeruginosa* twitching motility: Type IV pili in action. *Annu. Rev. Microbiol.* **2012**, *66*, 493–520.

47. Vlamakis, H.; Aguilar, C.; Losick, R.; Kolter, R. Control of cell fate by the formation of an architecturally complex bacterial community. *Genes Dev.* **2008**, *22*, 945–953.

48. Klausen, M.; Gjermansen, M.; Kreft, J.U.; Tolker-Nielsen, T. Dynamics of development and dispersal in sessile microbial communities: Examples from *Pseudomonas aeruginosa* and *Pseudomonas putida* model biofilms. *FEMS Microbiol. Lett.* **2006**, *261*, 1–11.

49. Reed, D.C.; Algar, C.K.; Huber, J.A.; Dick, G.J. Gene-centric approach to integrating environmental genomics and biogeochemical models. *Proc. Natl. Acad. Sci. USA* **2014**, *111*, 1879–1884.

50. Zhang, T.; Pabst, B.; Klapper, I.; Stewart, P.S. General theory for integrated analysis of growth, gene, and protein expression in biofilms. *PLoS ONE* **2013**, *8*, e83626.

51. Shipley, B.; Vile, D.; Garnier, É. From plant traits to plant communities: A statistical mechanistic approach to biodiversity. *Science* **2006**, *314*, 812–814.

52. Laughlin, D.C.; Joshi, C.; Bodegom, P.M.; Bastow, Z.A.; Fulé, P.Z. A predictive model of community assembly that incorporates intraspecific trait variation. *Ecol. Lett.* **2012**, *15*, 1291–1299.

53. Lennon, J.T.; Aanderud, Z.T.; Lehmkuhl, B.; Schoolmaster, D.R. Mapping the niche space of soil microorganisms using taxonomy and traits. *Ecology* **2012**, *93*, 1867–1879.

54. Miller, M.B.; Bassler, B.L. Quorum sensing in bacteria. *Annu. Rev. Microbiol.* **2001**, *55*, 165–199.

55. Waters, C.M.; Bassler, B.L. Quorum sensing: Cell-to-cell communication in bacteria. *Annu. Rev. Cell Dev. Biol.* **2005**, *21*, 319–346.

56. Williams, P. Quorum sensing, communication and cross-kingdom signalling in the bacterial world. *Microbiology* **2007**, *153*, 3923–3938.

57. Rice, S.; Koh, K.; Queck, S.; Labbate, M.; Lam, K.; Kjelleberg, S. Biofilm formation and sloughing in *Serratia marcescens* are controlled by quorum sensing and nutrient cues. *J. Bacteriol.* **2005**, *187*, 3477–3485.

58. Arciola, C.R.; Campoccia, D.; Speziale, P.; Montanaro, L.; Costerton, J.W. Biofilm formation in *Staphylococcus* implant infections. A review of molecular mechanisms and implications for biofilm-resistant materials. *Biomaterials* **2012**, *33*, 5967–5982.

59. McDougald, D.; Rice, S.A.; Barraud, N.; Steinberg, P.D.; Kjelleberg, S. Should we stay or should we go: Mechanisms and ecological consequences for biofilm dispersal. *Nat. Rev. Microbiol.* **2012**, *10*, 39–50.

60. Ward, J. Mathematical Modeling of Quorum-Sensing Control in Biofilms. In *Control of Biofilm Infections by Signal Manipulation*; Springer: Berlin/Heidelberg, Germany, 2008; pp. 79–108.

61. Pesci, E.C.; Pearson, J.P.; Seed, P.C.; Iglewski, B.H. Regulation of las and rhl quorum sensing in *Pseudomonas aeruginosa*. *J. Bacteriol.* **1997**, *179*, 3127–3132.

62. James, S.; Nilsson, P.; James, G.; Kjelleberg, S.; Fagerström, T. Luminescence control in the marine bacterium *Vibrio fischeri*: An analysis of the dynamics of lux regulation. *J. Mol. Biol.* **2000**, *296*, 1127–1137.

63. Dockery, J.D.; Keener, J.P. A mathematical model for quorum sensing in *Pseudomonas aeruginosa*. *Bull. Math. Biol.* **2001**, *63*, 95–116.

64. Horswill, A.R.; Stoodley, P.; Stewart, P.S.; Parsek, M.R. The effect of the chemical, biological, and physical environment on quorum sensing in structured microbial communities. *Anal. Bioanal. Chem.* **2007**, *387*, 371–380.

65. Kirisits, M.J.; Margolis, J.J.; Purevdorj-Gage, B.L.; Vaughan, B.; Chopp, D.L.; Stoodley, P.; Parsek, M.R. Influence of the hydrodynamic environment on quorum sensing in *Pseudomonas aeruginosa* biofilms. *J. Bacteriol.* **2007**, *189*, 8357–8360.

66. Nadell, C.D.; Xavier, J.B.; Levin, S.A.; Foster, K.R. The evolution of quorum sensing in bacterial biofilms. *PLoS Biol.* **2008**, *6*, e14.

67. Kreft, J.U.; Picioreanu, C.; Wimpenny, J.W.; van Loosdrecht, M.C. Individual-based modelling of biofilms. *Microbiology* **2001**, *147*, 2897–2912.

68. Xavier, J.B.; Picioreanu, C.; Van Loosdrecht, M. A framework for multidimensional modelling of activity and structure of multispecies biofilms. *Environ. Microbiol.* **2005**, *7*, 1085–1103.

69. Picioreanu, C.; Van Loosdrecht, M.C.; Heijnen, J.J. Effect of diffusive and convective substrate transport on biofilm structure formation: A two-dimensional modeling study. *Biotechnol. Bioeng.* **2000**, *69*, 504–515.

70. Sakuragi, Y.; Kolter, R. Quorum-sensing regulation of the biofilm matrix genes (pel) of *Pseudomonas aeruginosa*. *J. Bacteriol.* **2007**, *189*, 5383–5386.

71. Heithoff, D.M.; Mahan, M.J. Vibrio cholerae biofilms: Stuck between a rock and a hard place. *J. Bacteriol.* **2004**, *186*, 4835–4837.

72. Seressiotis, A.; Bailey, J.E. MPS: An artificially intelligent software system for the analysis and synthesis of metabolic pathways. *Biotechnol. Bioeng.* **1988**, *31*, 587–602.

73. Mavrovouniotis, M.L.; Stephanopoulos, G.; Stephanopoulos, G. Computer-aided synthesis of biochemical pathways. *Biotechnol. Bioeng.* **1990**, *36*, 1119–1132.

74. Schilling, C.H.; Schuster, S.; Palsson, B.O.; Heinrich, R. Metabolic pathway analysis: Basic concepts and scientific applications in the post-genomic era. *Biotechnol. Progress* **1999**, *15*, 296–303.

75. Orth, J.D.; Thiele, I.; Palsson, B.Ø. What is flux balance analysis? *Nat. Biotechnol.* **2010**, *28*, 245–248.

76. Harcombe, W.R.; Riehl, W.J.; Dukovski, I.; Granger, B.R.; Betts, A.; Lang, A.H.; Bonilla, G.; Kar, A.; Leiby, N.; Mehta, P.; et al. Metabolic resource allocation in individual microbes determines ecosystem interactions and spatial dynamics. *Cell Rep.* **2014**, *7*, 1104–1115.

77. Mahadevan, R.; Edwards, J.S.; Doyle, F.J. Dynamic flux balance analysis of diauxic growth in *Escherichia coli*. *Biophys. J.* **2002**, *83*, 1331–1340.

78. Phalak, P.; Chen, J.; Carlson, R.P.; Henson, M.A. Metabolic modeling of a chronic wound biofilm consortium predicts spatial partitioning of bacterial species. *BMC Syst. Biol.* **2016**, *10*, 90.

79. Sigurdsson, G.; Fleming, R.M.; Heinken, A.; Thiele, I. A systems biology approach to drug targets in *Pseudomonas aeruginosa* biofilm. *PLoS ONE* **2012**, *7*, e34337.

80. Oberhardt, M.A.; Puchałka, J.; Fryer, K.E.; Dos Santos, V.A.M.; Papin, J.A. Genome-scale metabolic network analysis of the opportunistic pathogen *Pseudomonas aeruginosa* PAO1. *J. Bacteriol.* **2008**, *190*, 2790–2803.

81. Ramkrishna, D.; Song, H.S. Dynamic models of metabolism: Review of the cybernetic approach. *AIChE J.* **2012**, *58*, 986–997.

82. James, G.; Beaudette, L.; Costerton, J. Interspecies bacterial interactions in biofilms. *J. Ind. Microbiol.* **1995**, *15*, 257–262.

83. Hansen, S.K.; Rainey, P.B.; Haagensen, J.A.; Molin, S. Evolution of species interactions in a biofilm community. *Nature* **2007**, *115*, 533–536.

84. Lee, K.; Periasamy, S.; Mukherjee, M.; Xie, C.; Kjelleberg, S.; Rice, S.A. Biofilm development and enhanced stress resistance of a model, mixed-species community biofilm. *ISME J.* **2014**, *8*, 894–907.

85. Fuhrman, J.A. Microbial community structure and its functional implications. *Nature* **2009**, *459*, 193–199.

86. Freilich, S.; Kreimer, A.; Meilijson, I.; Gophna, U.; Sharan, R.; Ruppin, E. The large-scale organization of the bacterial network of ecological co-occurrence interactions. *Nucleic Acids Res.* **2010**, *38*, 3857–3868.

87. Faust, K.; Raes, J. Microbial interactions: From networks to models. *Nat. Rev. Microbiol.* **2012**, *10*, 538–550.

88. De Smet, R.; Marchal, K. Advantages and limitations of current network inference methods. *Nat. Rev. Microbiol.* **2010**, *8*, 717–729.

89. Agrawal, R.; Imieliński, T.; Swami, A. Mining association rules between sets of items in large databases. In *ACM Sigmod Record*; ACM: New York, NY, USA, 1993; Volume 22, pp. 207–216.

90. Mounier, J.; Monnet, C.; Vallaeys, T.; Arditi, R.; Sarthou, A.S.; Hélias, A.; Irlinger, F. Microbial interactions within a cheese microbial community. *Appl. Environ. Microbiol.* **2008**, *74*, 172–181.

91. Faust, K.; Sathirapongsasuti, J.F.; Izard, J.; Segata, N.; Gevers, D.; Raes, J.; Huttenhower, C. Microbial co-occurrence relationships in the human microbiome. *PLoS Comput. Biol.* **2012**, *8*, e1002606.

92. Aitchison, J. *The Statistical Analysis of Compositional Data*; Chapman and Hall, Ltd.: London, UK, 1986.

93. Buehlmann, P. Boosting for high-dimensional linear models. *Ann. Stat.* **2006**, 559–583.

94. Human Microbiome Project Consortium. A framework for human microbiome research. *Nature* **2012**, *486*, 215–221.

95. Widder, S.; Besemer, K.; Singer, G.A.; Ceola, S.; Bertuzzo, E.; Quince, C.; Sloan, W.T.; Rinaldo, A.; Battin, T.J. Fluvial network organization imprints on microbial co-occurrence networks. *Proc. Natl. Acad. Sci. USA* **2014**, *111*, 12799–12804.

96. McNally, L.; Viana, M.; Brown, S.P. Cooperative secretions facilitate host range expansion in bacteria. *Nat. Commun.* **2014**, *5*, 4594.

97. Michealis, L.; Menten, M. The kinetics of invertase activity. *Biochem. Z.* **1913**, *49*, 333–369.

98. Desmond-Le Quéméner, E.; Bouchez, T. A thermodynamic theory of microbial growth. *ISME J.* **2014**, *8*, 1747.

99. Jin, Q.; Roden, E.E.; Giska, J.R. Geomicrobial kinetics: Extrapolating laboratory studies to natural environments. *Geomicrobiol. J.* **2013**, *30*, 173–185.

100. Bonachela, J.A.; Raghib, M.; Levin, S.A. Dynamic model of flexible phytoplankton nutrient uptake. *Proc. Natl. Acad. Sci. USA* **2011**, *108*, 20633–20638.

101. Alpkvist, E.; Klapper, I. A Multidimensional multispecies continuum model for heterogenous biofilm. *Bull. Math. Biol.* **2007**, *69*, 765–789.

102. Zhang, T.; Cogan, N.; Wang, Q. Phase field models for biofilms. II. 2-D numerical simulations of biofilm-flow interaction. *Commun. Comput. Phys.* **2008**, *4*, 72–101.

103. Dockery, J.D.; Klapper, I. Finger formation in biofilm layers. *SIAM. J. Appl. Math.* **2001**, *62*, 853–869.

104. Storck, T.; Picioreanu, C.; Virdis, B.; Batstone, D.J. Variable cell morphology approach for individual-based modeling of microbial communities. *Biophys. J.* **2014**, *106*, 2037–2048.

Dynamics of the Bacterial Community Associated with *Phaeodactylum tricornutum* Cultures

Fiona Wanjiku Moejes [1] , Antonella Succurro [2,3,*,†] , Ovidiu Popa [2,4,†], Julie Maguire [1] and Oliver Ebenhöh [2,4,*]

[1]   Bantry Marine Research Station, Gearhies, Bantry P75 AX07, Co. Cork, Ireland;
      fmoejes@bmrs.ie (F.W.M.); jmaguire@bmrs.ie (J.M.)
[2]   Cluster of Excellence on Plant Sciences (CEPLAS), Heinrich-Heine University, Universitätsstrasse 1,
      40225 Düsseldorf, Germany; ovidiu.popa@hhu.de
[3]   Botanical Institute, University of Cologne, Zülpicher Strasse 47b, 50674 Cologne, Germany
[4]   Institute of Quantitative and Theoretical Biology, Heinrich-Heine University, Universitätsstrasse 1,
      40225 Düsseldorf, Germany
*   Correspondence: a.succurro@uni-koeln.de (A.S.); oliver.ebenhoeh@hhu.de (O.E.)
†   These authors contributed equally to this work.

Abstract: The pennate diatom *Phaeodactylum tricornutum* is a model organism able to synthesize industrially-relevant molecules. Commercial-scale cultivation currently requires large monocultures, prone to bio-contamination. However, little is known about the identity of the invading organisms. To reduce the complexity of natural systems, we systematically investigated the microbiome of non-axenic *P. tricornutum* cultures from a culture collection in reproducible experiments. The results revealed a dynamic bacterial community that developed differently in "complete" and "minimal" media conditions. In complete media, we observed an accelerated "culture crash", indicating a more stable culture in minimal media. The identification of only four bacterial families as major players within the microbiome suggests specific roles depending on environmental conditions. From our results we propose a network of putative interactions between *P. tricornutum* and these main bacterial factions. We demonstrate that, even with rather sparse data, a mathematical model can be reconstructed that qualitatively reproduces the observed population dynamics, thus indicating that our hypotheses regarding the molecular interactions are in agreement with experimental data. Whereas the model in its current state is only qualitative, we argue that it serves as a starting point to develop quantitative and predictive mathematical models, which may guide experimental efforts to synthetically construct and monitor stable communities required for robust upscaling strategies.

Keywords: microbial communities; host-microbe interactions; mathematical modelling; diatoms; synthetic ecology; algal biotechnology

---

1. Introduction

Microalgae are photosynthesis-driven cells able to store light energy by converting carbon dioxide into carbohydrates, lipids, proteins, and other cellular components with potential biofuel, food, feed, and pharmaceutical and nutraceutical applications [1]. Novel applications also include the use of microalgae as an alternative sustainable development tool [2]. One such microalgae is the pennate diatom *Phaeodactylum tricornutum* that is able to synthesize a number of industrially relevant molecules applicable in: aquaculture as feed in e.g., bivalve, echinoderm, crustacean and fish hatcheries [3,4]; as biomass for biofuels [5,6]; pharmaceuticals and nutraceuticals [5,7–9]; and nanotechnology [10], and bioremediation industries [11]. To fully exploit the industrial potential of microalgal-derived products, substantial quantities of microalgal biomass is required, preferably obtained while maintaining low production costs. This is achieved by implementation of large-scale cultivation methods such as open raceway ponds and photobioreactors. The majority of conventional cultivation methods rely on keeping monocultures of the desired species, particularly if the final product is a bioactive molecule for human consumption [12]. Photobioreactors are closed systems that allow for the production of monoseptic cultures, fully isolated from potential contamination if cultivation protocols are followed correctly [13]. However, high operational costs of photobioreactors might not be sustainable. Another option is open raceway ponds, which are simple open-air cultivation systems that have been in use since the 1950s [1]. They are highly susceptible to contamination, and unless the desired species is a halophile or thermophile [14], it is hard to maintain monocultures. Irrespective of the cultivation method, the establishment of unwanted organisms such as amoeba, ciliates, rotifers, bacteria, viruses, and other photosynthetic organisms in microalgal cultures, is a serious obstacle for large-scale microalgae cultivation [15,16]. Although much research is carried out in the field of microalgal culture upscaling, very little is known about the true identity and characteristics of these invading organisms, responsible for microalgal culture "crashes" which lead to loss of biomass, and therefore, loss of revenue.

Microalgae are not found in monoculture in nature and it is not surprising that imposing such an artificial environment results in unstable large-scale cultures. By understanding rather than attempting to push out these micro-invaders, potential alternatives such as "synthetic ecology" as novel scaling up techniques should be explored [17]. This concept is based on the Competitive Exclusion Principle, or Gause's Law, which states that two species competing for the same natural resource scarcely ever occupy a similar niche [18,19]. By "synthesizing" a community of organisms that fills every niche in the ecosystem of the microalgal culture and support, rather than harm, the growth of the phototroph, we would automatically optimize the utilization of nutrients and prevent the establishment of other potentially harmful organisms [17]. In order for synthetic ecology to be a legitimate contender as a novel scaling up technique, a greater understanding of species-specific interactions is required, starting with the bacterial faction, which are present in all of the Earths' biomes [20], and arguably the key players in maintaining balance within a system. Theoretical ecology employs mathematical models to study the emergent patterns in ecosystems dynamics [21]. Because of the many industrial applications of microbial communities, current research has shown great interest in improving our understanding of such systems [22]. In particular, mathematical models and interdisciplinary approaches are fundamental to understanding the crucial underlying mechanisms that regulate community dynamics [23,24]. Since the same system can be inspected at different spatio-temporal scales and at a different degree of complexity, it is important to select the most suitable method to describe the biological phenomena under study in mathematical terms [25]. The first ecosystem models at the population-scale date back to the 1920s with the well-known Lotka-Volterra (LV) predator-prey model [26,27]. Since then, LV models have been extensively used to represent cooperation/competition population dynamics with a system of ordinary differential equations (ODEs). In generalized LV models (gLV) the system includes an arbitrary number of co-existing organisms and they directly represent positive/negative pairwise interactions as fixed parameters [28]. Today, a gLV model can be developed by inferring a co-occurrence network from a time series of metagenomics data [29]. This however requires a reasonable number of time resolved metagenomics data and will provide information on direct, one-on-one interactions only.

Diatoms and bacteria have co-evolved for more than 200 million years [30], and their co-existence is most likely based on a "biological barter trade system", where substances such as trace metals, vitamins, and nutrients (nitrate, phosphate, silicate, carbon) are exchanged. In this work, we built on previous research that investigated algae-bacterial interactions including Provasoli's work from 1958 where he suggested that bacteria can enhance the growth of algae [31], and subsequent species-specific studies that further corroborated his initial idea [32–35]. We first characterized the relative composition of the bacterial community in non-axenic P. tricornutum cultivated in the presence and absence of trace metals, vitamins and sodium metasilicate at different time points. Secondly, using critical peer-reviewed literature we defined the most likely functional roles of the bacterial factions and constructed a putative interaction network. Lastly, from the derived putative network of interactions, we built an ODE model with modified Verhulst equations [36] for microbial growth that included the direct effect of nutrient availability. Mortality rates were also introduced as dependent on specific bactericidal substances. The qualitative mathematical model, with parameters fitted to the available experimental data, served as a proof-of-concept that data as obtained here is sufficient to reconstruct a theoretical model that (a) reproduces the experimental observations, thus demonstrating consistency of our assumptions, and (b) allows for testing different hypotheses regarding the nature of the metabolic interactions underlying the ecosystem dynamics. It therefore represents a starting point to gain a deeper understanding of the principles of microbial community dynamics by an iterative experimental and theoretical approach.

2. Materials and Methods

2.1. Strains and Culture Conditions

All *P. tricornutum* strains were obtained from the Culture Collection of Algae and Protozoa (CCAP) based in Oban, Scotland [37]. All cultures were obtained non-axenic. Based on previous experimental evidence [38], the *P. tricornutum* strain CCAP1052/1B displayed optimal growth in 5L cultures. *P. tricornutum* was cultured in Guillard's medium for diatoms (F/2 + Si) in filtered natural seawater chemically sterilised using sodium hypochlorite and sodium thiosulphate pentahydrate. *P. tricornutum* was grown in two media conditions; (1) complete F/2 medium containing sources of nitrogen (NaNO$_3$) and phosphorus (NaH$_2$PO$_4$·2H$_2$O), as well as trace metals and vitamins with the addition of sodium metasilicate, as per Guillard and Ryther 1962 [39] and Guillard 1975 [40], and (2) minimal media which contained just sources of nitrogen (NaNO$_3$) and phosphorus (NaH$_2$PO$_4$·2H$_2$O) at the same concentration as in the F/2 medium recipe. All cultures were grown in hanging 5L polyethylene bags with a "V" shaped bottom prepared using a heat sealer (Supplementary Figure S1). All cultures had a modified aeration system provided by a 10 mL pipette attached to the main pressurised air supply via 0.2 μm sterile air filters. Cultures were kept at 18–20 °C and 24 h light at an average of 132.3 μmol m^{-2} s^{-1}. All cultures, irrespective of media condition, were inoculated with 250 mL from the same 5L stock culture of actively growing non-axenic *P. tricornutum*.

2.2. Growth Measurements

Growth was monitored every 24 to 48 h using a light microscope and carrying out cell counts of each culture in quadruplicate. During the cell counts the ratios of the four different morphotypes (oval, fusiform, triradiate and cruciform) were recorded, and descriptions of each culture noted. Samples of each culture were subsequently taken using a sterile 10 mL syringe and placed in 50 mL Falcon centrifuge tubes and placed in a −20 °C freezer.

2.3. Genomic DNA Extraction

All samples from days 1, 8, 15, and 22 were thawed in a water bath set at 25 °C. As per de Gouvion Saint Cyr et al. [41], samples were centrifuged for 5 min at 2000 x *g* to gather the *P. tricornutum* in the pellet while particles such as debris, other organisms, bacteria, and soluble substances remain in the supernatant. Because the bacteria might be attached to the *P. tricornutum* cells in the pellet, the pellet was washed with deionised water and then centrifuged for 5 min at 2000× *g*. This was repeated twice to ensure that majority of the bacteria attached to the pellet were released and were included in the community analysis. Genomic DNA extraction was carried out in the Aquaculture and Fisheries Development Centre, University College Cork. Mo Bio's PowerWater® DNA Isolation Kit (MO BIO Laboratories, Inc., Carlsbad, CA, USA, catalogue No. 14900-100-NF) was utilized to carry out the genomic DNA extraction. Presence of gDNA was detected by running a 1% agarose-ethidium bromide gel with 72 wells. The samples were sent on dry ice to Heinrich Heine University, Düsseldorf, for the V6 16S sequencing.

2.4. Barcoded 16S-V6-Next Generation Sequencing

Ion Torrent[TM] barcoded Next Generation Sequencing protocol (Thermo Fischer Scientific Inc., Waltham, MA, USA) was used to sequence the bacterial gDNA [42,43]. Amplification of the V6 hyper variable region of 16S rRNA with forward and reverse primers (Supplementary Table S1) was carried out. Ion Reporter[TM] software (Thermo Fischer Scientific Inc., Waltham, MA, USA) assembled all the raw sequencing data and sorted all the reads using the unique sample-specific barcode sequences and removed them from the reads. The outcome was raw FASTQ files which were ready for analysis using bioinformatics tools.

2.5. Bioinformatics Analysis

A total of 87,077,374 reads were identified. The smallest sample was just over 1 million reads; the largest sample was just under 10 million reads. The sequencing data was subjected to a pipeline adapted and modified from Pyho et al. [44]. Primers were trimmed with fastq-mcf (version 1.04.807) [45], the resulting sequences were quality filtered and clustered into OTUs with usearch (version 8.0.1517; 32Bit-opensource) [46,47]. Taxonomy assignment was done by QIIME (version 1.9.0) [48] with the implemented uclust classifier based on 97% sequence identity to the reference 16S sequences from SILVA 111 database [49]. Statistical analyses were performed in R [50]. The complete protocol containing all processing steps is available on GitHub (see Supplementary Materials).

2.6. Mathematical Model

Starting from our understanding of the organism-to-organism interactions, we developed a dynamic model consisting of 13 ordinary differential equations (ODEs) and including 56 (55 free) parameters (see Appendix B.1). The model was built from the following working hypotheses:

(1)   the growth rate γ of each population followed a standard Verhulst equation [36] parametrized with a carrying capacity and scaled by Monod-type terms [51] that describe the dependency on (micro)nutrients. These terms are in practice positive scaling factors <1.

(2)   the mortality rate of each population was inversely proportional to $(1 + \gamma)$, to account for the fact that cells during replication (high growth rate) were healthier;

(3)   additional contributions to population mortality was given by the presence in the environment of noxious elements like bactericidal substances;

(4)   changes in metabolite concentrations are in general directly proportional to the growth γ of the consumers and producers;

(5)   in the event of micronutrient scarcity (Iron and Vitamins in our model), *P. tricornutum* will secrete more organic carbons favored by those bacteria able to provide the needed micronutrients.

The initial conditions for simulations are different between minimal and complete media:

(1)   the initial quantity of Iron and Vitamins is 10 times higher in complete media;

(2)   the initial quantity of *P. tricornutum* biomass is matched to the first data point.

The parameters were fitted separately in minimal and complete media using a genetic algorithm [52] which was run in different steps to optimise the fit of *P. tricornutum* growth and/or the bacteria relative abundances to the experimental data in evolving system conditions (see Appendix B.2 and Supplementary Material 2). The model was written in Python (Python Software Foundation, https://www.python.org/) and is available on GitHub with instructions and scripts for running (see Supplementary Materials).

3. Results

3.1. Characteristics of Phaeodactylum tricornutum Growth

The media composition was shown to have a significant effect on the growth characteristics of *P. tricornutum*. *P. tricornutum* cultivated in minimal media exhibited a statistically significantly ($p = 0.042$, unpaired Wilcoxon signed rank) higher cell density (11.2×10^6 cells/mL) when compared to cultivation in complete media (9.3×10^6 cells/mL). The growth rates during the exponential phase in both cultures were $\mu_{\text{complete}} = 0.43 \pm 0.07$ d^{-1} and $\mu_{\text{minimal}} = 0.51 \pm 0.04$ d^{-1} respectively. In contrast, the death rates when the cultures "crash" are $\delta_{\text{complete}} = 0.09 \pm 0.02$ d^{-1} and $\delta_{\text{minimal}} = 0.08 \pm 0.04$ d^{-1} respectively.

3.2. Bacterial Community Profile of Phaeodactylum tricornutum Cultures

Bacterial gDNA analysis showed that most of the operational taxonomic units (OTUs) could be assigned to the genera level (Supplementary Figure S2). Of the 9727 OTUs identified, 8109 corresponded to known sequences in the SILVA database (v.118) [49]. The OTU abundance at the phylum level showed that 99.97% of all OTUs belonged to Proteobacteria, Bacteroidetes, Actinobacteria and Firmicutes (Figure 1a). A comparison of the number of individual reads to the number of unique OTUs showed that the high number of reads per phyla was not the result of a single OTU (Supplementary Figure S3). OTUs with hits to known 16S *P. tricornutum* sequences were discarded.

Rarefaction curves were used to evaluate the alpha diversity in the different media conditions as well as at the different time points (Supplementary Figure S4). Species richness in both minimal and complete media was ~3000. Species richness over time remained between ~2400 and ~2600, with reduced species richness (~1300) on day 8 (both minimal and complete media) possibly due to elevated levels of 16S *P. tricornutum* chloroplast reads which had to be omitted. Greatest species richness (~3000) was shown on day 22. All datasets showed a diminished increase in the number of unique species as the sample size increased, confirming adequate species richness in all culture conditions. To compare the species composition between the different samples (days/media) we used a non-metric multidimensional scaling (NMDS) function based on generalized UniFrac distances [53]. This allowed us to characterize the relationship between the particular samples on a visual level by displaying the information contained in the distance matrix. Therefore, similar samples would be placed together in an N-dimensional space. Here we observed a clear gradient of similarity between the bacterial samples from the different time points. The ordination based on the sampling day indicated that the bacterial community was dynamic with a clear divergence visible between day 1 and the other three sampling days. Interestingly the similarity between the different time samples showed the evolving processes of the community over time (overlaps between day 8 with day 15 and day 15 with day 22) and the recovery to the original one (overlap day 22 and day 1) (Figure 1b).

The existence of one dominant family at each investigated time point was a particularly interesting observation. In minimal media (Figure 2a), the lag phase of *P. tricornutum* growth was dominated by Pseudoalteromonadaceae (85%). However, during the log phase, a wide diversity of bacterial families was observed, with members of the Alteromonadaceae family (21%) beginning to dominate. During the stationary phase, a clear dominance of Alteromonadaceae species (55%) in the community was observed. The decline phase, however, showed the Pseudomonadaceae (39%) as a dominant family, with Pseudoalteromonadaceae species (37%) increasing in abundance again. In complete media (Figure 2b), the lag phase was also dominated by Pseudoalteromonadaceae (63%). During the log phase, 50% of the community was composed of members of the Flavobacteriaceae family, with the other 50% distributed among a number of different families. Flavobacteriaceae (46%) remained high in abundance during the stationary phase, with Pseudoalteromonadaceae species (44%) beginning to increase in abundance again. As for minimal media, Pseudoalteromonadaceae (57%) showed a clear dominance of the community during the decline phase.

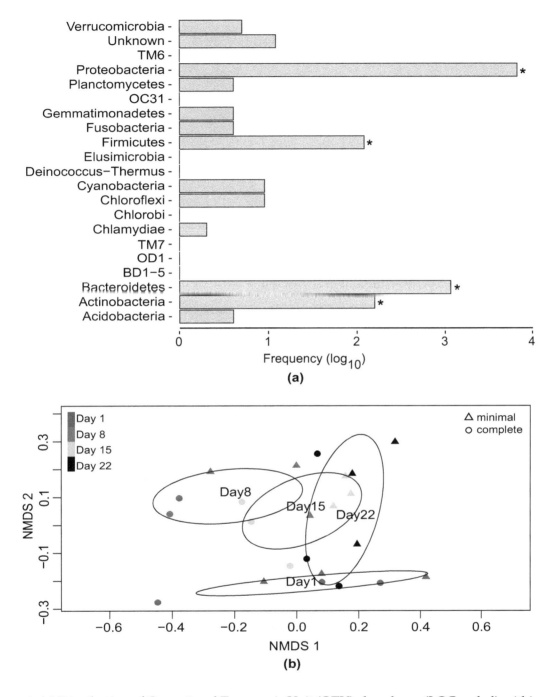

Figure 1. (**a**) Distribution of Operational Taxonomic Unit (OTU) abundance (LOG scaled) within phyla from complete data set. The bins marked with asterisks (*) correspond to 99.97% of all which belong to Proteobacteria, Bacteriodetes, Actinobacteria and Firmicutes. (**b**) Ordination plot of bacterial community in the two media conditions for all sampling points. Triangles and circles correspond to minimal media and complete media conditions, respectively. Blue represents day 1. Red day 8. Green day 15. Black day 22. The ellipses correspond to the 99% confidence interval to each group centroid.

An adapted version of PermanovaG was used to carry out permutational multivariate analysis of variance using multiple distance matrices which were previously calculated based on the generalized UniFrac distance [53]. The significance for the test was assessed by 5000 permutations. The results of the PermanovaG tests support the NMDS ordination, confirming a statistically significant effect in the bacterial community profile at the different sampling points and in the two media conditions whereas no significant effect was found in the experimental replicates (Figure A1).

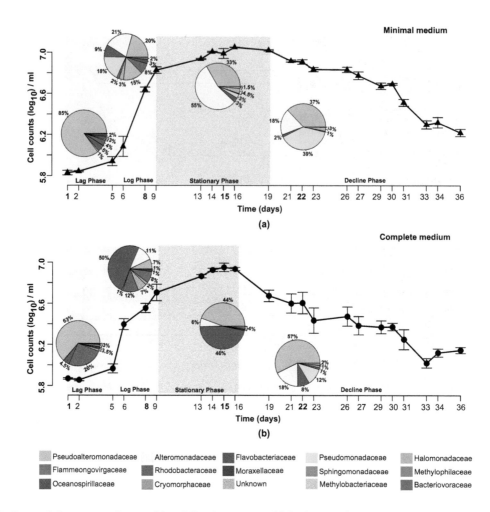

Figure 2. Bacterial community profile of *P. tricornutum* (CCAP 1052/1B) over a 36 day period in culture conditions: (**a**) minimal media; (**b**) complete media. The growth curves are partitioned into lag (green), log (blue), stationary (red), and decline (yellow) phases. The abundance (%) of the "Top Ten" bacterial families (corresponding colors described in the key) is depicted in pie charts on days 1, 8, 15 and 22 in both media conditions.

3.3. *Effect of Temporal Evolution and Media Composition on the Bacterial Community Profile*

We compared the bacterial community profiles over time and in the different media conditions at the family level to avoid diluting the signal of the less abundant genera. Supplementary Figures S5 and S6 show no dynamical difference within the genera that cannot be observed at the family level. By investigating the bacterial community dynamics at the family level, we also included taxonomical information that is unavailable at the genus level. The families over-represented in all samples were Pseudoalteromonadaceae, Alteromonadaceae, Flavobacteriaceae and Pseudomonadaceae. Figure 2 illustrates the temporal evolution of the bacterial community in both minimal and complete media with a unique composition at each time point. A remarkable feature is that at all investigated time points there exist one or two dominant families.

In complete media, members of the Pseudoalteromonadaceae family were highly abundant when *P. tricornutum* cell densities were low (63% and 57% on day 1 and day 22, respectively). Flavobacteriaceae species dominated (50%) when the *P. tricornutum* culture was growing exponentially (day 8). Day 15, when *P. tricornutum* cell densities were at their highest, showed co-dominance of both Flavobacteriaceae (46%) and Pseudoalteromonadaceae (44%).

In minimal media, members of the Pseudoalteromonadaceae family were highly abundant when *P. tricornutum* cell densities were low. On day 22 Pseudomonadaceae (39%) and Pseudoalteromonadaceae

(37%) were both overrepresented. When the *P. tricornutum* culture was in its exponential growth phase (day 8), a cluster of Families dominated; namely Alteromonadaceae (21%), Pseudoalteromonadaceae (20%), Pseudomonadaceae (18%), Halomonadaceae (15%) and Flavobacteriaceae (9%). When the cell density of *P. tricornutum* peaked (day 15), the Alteromonadaceae species took over (55%).

The bacterial communities within the two media conditions on day 1 were more closely related than the communities on days 8 and 15 (see Supplementary Table S2 for generalized UniFrac distances). As the cultures begin to "crash" (day 22), the bacterial communities in the two media conditions increased in similarity again. In general, the main families identified showed a distinct pattern of disappearance and regeneration within the bacterial community. In the complete media, Pseudoalteromonadaceae species started at 63% (day 1), dropped in abundance to 7% (day 8) then recovered to 57% (day 22). Flavobacteriaceae species, in complete media, started at 4.5% (day 1), increased in abundance to 50% (day 8), and then fell back to 8% (day 22). In the minimal media, Alteromonadaceae species had an abundance of only 1% (day 1), peaked at 55% (day 15), and decreased down to 18% (day 22).

3.4. Network of Putative Interactions between Phaeodactylum tricornutum and Identified Bacterial Families

The putative roles of each of the dominant families are illustrated in Figure 3. Based on an extensive literature review, five metabolites were identified as playing a crucial role in the interactions between *P. tricornutum* and the identified bacterial families. These are: bactericidal metabolites; iron; vitamins; dissolved organic carbons; dissolved organic phosphates.

- *Bactericidal metabolites.* Several species of the Pseudoalteromonadaceae family have been reported to possess bactericidal effects [54]. This ability to suppress the growth of competing bacteria could explain the dominance of Pseudoalteromonadaceae in almost all cultures irrespective of media composition. *P. tricornutum* also demonstrates bactericidal properties by excreting fatty acids (such as eicosapentaenoic acid or EPA), nucleotides, peptides, and pigment derivatives [55].
- *Iron.* Iron acquisition is essential for biological processes such as photosynthesis, respiration and nitrogen fixation. Bacteria produce and excrete siderophores, which scavenge iron. Diatoms are not known to produce siderophores, but genome sequence analyses identified the presence of a gene orthologue of a bacterial ferrichrome binding protein that suggests the possibility of iron (III)-siderophore utilization by *P. tricornutum* [56,57]. Furthermore, it was shown that *P. tricornutum* was able to uptake siderophores ferrioxamines B and E [58].
- *Vitamins.* Prokaryotes are thought to be the main producers of B vitamins [59,60]. Although *P. tricornutum* does not require cobalamin, thiamine and biotin [61], production of organic compounds such as EPA can be considerably enhanced by the bioavailability of co-factors such as cobalamin [62]. This provides the basis for potential mutualistic interactions. For example, Alteromonadales, dominant in our cultures, are thought to be capable of producing B vitamins [63].
- *Dissolved Organic Carbon (DOC).* It is estimated that up to 50% of carbon fixed via phytoplankton-mediated photosynthesis is utilized by marine bacteria [64], mainly as DOC compounds, defined as the organic material <0.7 μm in size [65]. DOC from diatoms originates either from live cells or recently lysed or grazed cells, which determines the type of DOCs available, and therefore are likely to influence the bacterial consortia associated with the diatom [30].
- *Dissolved Organic Phosphate (DOP).* Both diatoms and bacteria primarily utilize orthophosphate as a source of phosphorus. However, to access phosphate from DOP compounds, both diatoms and bacteria developed mechanisms to release orthophosphate (PO_4^{3-}) from DOP. The mechanism is not species-specific, which consequently means the "free" orthophosphates can be acquired by any organism [66].

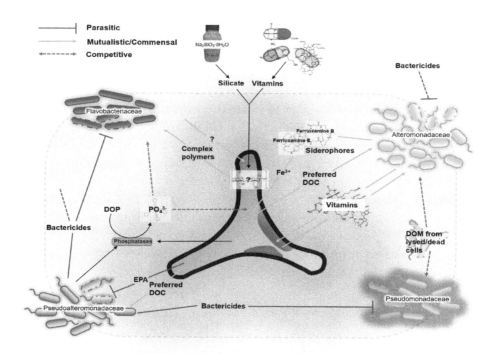

Figure 3. Network of putative interactions between *Phaeodactylum tricornutum* and identified bacterial families. The dotted grey line depicts the "phycosphere"; a term coined by Bell and Mitchell in 1972 as an aquatic equivalent of the "rhizosphere", denoting the region extending outwards from the algal cell in which bacterial growth is stimulated by extracellular products of the alga [67].

3.5. Mathematical Model Simulations

Based on the network of putative interactions between diatoms, bacteria, and the environment, we constructed a dynamic mathematical model, based on generalized Verhulst growth-laws [36] extended with Monod-type terms [51] to reflect the dependencies on metabolites (see Materials and Methods and Appendix B.1). Figure 4 presents results of the model simulations after the model parameters were fitted to the data in minimal and complete media conditions, respectively (see Appendix B.1). Experimental data are superimposed. The top panel shows biomasses of the five organisms (data available only for the diatom), the bottom panel shows relative bacteria abundance versus time (individual biomass divided by total bacterial biomass). Because of the qualitative nature of the model, units are arbitrary. The figures show that the model is able to reproduce the main features of the bacterial community dynamics, such as the disappearance and return of Pseudoalteromonadaceae in complete media and the peak of Alteromonadaceae at the end of the exponential growth phase of *P. tricornutum* in minimal media. Supplementary Figure S7b and S7d show the dynamics of metabolite concentrations, for which no data are available.

Due to the large number of free parameters, the fit was certainly not unique. Supplementary Material 2 presents additional checks we performed on the parameter fitting procedures. The parameter space could be in principle reduced to 43 free parameters, but this did not change the results. It was not possible to find a unique parameter set valid in both minimal and complete media conditions. This is however consistent with the fact that the model was not constructed to capture effects like metabolic re-adjustments, something that would be observed e.g., as a different parameter value for growth rate or metabolite consumption. With the data available, it was not possible to make any quantitative statements about the actual interaction parameters, neither could it be assumed that simulation results are of general validity. Despite these limitations, the model did represent a possible configuration of diatom-bacteria-environment interactions, which was in agreement with the experimentally observed bacterial dynamics.

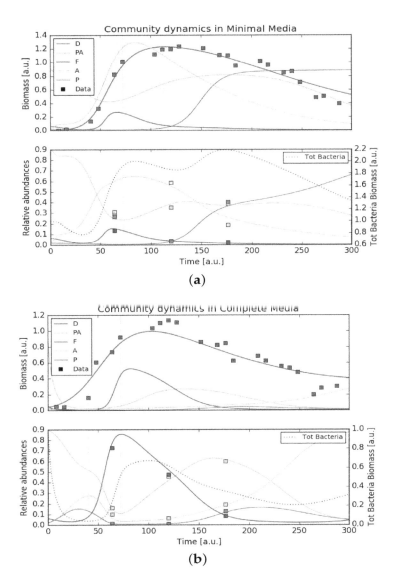

Figure 4. Simulation results (lines) and experimental data (squares) for communities of *P. tricornutum* (*D*), Pseudoalteromonadaceae (*PA*), Flavobacteriaceae (*F*), Alteromonadaceae (*A*) and Pseudomonadaceae (*P*) in (**a**) minimal media and (**b**) complete media conditions. The top panel shows the biomass time course (arbitrary units) for the five organisms and the rescaled data points (squares) for the *P. tricornutum*. The bottom panel shows the variations in relative abundances of the four bacteria (single bacteria biomass/total bacteria biomass) over time and the three sets of data points from the sequencing analysis (the first data point is used as starting condition at time 0). Also shown in the bottom plot (dotted line, right *y*-axis) is the total bacterial biomass in arbitrary units.

In order to test the stability of the bacterial community and how it supports the growth of *P. tricornutum* we ran simulations using the same set of parameters (either minimal or complete media conditions) and varied the initial community composition removing one bacteria per simulation. In complete media the simulated growth of *P. tricornutum* still fit the experimental data, rather independently from the bacterial community (Figure 5). Also under axenic conditions, diatom growth was predicted to be largely unperturbed. This situation was different in minimal media. While under these conditions diatom growth was also unaffected upon removal of the three bacterial families Pseudoalteromonadaceae, Flavobacteriaceae and Pseudomonadaceae, removing Alteromonadaceae from the community resulted in the total absence of *P. tricornutum* growth (Figure 6). This behavior was expected from the hypothesized central role of Alteromonadaceae in supplying the diatom with micronutrients. Surprisingly, the removal of a single bacteria from the community in both media conditions still gave, in general, a good fit of the (recomputed) relative abundances, except when removing Alteromonadaceae in minimal media (as direct consequence of what stated previously) and when removing Pseudoalteromonadaceae in both media conditions. This hinted to a relevant role of Pseudoalteromonadaceae in regulating the community composition through its predatory strategy of releasing bactericidal substances. Finally, the community composition at the last time point was overall better captured. This suggests that the mathematical model consistently captured the general interactions leading one bacterial family to dominate over the others on the long term.

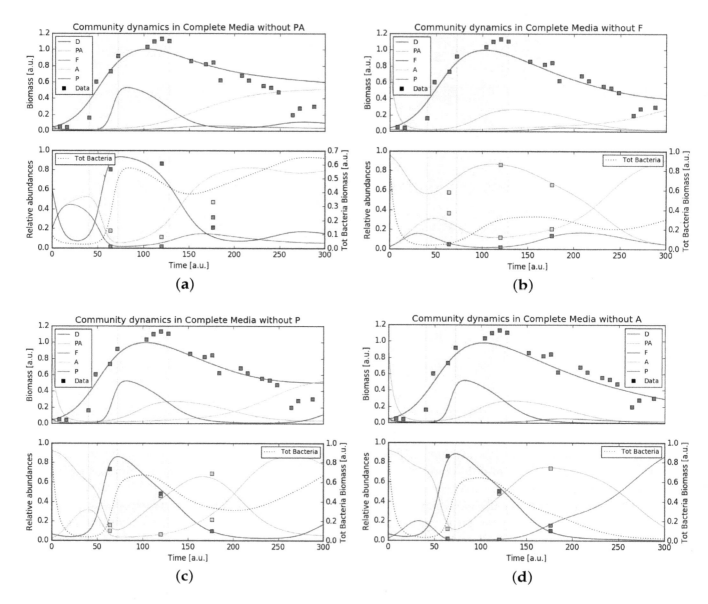

Figure 5. Reduced community simulation results (lines) and experimental data (squares) for communities of *P. tricornutum* (*D*), Pseudoalteromonadaceae (*PA*), Flavobacteriaceae (*F*), Alteromonadaceae (*A*) and Pseudomonadaceae (*P*) in complete media conditions. Simulations are run removing from the bacterial community one member: (**a**) *PA*; (**b**) *F*; (**c**) *P*; (**d**) *A*. The top panel shows the biomass time course (arbitrary units) for the four organisms and the rescaled data points (squares) for the *P. tricornutum*. The bottom panel shows the variations in relative abundances of the three bacteria (single bacteria biomass/total bacteria biomass) over time and the three sets of data points from the sequencing analysis (the first data point is used as starting condition at time 0). Also shown in the bottom plot (dotted line, right *y*-axis) is the total bacterial biomass in arbitrary units.

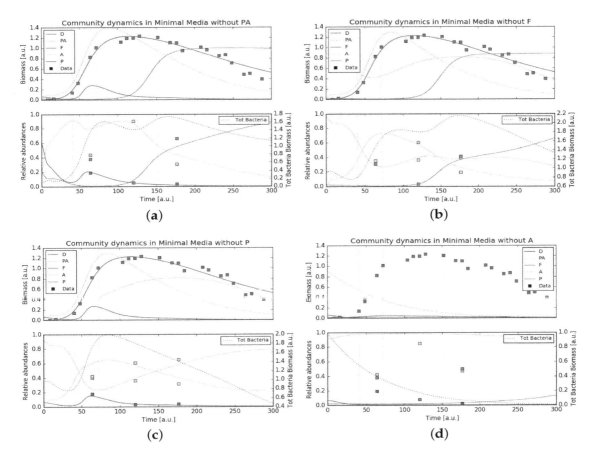

Figure 6. Reduced community simulation results (lines) and experimental data (squares) for communities of *P. tricornutum* (*D*), Pseudoalteromonadaceae (*PA*), Flavobacteriaceae (*F*), Alteromonadaceae (*A*) and Pseudomonadaceae (*P*) in minimal media conditions. Simulations are run removing from the bacterial community one member: (**a**) *PA*; (**b**) *F*; (**c**) *P*; (**d**) *A*. The top panel shows the biomass time course (arbitrary units) for the four organisms and the rescaled data points (squares) for the *P. tricornutum*. The bottom panel shows the variations in relative abundances of the three bacteria (single bacteria biomass/total bacteria biomass) over time and the three sets of data points from the sequencing analysis (the first data point is used as starting condition at time 0). Also shown in the bottom plot (dotted line, right *y*-axis) is the total bacterial biomass in arbitrary units.

4. Discussion

In nature, *P. tricornutum* does not exist as an isolated entity. In fact, it is part of a complex ecosystem whose complete network of interactions with both its environment and other organisms remains poorly understood. Microbial ecosystems are of high interest for a wide range of applications in fields, such as medicine, renewable energy, and agriculture. Within the scope of this project, the complexity of a natural, variable system was reduced by investigating the batch growth of non-axenic laboratory strains of *P. tricornutum* from a culture collection. The cultivation method we developed was designed to compromise between highly controlled small-scale laboratory conditions and a large-scale industrial set-up. The bacterial community was characterized and the community dynamics investigated in two conditions: minimal and complete media. The data was then complimented with an extensive study of existing, peer-reviewed literature to identify the putative role of the dominant bacterial families associated with *P. tricornutum*. We then validated the derived network of interactions by developing a mathematical model which could reproduce the observed dynamics. The presented approach, integrating experiments, bioinformatics and mathematical modeling, illustrates a possible way towards the development of a monitoring pipeline for non-axenic microalgae cultures.

4.1. Experimental Observation of the Dynamics of the Bacterial Community Associated to Phaeodactylum tricornutum

The growth dynamics of *P. tricornutum* in the two media conditions showed an accelerated "culture crash" in the complete media compared to the minimal media, which indicated a more stable culture in the minimal media (Figure 2). This also suggested that non-axenic cultures of *P. tricornutum* might not require expensive trace metals and vitamins for optimal growth under the conditions provide, an observation crucial to the large-scale industrial cultivation of *P. tricornutum* as this would drastically decrease the production costs. Simultaneously, the dynamics of the bacterial community revealed that the community in the minimal media increased in complexity over time. The link between ecosystem complexity and stability based on theoretical and experimental data has been debated by ecologists for over half a century [68–71]. Our observations are in agreement with more recent hypotheses indicating that diversity generally increases the stability of an ecosystem [72].

4.2. Literature-Based Assessment of the Putative Role of Each Bacterial Family

The bioinformatics analysis of bacterial gDNA abundance showed clear dominance of four bacterial families: Pseudoalteromonadaceae, Alteromonadaceae, Flavobacteriaceae and Pseudomonadaceae. These bacteria were over-represented in all samples and their relative abundances showed different temporal dynamics among the two *P. tricornutum* growth conditions. In order to understand the putative functional role of these bacteria an extensive study of peer-reviewed literature was carried out.

Pseudoalteromonadaceae. Members of Pseudoalteromonadaceae family have been isolated from coastal, open and deep-sea waters, sediments, marine invertebrates, as well as marine fish and algae [73]. The Pseudoalteromonadaceae family has three genera, namely *Pseudoalteromonas*, *Algicola* and *Psychrosphaera* [74]. Several species of Pseudoalteromonadaceae are reported to possess antibiotic properties with bactericidal effects [54]. For example, concentrated supernatant of a marine bacterium *Pseudoalteromonas* sp. strain A28 contained various enzymes including proteases, DNAses, cellulases, and amylases, capable of causing the lysis of the diatom *Skeletonema costatum* [75]. Species of Pseudoalteromonadaceae are also capable of producing cold-adapted enzymes [76–81]. Pseudoalteromonadaceae species can produce extracellular polymeric substances allowing them to colonise surfaces, enhancing nutrient uptake whilst limiting diffusion of particular substances across the cell membrane [82]. The ability of Pseudoalteromonadaceae species to suppress the growth of competing bacteria could explain the dominance of Pseudoalteromonadaceae in almost all cultures irrespective of media composition, particularly when *P. tricornutum* abundance is limited (Figure 2, days 1 and 22). *P. tricornutum* on the other hand, may protect other bacterial community members from the bacteriolytic ability of Pseudoalteromonadaceae by producing specific antibacterial compounds themselves. Desbois et al. showed that *P. tricornutum* excreted bacteriolytic fatty acids such as eicosapentaenoic acid (EPA; 20:5n-3), nucleotides, peptides, and pigment derivatives that can eliminate unwanted competition for nutrients such as organic phosphates from certain bacteria [55].

Alteromonadaceae. The Alteromonadaceae family consists of 16 (yet annotated) named genera found predominantly in marine environments [74]. Members of this family were isolated from nutrient-rich environments such as coastal, open, and deep-sea waters, sediments, marine invertebrates and vertebrates, algae, and temperate and Antarctic marine environments [83]. They are able to utilize a vast array of compounds as carbon sources; from glucose to glycerol [74]. Members of this family are known siderophore producers [57,84,85]. Greek for "iron carrier", siderophores are a group of iron scavengers that act by chelating iron (III) that are produced and excreted by bacteria, and some cyanobacteria, which then reuptake the siderophores with bound iron (III) via outer-membrane transporters that are siderophore-specific [86]. Most bioactive trace metals, including iron, exist at nanomolar to picomolar concentrations in our oceans, approximately one-millionth of the intracellular concentration in diatoms [87,88]. No trace metals, including iron (III), were provided to minimal media cultures. However, natural seawater may contain minute traces of bioactive trace metals. The high abundance of Alteromonadaceae in the minimal media suggests a potential supportive role in sequestering traces of iron (III) that may be present in the sterile natural seawater to the *P. tricornutum* (Figure 2). This is further supported by the very low level of Alteromonadaceae in the complete media (11% in complete media compared to 55% in minimal media, both on day 15) where the culture has been supplied with 11.7 µM of iron (III) chloride hexahydrate.

Flavobacteriaceae. Flavobacteriaceae are members of the Bacteroidetes phylum and include over 120 genera found in soil, sediments and seawater (see [89] for further references). Flavobacteriaceae belong within the *Cytophaga-Flavobacterium* cluster which has been shown to account for more than 10% of the total bacterial community in coastal and offshore waters [90–92]. Members of Flavobacteriaceae can proficiently degrade various biopolymers such as cellulose, chitin and pectin [93,94]. They were shown to be omnipresent during phytoplankton blooms, and their preference for consuming more complex polymers rather than monomers suggests an active role in the processing of organic matter during these blooms [95,96]. Although the exact mechanisms behind them are not perfectly understood, algal blooms are a consequence of exponential growth of phytoplankton [97]. In this respect, the phase of exponential growth of *P. tricornutum* in complete media, when our results showed highest abundance of Flavobacteriaceae, is the artificial equivalent of an algal bloom of *P. tricornutum* (Figure 2). In the minimal media, the abundance of Flavobacteriaceae remains very low; at its maximum on day 8 it only accounts for 9% of the total bacterial community. Members of the Flavobacteriaceae family could be more demanding than other bacteria that require lower nutrient levels to thrive.

Pseudomonadaceae. Pseudomonadaceae are an extraordinarily diverse family of bacteria found in almost all habitats on Earth; in soils, freshwater as well as marine environments, as well as plant and animal-associated pathogens [98]. Species from the *Pseudomonas* genus are the best studied of the Pseudomonadaceae family, whose sheer genetic diversity explains the ability to thrive in such a wide range of environments [99]. Marine isolates from the *Pseudomonas* genus have been shown to produce a wide range of bioactive compounds, many of which exhibit antibacterial as well as antiviral properties (see [100] for further references). Our results, indeed show an elevated level of Pseudomonadaceae OTUs evident on day 22 of the complete media cultures, and on days 8 and 22 of the minimal media cultures. The increased presence of Pseudomonadaceae when the *P. tricornutum* culture has "crashed" could be attributed to its ability to produce antibacterial compounds allowing members of this family to begin to thrive in the community through inhibition of its competitors. Given its exceptional genetic diversity, and thus, its metabolic versatility, allows for members of Pseudomonadaceae to be truly saprophytic; providing a hypothetical explanation of its abundance we could measure when the *P. tricornutum* cultures crash (Figure 2, day 22 in both media conditions).

4.3. Putative Network of Interactions and Validation with a Qualitative Mathematical Model

The literature review work revealed interesting insights into the possible metabolic exchanges going on and allowed to infer interaction links among *P. tricornutum* and its associated bacterial community. We critically considered which metabolites were most relevant for survival (organic carbons for the bacteria, iron, vitamins and phosphates for the diatom) and which ones could play a role in competition and predation among the microbes (bactericidal metabolites). From these considerations we designed a putative network of interactions that was then translated into a mathematical model. In particular we chose, besides the five microbes' biomasses, a total of eight metabolites as the variables that directly and specifically influence the interactions among the different organisms. These were: four different possible sources of organic carbons, each preferred by a different bacterial family [30]; two bactericidal substances, one produced by Pseudoalteromonadaceae and affecting all other bacteria [54], the other produced by *P. tricornutum* and targeting specifically Pseudoalteromonadaceae [55]; vitamins produced by Alteromonadaceae and needed by *P. tricornutum* [63]; bio-available iron that is chelated by Alteromonadaceae and efficiently absorbed by *P. tricornutum* [57]. For the scope of the model, we ignored other free iron forms that can be uptaken by all bacteria as well as phosphates that are not species-specific and are present in both minimal and complete media.

Direct metabolic exchanges are known to be central in microbial community interactions [101], but usually population dynamics models like gLV [28] do not include this information. This work, therefore, modified the standard formulation of the Verhulst equation [36] for bacterial growth to include organism-to-organism interactions depending on the production/consumption of metabolites, modeled as Monod-type terms [51]. Nutrients availability can indeed drastically change the "metabolic state" of an organism, inducing a reprogramming of resource allocation to face nutrient scarcity [102]. This was shown at the gene expression level for example for *Escherichia coli* [103] and *Shewanella oneidensis* [104] grown in minimal and rich media condition. Micronutrients can as well affect microbial gene expression, as is for example the case with vitamin B_{12}, whose presence induces the expression of the cofactor-dependent methionine synthase enzyme METH, while in its absence the cofactor-independent methionine synthase enzyme METE is expressed [105]. An ODE model at the population level cannot, of course, capture mechanisms such as metabolic shifts caused by changes in the environment such as the supplementation of minimal or complete media [25]. Therefore, we did not expect to find a unique set of parameters for the model in the two conditions. However, the parameters fitted to the data of *P. tricornutum* with four bacterial families still provide good fits in simulations with altered community composition. Even though the parameter values could not directly be interpreted biologically, we could use the simulation results on metabolites dynamics (an information absent in data) to speculate about the reason for the lower cell count of *P. tricornutum* in complete media with respect to minimal media (Supplementary Material 3). In minimal media, Alteromonadaceae maintained constant iron levels and the fitted values for parameters characterizing *P. tricornutum*'s sensitivity to micronutrients levels were significantly lower. This would suggest a key role of Alteromonadaceae in supporting *P. tricornutum* growth combined with the diatom's adaptation to scarce micronutrients availability.

Considering the limited information that can be extracted from the current experimental data available, the model we proposed is purely qualitative and provides a proof-of-concept that a quantitative model can, in principle, be constructed if dedicated experiments are designed for calibration. The current qualitative model provides therefore a preliminary validation of our putative network of interactions, and serves as motivation for further research bringing the model to a quantitative, predictive level. Indeed, starting with systematic measurements of model parameters in co-cultivation experiments, the simulations can gain predictive power and become a powerful tool towards the goal of synthetic community design and control.

5. Conclusions and Outlook

This study demonstrated that the bacterial community associated with non-axenic laboratory strains of *P. tricornutum* is not randomly assembled but follows dynamics that can be reproduced. We postulate that a role within the community can be filled by a number of bacterial species capable of carrying out a certain function (guilds) rather by one specific species of bacteria. Which bacteria fill the role is dependent upon the environmental characteristics and the prevailing needs of the community as a whole at any given time. Unfilled niches will be seized by bacteria with the ideal metabolic functionality. The absence of certain micronutrients creates a new niche that can be filled by a certain unique bacterial faction. Further work is necessary to explore the hypotheses postulated and to further develop the qualitative mathematical model to understand the specific community roles and the ecological niches. In the context of fundamental research, one approach would be to carry out systematic time-resolved omics studies, which provide a holistic view of the genes (genomics) and metabolites (metabolomics) in a specific biological sample in a non-targeted and non-biased manner [106], and use them to develop an "expanded gLV" mathematical model where the species specific interaction terms depend on the metabolite concentrations. This would allow to derive a network of interactions independent of *a priori* hypotheses, and thus represent a significant step forward in understanding community dynamics based on metabolic exchanges. In the context of industrial scale-up, systematic co-culture experiments with culturable members of the bacterial families of interest, chosen based on desired functional roles, could be used to parametrize a mathematical model like the one we presented and develop it into a powerful predictive tool for culture monitoring. For example, samplings assessing the community composition can be used to predict the harvesting point and avoid "culture crash". The development of novel co-cultivation strategies for scale-up is extremely relevant for pharma- and nutraceutical, as well as animal feed industries. Therefore there will be increasing interest in further research into co-cultivation approaches and in general in the field of synthetic ecology.

Acknowledgments: This work was supported by the European Commission Seventh Framework Marie Curie Initial Training Network project 'AccliPhot' (grant agreement number PITN-GA-2012-316427) to F.W.M. and A.S.; and the Deutsche Forschungsgemeinschaft, Cluster of Excellence on Plant Sciences CEPLAS (EXC 1028) to O.P., A.S. and O.E. Genomic DNA extraction was carried out at the Aquaculture and Fisheries Development Centre, University College Cork, Ireland (funded by Beaufort Marine Research Award in Fish Population Genetics funded by the Irish Government under the Sea Change Programme). Barcoded 16S-V6-Next Generation Sequencing was carried out by the Genomics and Transcriptomics Laboratory at Heinrich-Heine University, Düsseldorf, Germany.

Author Contributions: F.W.M., O.E. and J.M. conceived and designed the experiments; F.W.M. performed the experiments; O.P. performed the bioinformatics and statistical data analysis; A.S. developed the mathematical model and performed simulations; all authors contributed to the interpretation of the results and wrote the paper.

Appendix A. Data Analysis

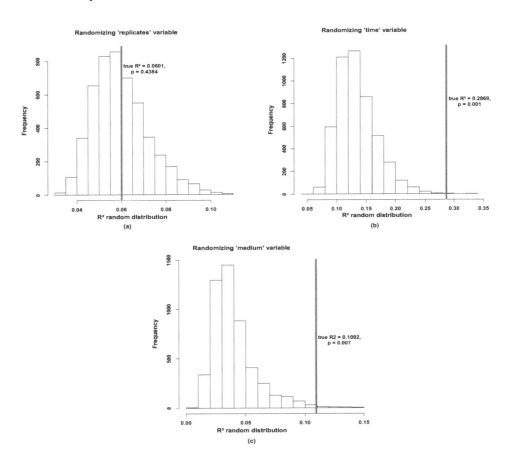

Figure A1. Beta diversity. A modified version of PermanovaG was used to carry out permutational multivariate analysis of variance using multiple distance matrices. The distance matrices [24 × 24] were previously calculated based on the generalized UniFrac distance [53], weighted UniFrac and unweighted UniFrac [107] distance. The significance for the test was assessed by 5000 permutations. (**a**) shows no significant effect between the replicates (*p*-value of 0.4384). (**b**) shows a significant effect for the time variable (*p*-value of 0.001). (**c**) shows also shows a significant effect for the medium variable (*p*-value of 0.007).

Appendix B. Mathematical Model Additional Material

The mathematical model is a system of 13 ODEs describing the variation in time of the populations (cell counts) of

- *P. tricornutum (D)*;
- Pseudoalteromonadaceae (*PA*);
- Flavobacteriaceae (*F*);
- Alteromonadaceae (*A*);
- Pseudomonadaceae (*P*);

and the production and consumption of the metabolites we consider as mainly contributing to drive the community dynamics:

- the dissolved organic carbons of preference for *PA* and *A* (DOC_{PA} and DOC_A, respectively);
- the complex polymers (COP) consumed by *F*;
- generic vitamins (Vit) and iron (Fe) needed by *D* and produced by *A*;
- bactericidial molecules (EPA and Bac, produced by *D* and by *PA* respectively);
- the dissolved organic matter (DOM).

The model has 55 unknown free parameters:

- 5 carrying capacities *CC*;
- 34 maximal rates *v*;
- 15 Monod-type coefficients *K*;
- the fraction of DOC_A-dependent growth of *A*, ϵ_{DOC_A}.

Appendix B.1. ODEs System

Five ODEs describe the variation in time of the populations of organism *O*, with γ^O and δ^O being its growth and death rate:

$$\frac{dD}{dt} = \gamma^D D - \delta^D D \tag{A1}$$

$$\frac{dPA}{dt} = \gamma^{PA} PA - \delta^{PA} PA \tag{A2}$$

$$\frac{dF}{dt} = \gamma^F F - \delta^F F \tag{A3}$$

$$\frac{dA}{dt} = \gamma^A A - \delta^A A \tag{A4}$$

$$\frac{dP}{dt} = \gamma^P P - \delta^P P \tag{A5}$$

Eight ODEs describe the variation in time of the metabolites *J*, with $v_J^{\text{prod/cons}(O)}$ being the maximal production/consumption rate of *J* by organism *O*:

$$\frac{d\text{Vit}}{dt} = v_{\text{Vit}}^{\text{prod}(A)} \gamma^A A - v_{\text{Vit}}^{\text{cons}(D)} \gamma^D D \tag{A6}$$

$$\frac{d\text{Fe}}{dt} = v_{\text{Fe}}^{\text{prod}(A)} \gamma^A A - v_{\text{Fe}}^{\text{cons}(D)} \gamma^D D \tag{A7}$$

$$\frac{d\text{DOC}_{PA}}{dt} = v_{\text{DOC}_{PA}}^{\text{prod}(D)} \gamma^D D - v_{\text{DOC}_{PA}}^{\text{cons}(PA)} \gamma^{PA} PA \tag{A8}$$

$$\frac{d\text{DOC}_A}{dt} = (v_{\text{DOC}_A}^{\text{prod}(D)} + \phi) \gamma^D D - v_{\text{DOC}_A}^{\text{cons}(A)} \gamma^A A \tag{A9}$$

$$\frac{d\text{COP}}{dt} = (v_{\text{COP}}^{\text{prod(D)}} + \psi)\gamma^D D - v_{\text{COP}}^{\text{cons(F)}}\gamma^F F \tag{A10}$$

$$\frac{d\text{EPA}}{dt} = v_{\text{EPA}}^{\text{prod(D)}}\gamma^D D - v_{\text{EPA}}^{\text{deg}}\text{EPA} \tag{A11}$$

$$\frac{d\text{Bac}}{dt} = v_{\text{Bac}}^{\text{prod(PA)}}\gamma^{PA} PA - v_{\text{Bac}}^{\text{deg}}\text{Bac} \tag{A12}$$

$$\frac{d\text{DOM}}{dt} = v_{\text{DOM}}^{\text{prod(D)}}\delta^D D - v_{\text{DOM}}^{\text{cons(A)}}\gamma^A A - v_{\text{DOM}}^{\text{cons(P)}}\gamma^P P \tag{A13}$$

ϕ and ψ are additional terms for DOC_A and COP production respectively (see Appendix B.1.1). v_J^{deg} is the degradation rate of the bactericidal substances. Organism O growth and death rates depend in general on carrying capacity CC^O, maximal rates $v_{\gamma/\delta}^O$ and on metabolites concentrations J with Monod-type coefficient K_J^O and eventually maximal rates v_J^O:

$$\gamma^D = v_\gamma^D \cdot \frac{\text{Vit}}{\text{Vit} + K_{\text{Vit}}^D} \frac{\text{Fe}}{\text{Fe} + K_{\text{Fe}}^D}(1 - \frac{D}{CC^D}) \tag{A14}$$

$$\delta^D = v_\delta^D \frac{1}{1 + \gamma^D} \tag{A15}$$

$$\gamma^{PA} = v_\gamma^{PA} \frac{\text{DOC}_{PA}}{\text{DOC}_{PA} + K_{\text{DOC}_{PA}}}(1 - \frac{PA}{CC^{PA}}) \tag{A16}$$

$$\delta^{PA} = v_\delta^{PA}(1 + \frac{v_{\text{EPA}}^{PA} \cdot \text{EPA}}{\text{EPA} + K_{\text{EPA}}})\frac{1}{1 + \gamma^{PA}} \tag{A17}$$

$$\gamma^F = v_\gamma^F \frac{\text{COP}}{\text{COP} + K_{\text{COP}}}(1 - \frac{F}{CC^F}) \tag{A18}$$

$$\delta^F = v_\delta^F(1 + \frac{v_{\text{Bac}}^F \cdot \text{Bac}}{\text{Bac} + K_{\text{Bac}}^F})\frac{1}{1 + \gamma^F} \tag{A19}$$

$$\gamma^A = \gamma_{\text{DOC}_A}^A + \gamma_{\text{DOM}}^A \tag{A20}$$

$$\gamma_{\text{DOC}_A}^A = v_\gamma^A \frac{\epsilon_{\text{DOC}_A} \cdot \text{DOC}_A}{\text{DOC}_A + K_{\text{DOC}_A}^A}(1 - \frac{A}{CC^A}) \tag{A21}$$

$$\gamma_{\text{DOM}}^A = v_\gamma^A \frac{(1 - \epsilon_{\text{DOC}_A}) \cdot \text{DOM}}{\text{DOM} + K_{\text{DOM}}^A}(1 - \frac{A}{CC^A}) \tag{A22}$$

$$\delta^A = v_\delta^A(1 + \frac{v_{\text{Bac}}^A \cdot \text{Bac}}{\text{Bac} + K_{\text{Bac}}^A})\frac{1}{1 + \gamma^A} \tag{A23}$$

$$\gamma^P = v_\gamma^P \frac{\text{DOM}}{\text{DOM} + K_{\text{DOM}}^P}(1 - \frac{P}{CC^P}) \tag{A24}$$

$$\delta^P = v_\delta^P(1 + \frac{v_{\text{Bac}}^P \cdot \text{Bac}}{\text{Bac} + K_{\text{Bac}}^P})\frac{1}{1 + \gamma^P} \tag{A25}$$

For example in Equation (A14), describing the growth rate of the diatom, the Verhulst growth equation $dD/dt = v_\gamma^D(1 - D/CC^D)$ describes a standard logistic growth, while adding the Monod-type coefficients of the form $X/(X + K_X)$ introduce a dependency on the micronutrients Vit and Fe, in practice scaling down the effective growth rate if micronutrients are scarce. In the case of A, where growth is thought to be sustained by two different complementary nutrients, the final growth γ can be represented as the sum of two terms $\gamma_{\text{DOC}_A}^A$ and γ_{DOM}^A (Equations (A21) and (A22)), with the parameter $0 < \epsilon_{\text{DOC}_A} < 1$.

Appendix B.1.1. DOC_A and COP Production

When D is grown in minimal media conditions, the emergence of A is observed over F. From this observation we hypothesise that D can produce extra organic carbons for either A or F depending on the scarcity of micronutrients to favor the growth of A if more Vit or Fe is needed. We model the production of DOC_A and COP (Equations (A9) and (A10)) introducing the functions ϕ and ψ defined as:

$$\phi = v^D_{\text{DOC}_A\text{COP}} \cdot (1 - \xi) \tag{A26}$$

$$\psi = v^D_{\text{DOC}_A\text{COP}} \cdot \xi \tag{A27}$$

$$\xi = \frac{\text{Vit}^4}{\text{Vit}^4 + K'^D_{\text{Vit}}} \frac{\text{Fe}^4}{\text{Fe}^4 + K'^D_{\text{Fe}}} \tag{A28}$$

where $v^D_{\text{DOC}_A\text{COP}}$ is the maximal additional production rate and $0 < \xi < 1$ depends on Vit and Fe with fourth order Hill equations terms parametrised with K'^D_{Vit} and K'^D_{Fe} (see Figure A2).

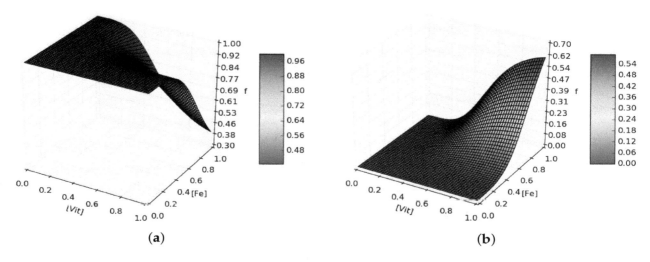

Figure A2. Example for DOC_A ((a), $1 - \xi$) and COP ((b), ξ) additional production rates dependent on Vit and Fe availability in the media. Here $K'^D_{\text{Vit}} = 0.1$, $K'^D_{\text{Fe}} = 0.5$.

Appendix B.2. Parameter Fitting

The model has 56 parameters, of which 55 are free parameters (see Table A1). Being a qualitative model, we do not aim at interpreting the absolute parameter values in a biological sense.

Table A1. Total number of parameters for each parameter set. The dependent parameter is $\epsilon_{\text{DOM}} = 1 - \epsilon_{\text{DOC}_A}$ in the sub-set of A parameters $\mathcal{P}(A)$.

Parameter Sub-Set	$\mathcal{P}(D)$	$\mathcal{P}(PA)$	$\mathcal{P}(A)$	$\mathcal{P}(F)$	$\mathcal{P}(P)$	Degradation
Sub-set size	15	9	14	8	8	2

The available data that can be used to fit the model parameters are the diatom biomass growth in two media conditions and four time points with bacteria relative abundances again in two media conditions. We can therefore fit the diatom biomass D evolution and the four relative bacteria i abundances $B_i / \sum_j B_j$ time-course.

We implement as general strategy a genetic algorithm, where an "individual" i is a full set of 56 parameters $\mathcal{P}_i$, a "population" is an *ensemble* of parameter sets $\{\mathcal{P}_i\}$, a population at a certain evolution step is a "generation" and "evolution" goes as:

(1)   the first generation $\{\mathcal{P}_i\}^0$ is populated by extracting the parameters from random uniform distributions within user-chosen ranges;

(2)   for each $\mathcal{P}_i$ the ODE system is solved and a fitness score (see Appendix B.2.1) is computed;

(3)   the most fit 10% individuals are retained as parents for the next generation;
(4)   the remaining individuals have a probability $p = 0.05$ to be also selected as parents;
(5)   parents are crossed to obtain enough children to reach the original population size;
(6)   crossing means randomly pick a parameter sub-set from one parent or the other;
(7)   each children has a probability $p = 0.3$ to randomly mutate one parameter;
(8)   the process is repeated from step 2. until generation $\{\mathcal{P}_i\}^{G_{max}}$.

Appendix B.2.1. Fitness Score

Fitness scores are computed in a different way when fitting the diatom growth or the bacteria relative abundances. When fitting to the diatom biomass data we compute the score as a simple euclidean distance:

$$s = \sqrt{\sum_t (x_t - X_t)^2} \tag{A29}$$

where the sum over time extends over 22 time points, x_t is the D biomass at time t and X_t is the biomass data at time t. The lower s, the better the fit. This score definition works well to fit the measurements of diatom biomass, but presents a big problem when used with bacteria relative abundances. A relative abundance is a number between 0 and 1, and we observe high variations including bacteria population going from very close to 0 to high abundance. Having only three time points to fit (the first 16S measurement is used as initial point), it can happen that constantly low abundant population are kept by the algorithm. We therefore define for the fit of bacteria relative abundances the following score:

$$s = \sum_t \sqrt{\sum_o \left(1 - e^{\frac{r_{o_t} - R_{o_t}}{r_{o_t}}}\right)^2} \tag{A30}$$

where the sum over time extends over 3 time points and the sum over organisms over the 4 bacterial species, r_{o_t} is the relative abundance from the ODEs system solution for organism o at time t and R_{o_t} is it the corresponding experimental relative abundance. This score definition allows to penalize the event of population extinction: when r is 0, the exponential term is 0 and the score is 1, while when $r = R$ the exponential term is 1 and the score is 0.

Table A2. Datasets used to fit diatom growth in minimal and complete media (MM and CM respectively). Time is scaled (1/3 of a day) to fit reasonably the growth phases (lag-log-exp-decay) using parameters $\mathcal{O}(1)$. For the same reason cell counts are scaled to bring the lower count close to 0, but not feature-scaled to avoid loosing information on differences among MM and CM conditions. Only average values, and not experimental errors, are taken into account.

T	8	16	40	48	64	72	104	112	120	128	152
MM	0.004	0.021	0.133	0.325	0.820	1.012	1.121	1.187	1.192	1.233	1.209
CM	0.050	0.044	0.162	0.605	0.733	0.919	1.037	1.099	1.134	1.108	0.859

T	168	176	184	208	216	232	240	248	264	272	288
MM	1.104	1.096	0.951	1.015	0.965	0.851	0.869	0.704	0.481	0.504	0.394
CM	0.821	0.844	0.624	0.682	0.624	0.556	0.535	0.478	0.199	0.282	0.303

Table A3. Relative abundances of the four bacterial families at three intermediate time points (days 8, 15 and 22). The abundances were scaled from the experimental values (where more families were present) to add to unity.

	Complete Media				Minimal Media			
t	PA	F	A	P	PA	F	A	P
64	0.101	0.724	0.159	0.014	0.294	0.132	0.308	0.264
120	0.453	0.474	0.061	0.010	0.351	0.031	0.585	0.031
176	0.600	0.084	0.189	0.126	0.385	0.020	0.187	0.406

Appendix B.2.2. Results of the Genetic Algorithm

The chosen population size is 200 and the algorithm stops either after non significant increase in fitness or at generation number 50. The algorithm can be run to fit six scenarios:

- D-MM: D Biomass in Minimal Media;
- D-CM: D Biomass in Complete Media;
- B-MM: Bacteria relative abundances in Minimal Media;
- B-CM: Bacteria relative abundances in Complete Media;
- D*B-MM: D Biomass and Bacteria relative abundances in Minimal Media;
- D*B-CM: D Biomass and Bacteria relative abundances in Complete Media;

For D-type fits, the fitness score of Equation (A29) is used. For B-type fits, the fitness score of Equation (A30) is used. For D*B-type fits, the fitness score is the product of the two scores. We will refer to D-fit, B-fit and D*B-fit in the following if media is not to be specified.

Considering the fact that a simple ODE model cannot capture metabolic readjustment, we do not expect to obtain the same parameters for CM and MM conditions. The fitting is therefore performed separately in the two conditions and in the following steps:

1. B-fit is run 20 times varying all 55 parameters in $\mathcal{O}(1)$ ranges
2. The parameters from the best B-fits are kept ($\mathcal{P}_{MM1}$ and $\mathcal{P}_{CM1}$)
3. After checking the effect of varying the different parameters sets, different variation ranges are chosen to perform refits
4. D*B-CM is run 5 times varying $\mathcal{P}(D, deg)_{CM1} \pm 50\%$, $\mathcal{P}(A, F, P)_{CM1} \pm 20\%$, $\mathcal{P}(PA)_{CM1} \pm 10\%$
5. D*B-MM is run 5 times varying $\mathcal{P}_{MM1} \pm 50\%$, and the best parameters are kept ($\mathcal{P}_{MM2}$)
6. D*B-MM is run again 5 times varying $\mathcal{P}(D)_{MM2} \pm 5\%$, $\mathcal{P}(A, F, P, PA, deg)_{MM2} \pm 80\%$

The last rounds of fitting were run on different sets of parameters considered equally good. The final parameter sets $\mathcal{P}$ are presented in Table A4.

Table A4. Final parameter sets used for simulation in CM ($\mathcal{P}_{CM}$) and MM ($\mathcal{P}_{MM}$). Also reported are the overall average and standard deviation values from all the last rounds of fitting.

	$\mathcal{P}_{CM}$	$\mu(\mathcal{P}_{CM})$	$\sigma(\mathcal{P}_{CM})$	$\mathcal{P}_{MM}$	$\mu(\mathcal{P}_{MM})$	$\sigma(\mathcal{P}_{MM})$
K_{Bac}^A	0.562780	0.476829	0.235751	0.23821	0.281329	0.257813
$K_{DOC_A}^A$	0.463690	0.249183	0.152283	0.02253	0.332969	0.365116
K_{DOM}^A	1.043490	0.526842	0.339699	0.82552	0.671145	0.475618
v_{Bac}^A	1.884690	1.433304	0.495578	0.94702	1.085873	0.889537
CC^A	1.230920	1.112280	0.351905	2.74047	1.564099	0.668637
v_δ^A	0.036310	0.067159	0.085177	0.01697	0.058289	0.113161
$v_{DOC_A}^{cons(A)}$	0.504220	0.504324	0.149340	1.30073	0.806015	0.315439
ϵ_{DOC_A}	0.204470	0.464587	0.226651	0.99257	0.670187	0.250496
$v_{DOM}^{cons(A)}$	0.186340	0.493623	0.298676	0.74598	0.349684	0.246408
ϵ_{DOM}	0.795530	0.508826	0.229097	0.00743	0.323377	0.248408
$v_{DOM}^{prod(A)}$	0.030470	0.052871	0.094576	0.06702	0.130514	0.176652
$v_{Fe}^{prod(A)}$	0.134290	0.118939	0.028298	0.19948	0.206115	0.091565
v_γ^A	0.329520	0.434683	0.340005	0.34841	1.037434	0.510106
$v_{Vit}^{prod(A)}$	0.954240	0.644251	0.317607	1.09226	1.241136	0.618984
v_{Bac}^{deg}	0.108110	0.409152	0.213390	0.07769	0.263422	0.330743
v_{EPA}^{deg}	0.350050	0.373353	0.117858	0.57995	0.784058	0.412260

Table A4. *Cont.*

	$\mathcal{P}_{CM}$	$\mu(\mathcal{P}_{CM})$	$\sigma(\mathcal{P}_{CM})$	$\mathcal{P}_{MM}$	$\mu(\mathcal{P}_{MM})$	$\sigma(\mathcal{P}_{MM})$
K_{Fe}^{D}	0.488680	0.583597	0.358354	0.02979	0.124157	0.134217
$K_{Fe}'^{D}$	1.199730	1.048777	0.343732	0.33321	0.486298	0.256807
K_{Vit}^{D}	0.844900	0.645544	0.226955	0.46274	0.346839	0.088571
$K_{Vit}'^{D}$	0.469600	0.903463	0.501364	0.09782	0.339058	0.356208
$v_{DOC_A COP}^{D}$	0.314330	0.853140	0.557108	0.36480	0.546552	0.330353
CC^{D}	1.875200	1.584920	0.515701	1.57897	1.427444	0.420945
$v_{COP}^{prod(D)}$	1.005110	1.268507	0.463020	0.70666	0.736754	0.370734
v_{δ}^{D}	0.007180	0.016960	0.051891	0.00681	0.013765	0.049375
$v_{DOC_A}^{prod(D)}$	1.770740	0.987437	0.490546	1.65657	1.350378	0.421454
$v_{DOC_{PA}}^{prod(D)}$	1.055270	0.990547	0.415673	0.83897	1.081959	0.591094
$v_{DOM}^{prod(D)}$	0.135980	0.653181	0.351953	0.54133	0.565364	0.231813
$v_{EPA}^{prod(D)}$	1.214350	0.899207	0.360058	1.28659	1.070999	0.478554
$v_{Fe}^{cons(D)}$	0.665740	0.755699	0.241111	0.31684	0.363974	0.099207
v_{γ}^{D}	0.194310	0.200395	0.069030	0.52737	0.562459	0.149546
$v_{Vit}^{cons(D)}$	0.367880	0.566566	0.416514	1.78450	0.909564	0.404000
K_{Bac}^{F}	0.936420	0.583317	0.263447	0.16731	0.299761	0.198921
K_{COP}	0.477700	0.588674	0.234155	0.74525	0.451922	0.315203
v_{Bac}^{F}	0.184360	0.311845	0.105780	0.23234	1.237169	1.012458
CC^{F}	1.351050	1.206888	0.384417	0.54187	1.074951	0.758107
$v_{COP}^{cons(F)}$	0.139320	0.175972	0.045416	0.57005	0.330531	0.127086
v_{δ}^{F}	0.382820	0.318181	0.144775	0.18005	0.200895	0.150824
$v_{DOM}^{prod(F)}$	0.092860	0.080066	0.074871	0.00984	0.135875	0.283818
v_{γ}^{F}	0.765450	0.726578	0.223690	1.50156	0.888556	0.545041
K_{Bac}^{P}	0.020100	0.148399	0.132143	0.16823	0.326145	0.294570
K_{DOM}^{P}	0.609800	0.560853	0.171693	1.12080	0.688621	0.413129
v_{Bac}^{P}	1.009740	1.238831	0.430709	2.11081	1.419892	0.958683
CC^{P}	1.301320	1.277678	0.407513	1.17750	2.585117	0.869802
v_{δ}^{P}	0.020440	0.069033	0.167148	0.01591	0.036821	0.150349
$v_{DOM}^{cons(P)}$	0.698330	0.523151	0.203572	0.17625	0.124345	0.136063
$v_{DOM}^{prod(P)}$	0.195450	0.189091	0.103414	0.03107	0.116528	0.203261
v_{γ}^{P}	0.820720	0.440066	0.249502	0.57938	0.527859	0.284980
$K_{DOC_{PA}}$	0.245720	0.351941	0.221873	0.42128	0.564689	0.489764
K_{EPA}	0.755570	0.541606	0.297743	0.05329	0.404319	0.414359
v_{EPA}^{PA}	1.577050	1.484135	0.474372	2.65508	1.368551	0.892632
$v_{Bac}^{prod(PA)}$	0.819580	0.848959	0.264185	0.28618	0.568550	0.438164
CC^{PA}	0.995130	1.029216	0.323852	1.28138	1.477045	0.533872
v_{δ}^{PA}	0.221040	0.284309	0.181709	0.01861	0.052638	0.102986
$v_{DOC_{PA}}^{cons(PA)}$	0.236820	0.254966	0.144860	0.41130	0.249994	0.182493
$v_{DOM}^{prod(PA)}$	0.130620	0.110548	0.125387	0.01816	0.108930	0.154334
v_{γ}^{PA}	0.327430	0.468045	0.287832	0.12769	0.350329	0.210662

Appendix B.2.3. Sanity Checks of the Parameter Fits

The parameters of the algorithm were chosen to obtain a satisfactory convergence of the fit (Figure A3).

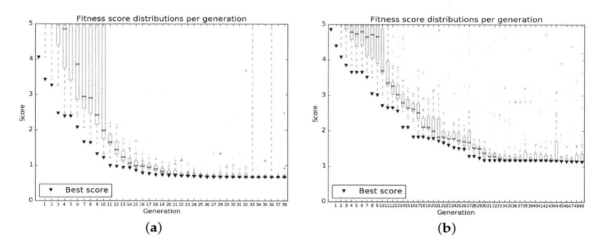

(a)   (b)

Figure A3. Distribution of fitness scores in populations over generations for the genetic algorithm runs chosen to perform the last fitting iteration in minimal (**a**) and complete (**b**) media conditions.

We checked the effect of varying the parameters δ^A, $v_{\mathrm{DOC}_A}^{\mathrm{cons}(A)}$ and $v_{\mathrm{Fe}}^{\mathrm{prod}(A)}$ (the only bacterial parameters observed to influence the biomass growth curve in CM) by $\pm10\%$ and $\pm50\%$. The diatom growth is almost insensitive to these variations in CM (Figure A4), while it shows greater effects in MM (Figure A5).

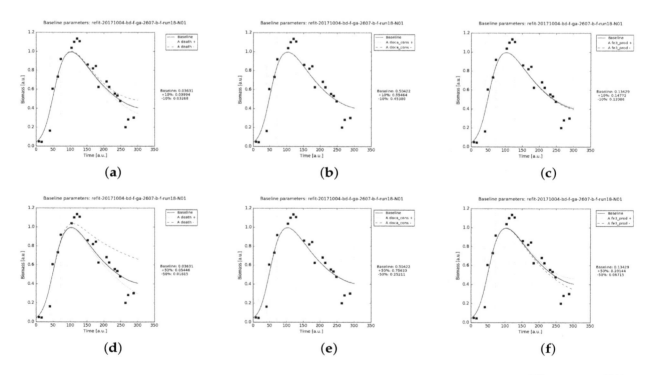

Figure A4. Diatom growth in CM simulation results. The parameters δ^A, $v_{\mathrm{DOC}_A}^{\mathrm{cons}(A)}$ and $v_{\mathrm{Fe}}^{\mathrm{prod}(A)}$, are varied by $\pm10\%$ ((**a**–**c**) respectively) and by $\pm50\%$ ((**d**–**f**) respectively).

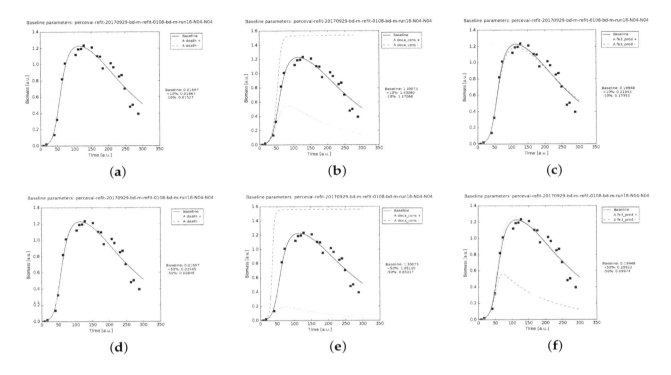

Figure A5. Diatom growth in MM simulation results. The parameters δ^A, $v_{DOC_A}^{cons(A)}$ and $v_{Fe}^{prod(A)}$, are varied by $\pm 10\%$ ((**a**–**c**) respectively) and by $\pm 50\%$ ((**d**–**f**) respectively).

Parameter profiling shows that the algorithm correctly converges towards local minima and that in general those minima are rather stable to perturbation $p \pm 50\%$. Figure A6 shows examples of the most unstable profiles from this first set of fits. Additional information is provided in the Supplementary Material 2.

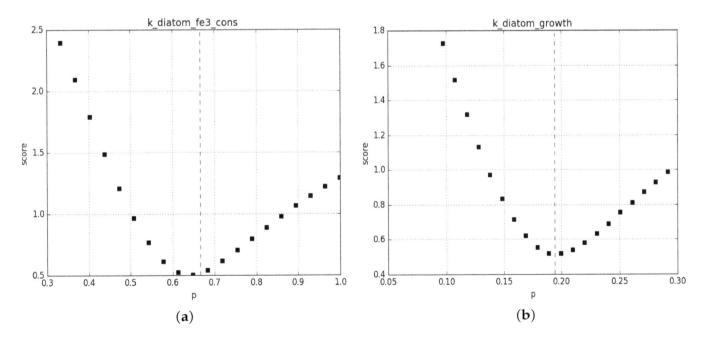

Figure A6. *Cont.*

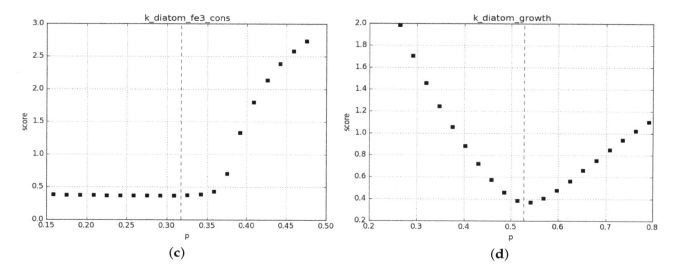

Figure A6. Profiling of the parameters $v_{Fe}^{cons(D)}$ in CM (**a**) and MM (**c**) and γ^D in CM (**b**) and MM (**d**). The red line shows the value chosen by the fit.

References

1. Chisti, Y. Biodiesel from microalgae. *Biotechnol. Adv.* **2007**, *25*, 294–306.
2. Moejes, F.W.; Moejes, K.B. Algae for Africa: Microalgae as a source of food, feed and fuel in Kenya. *Afr. J.Biotechnol.* **2017**, *16*, 288–301.
3. Ryther, J.; Goldman, J. Microbes as food in mariculture. *Annu. Rev. Microbiol.* **1975**, *29*, 429–443.
4. Tredici, M.R.; Biondi, N.; Ponis, E.; Rodolfi, L.; Zittelli, G.C.; Burnell, G.; Allan, G. Advances in microalgal culture for aquaculture feed and other uses. In *New Technologies in Aquaculture: Improving Production Efficiency, Quality and Environmental Management*; Burnell, G., Allan, G., Eds.; Woodhead Publishing Ltd.: Cambridge, UK, 2009; pp. 610–676.
5. Rebolloso-Fuentes, M.; Navarro-Pérez, A.; Ramos-Miras, J.; Guil-Guerrero, J. Biomass nutrient profiles of the microalga Phaeodactylum tricornutum. *J. Food Biochem.* **2001**, *25*, 57–76.
6. Kates, M.; Volcani, B.E. Lipid components of diatoms. *Biochim. Biophys. Acta* **1966**, *116*, 264–278.
7. Fajardo, A.R.; Cerdán, L.E.; Medina, A.R.; Fernández, F.G.A.; Moreno, P.A.G.; Grima, E.M. Lipid extraction from the microalga Phaeodactylum tricornutum. *Eur. J. Lipid Sci. Technol.* **2007**, *109*, 120–126.
8. Siron, R.; Giusti, G.; Berland, B. Changes in the fatty acid composition of Phaeodactylum tricornutum and Dunaliella tertiolecta during growth and under phosphorus deficiency. *Mar. Ecol. Prog. Ser.* **1989**, *55*, 95–100.
9. Yashodhara, B.M.; Umakanth, S.; Pappachan, J.M.; Bhat, S.K.; Kamath, R.; Choo, B.H. Omega-3 fatty acids: A comprehensive review of their role in health and disease. *Postgrad. Med. J.* **2009**, *85*, 84–90.
10. Francius, G.; Tesson, B.; Dague, E.; Martin-jézéquel, V.; Dufrêne, Y.F. Nanostructure and nanomechanics of live Phaeodactylum tricornutum morphotypes. *Environ. Microbiol.* **2008**, *10*, 1344–1356.
11. Santaeufemia, S.; Torres, E.; Mera, R.; Abalde, J. Bioremediation of oxytetracycline in seawater by living and dead biomass of the microalga Phaeodactylum tricornutum. *J. Hazard. Mater.* **2016**, *320*, 315–325.
12. Mata, T.; Martins, A.; Caetano, N. Microalgae for biodiesel production and other applications: A review. *Renew. Sustain. Energy Rev.* **2010**, *14*, 217–232.
13. Grima, E.; Fernández, F. Photobioreactors: Light regime, mass transfer, and scaleup. *J. Biotechnol.* **1999**, *70*, 231–247.
14. Parmar, A.; Singh, N.; Pandey, A. Cyanobacteria and microalgae: A positive prospect for biofuels. *Bioresour. Technol.* **2011**, *102*, 10163–10172.
15. Day, J.G.; Thomas, N.J.; Achilles-Day, U.E.M.; Leakey, R.J.G. Early detection of protozoan grazers in algal biofuel cultures. *Bioresour. Technol.* **2012**, *114*, 715–719.
16. Wang, H.; Zhang, W.; Chen, L.; Wang, J.; Liu, T. The contamination and control of biological pollutants in mass cultivation of microalgae. *Bioresour. Technol.* **2013**, *128*, 745–750.

17. Kazamia, E.; Aldridge, D.C.; Smith, A.G. Synthetic ecology—A way forward for sustainable algal biofuel production? *J. Biotechnol.* **2012**, *162*, 163–169.

18. Gause, G. *The Struggle for Existence*; The Williams & Wilkins company: Baltimore, MD, USA, 1934; p. 163.

19. Hardin, G. The competitive exclusion principle. *Science* **1960**, *131*, 1292–1297.

20. Dykhuizen, D. Santa Rosalia revisited: Why are there so many species of bacteria? *Antonie Van Leeuwenhoek* **1998**, *73*, 25–33.

21. Hagstrom, G.I.; Levin, S.A. Marine Ecosystems as Complex Adaptive Systems: Emergent Patterns, Critical Transitions, and Public Goods. *Ecosystems* **2017**, *20*, 458–476.

22. Widder, S.; Allen, R.J.; Pfeiffer, T.; Curtis, T.P.; Wiuf, C.; Sloan, W.T.; Cordero, O.X.; Brown, S.P.; Momeni, B.; Shou, W.; et al. Challenges in microbial ecology: Building predictive understanding of community function and dynamics. *ISME J.* **2016**, *10*, 2557–2568.

23. Song, H.S.; Cannon, W.; Beliaev, A.; Konopka, A. Mathematical Modeling of Microbial Community Dynamics: A Methodological Review. *Processes* **2014**, *2*, 711–752.

24. Zomorrodi, A.R.; Segrè, D. Synthetic Ecology of Microbes: Mathematical Models and Applications. *J. Mol. Biol.* **2015**, *428*, 837–861.

25. Succurro, A.; Moejes, F.W.; Ebenhöh, O. A Diverse Community To Study Communities: Integration of Experiments and Mathematical Models To Study Microbial Consortia. *J. Bacteriol.* **2017**, *199*, doi:10.1128/JB.00865-16.

26. Lotka, A. *Elements of Physical Biology*; Williams & Wilkins Company: Philadelphia, PA, USA, 1925.

27. Volterra, V. Fluctuations in the abundance of a species considered mathematically. *Nature* **1926**, *118*, 558–560.

28. Hofbauer, J.; Hutson, V.; Jansen, W. Coexistence for systems governed by difference equations of Lotka-Volterra type. *J. Math. Biol.* **1987**, *25*, 553–570.

29. Berry, D.; Widder, S. Deciphering microbial interactions and detecting keystone species with co-occurrence networks. *Front. Microbiol.* **2014**, *5*, 219.

30. Amin, S.A.; Parker, M.S.; Armbrust, E.V. Interactions between diatoms and bacteria. *Microbiol. Mol. Biol. Rev.* **2012**, *76*, 667–684.

31. Provasoli, L. Nutrition and ecology of Protozoa and Algae. *Annu. Rev. Microbiol.* **1958**, *12*, 279–308.

32. Delucca, R.; McCracken, M.D. Observations on interactions between naturally-collected bacteria and several species of algae. *Hydrobiologia* **1977**, *55*, 71–75.

33. Suminto; Hirayama, K. Application of a growth-promoting bacteria for stable mass culture of three marine microalgae. *Hydrobiologia* **1997**, *358*, 223–230.

34. Bruckner, C.G.; Rehm, C.; Grossart, H.P.; Kroth, P.G. Growth and release of extracellular organic compounds by benthic diatoms depend on interactions with bacteria. *Environ. Microbiol.* **2011**, *13*, 1052–1063.

35. Amin, S.A.; Hmelo, L.R.; van Tol, H.M.; Durham, B.P.; Carlson, L.T.; Heal, K.R.; Morales, R.L.; Berthiaume, C.T.; Parker, M.S.; Djunaedi, B.; et al. Interaction and signalling between a cosmopolitan phytoplankton and associated bacteria. *Nature* **2015**, *522*, 98–101.

36. Verhulst, P. Notice sur la loi que la population suit dans son accroissement. *Corresp. Math. Phys. l'Obs. Brux.* **1838**, *10*, 113–121.

37. Gachon, C.M.; Day, J.G.; Campbell, C.N.; Pröschold, T.; Saxon, R.J.; Küpper, F.C. The Culture Collection of Algae and Protozoa (CCAP): A biological resource for protistan genomics. *Gene* **2007**, *406*, 51–57.

38. Moejes, F.W. Dynamics of the bacterial community associated with Phaeodactylum tricornutum cultures: A novel approach to scaling up microalgal cultures. Ph.D. Thesis, Heinrich Heine University Dusseldorf, Düsseldorf, Germany, 2016.

39. Guillard, R.; Ryther, J. Studies of marine planktonic diatoms: I. Cyclotella nana Hutedt, and Detonula confervacea (Cleve) Gran. *Can. J. Microbiol.* **1962**, *8*, 229–239.

40. Guillard, R. Culture of phytoplankton for feeding marine invertebrates. In *Culture of Marine Invertebrate Animals*; Smith, W.L., Chanley, M.H., Eds.; Plenum Press: New York, NY, USA, 1975; pp. 29–60.

41. de Gouvion Saint Cyr, D.; Wisniewski, C.; Schrive, L.; Farhi, E.; Rivasseau, C. Feasibility study of microfiltration for algae separation in an innovative nuclear effluents decontamination process. *Sep. Purif. Technol.* **2014**, *125*, 126–135.

42. Quail, M.; Smith, M.E.; Coupland, P.; Otto, T.D.; Harris, S.R.; Connor, T.R.; Bertoni, A.; Swerdlow, H.P.; Gu, Y. A tale of three next generation sequencing platforms: Comparison of Ion torrent, pacific biosciences and illumina MiSeq sequencers. *BMC Genom.* **2012**, *13*, 341.

43. Grada, A.; Weinbrecht, K. Next-Generation Sequencing: Methodology and Application. *J. Investig. Dermatol.* **2013**, *133*, e11.

44. Pylro, V.S.; Roesch, L.F.W.; Morais, D.K.; Clark, I.M.; Hirsch, P.R.; Tótola, M.R. Data analysis for 16S microbial profiling from different benchtop sequencing platforms. *J. Microbiol. Methods* **2014**, *107*, 30–37.

45. Aronesty, E. ea-utils: Command-Line Tools for Processing Biological Sequencing Data, Durham, NC, USA, 2011. Available online: https://github.com/ExpressionAnalysis/ea-utils (accessed on 30 November 2017).

46. Edgar, R.C. Search and clustering orders of magnitude faster than BLAST. *Bioinformatics* **2010**, *26*, 2460–2461.

47. Edgar, R.C. UPARSE: Highly accurate OTU sequences from microbial amplicon reads. *Nat. Methods* **2013**, *10*, 996–998.

48. Caporaso, J.G.; Kuczynski, J.; Stombaugh, J.; Bittinger, K.; Bushman, F.D.; Costello, E.K.; Fierer, N.; Peña, A.G.; Goodrich, J.K.; Gordon, J.I.; et al. QIIME allows analysis of high-throughput community sequencing data. *Nat. Methods* **2010**, *7*, 335–336.

49. Quast, C.; Pruesse, E.; Yilmaz, P.; Gerken, J.; Schweer, T.; Yarza, P.; Peplies, J.; Glockner, F.O. The SILVA ribosomal RNA gene database project: Improved data processing and web-based tools. *Nucleic Acids Res.* **2013**, *41*, D590–D596.

50. R Development Core Team. R: A Language and Environment for Statistical Computing. R Foundation for Statistical Computing, Vienna, Austria. 2015. Available online: http://www.R-project.org/ (accessed on 30 November 2017).

51. Monod, J. The Growth of Bacterial Cultures. *Annu. Rev. Microbiol.* **1949**, *3*, 371–394.

52. Mitchell, M. *An Introduction to Genetic Algorithms*; MIT Press: Cambridge, MA, USA, 1996.

53. Chen, J.; Bittinger, K.; Charlson, E.S.; Hoffmann, C.; Lewis, J.; Wu, G.D.; Collman, R.G.; Bushman, F.D.; Li, H. Associating microbiome composition with environmental covariates using generalized UniFrac distances. *Bioinformatics* **2012**, *28*, 2106–2113.

54. Bowman, J. Bioactive compound synthetic capacity and ecological significance of marine bacterial genus Pseudoalteromonas. *Mar. Drugs* **2007**, *5*, 220–241.

55. Desbois, A.P.; Mearns-Spragg, A.; Smith, V.J. A Fatty Acid from the Diatom Phaeodactylum tricornutum is Antibacterial Against Diverse Bacteria Including Multi-resistant Staphylococcus aureus (MRSA). *Mar. Biotechnol.* **2009**, *11*, 45–52.

56. Soria-Dengg, S.; Reissbrodt, R.; Horstmann, U. Siderophores in marine coastal waters and their relevance for iron uptake by phytoplankton: Experiments with the diatom Phaeodactylum tricornutum. *Mar. Ecol. Prog. Ser.* **2001**, *220*, 73–82.

57. Amin, S.A.; Green, D.H.; Hart, M.C.; Küpper, F.C.; Sunda, W.G.; Carrano, C.J. Photolysis of iron, siderophore chelates promotes bacterial, algal mutualism. *Proc. Natl. Acad. Sci. USA* **2009**, *106*, 17071–17076.

58. Soria-Dengg, S.; Horstmann, U. Ferrioxamines B and E as iron sources for the marine diatom Phaeodactylum tricornutum. *Mar. Ecol. Prog. Ser.* **1995**, *127*, 269–277.

59. Provasoli, L. Organic regulation of phytoplankton fertility. In *The Sea: Ideas and Observations on Progress in the Study of the Seas*; Hill, M., Ed.; Wiley-Interscience: New York, NY, USA, 1963; pp. 165–219.

60. Provasoli, L.; Carlucci, A. Vitamins and growth regulators. In *Algal Physiology and Biochemistry, Botanical Monographs, 10*; Stewart, W., Ed.; Blackwell Scientific Publications: San Diego, CA, USA, 1974; pp. 741–787.

61. Croft, M.T.; Warren, M.J.; Smith, A.G. Algae need their vitamins. *Eukaryot. Cell* **2006**, *5*, 1175–1183.

62. Yongmanitchai, W.; Ward, O.P. Growth of and omega-3 fatty acid production by Phaeodactylum tricornutum under different culture conditions. *Appl. Environ. Microbiol.* **1991**, *57*, 419–425.

63. Sañudo-Wilhelmy, S.A.; Gómez-Consarnau, L.; Suffridge, C.; Webb, E.A. The Role of B Vitamins in Marine Biogeochemistry. *Annu. Rev. Mar. Sci.* **2014**, *6*, 339–367.

64. Azam, F.; Fenchel, T.; Field, J.; Gray, J.; Meyer-Reil, L.; Thingstad, F. The Ecological Role of Water-Column Microbes in the Sea. *Mar. Ecol. Prog. Ser.* **1983**, *10*, 257–263.

65. Stocker, R. Marine Microbes See a Sea of Gradients. *Science* **2012**, *338*, 628–633.

66. Persson, G.; Jansson, M.; Kluwer, C. Phosphate uptake and utilizaton by bacteria and algae. *Hydrobiologia* **1988**, *170*, 177–189.

67. Bell, W.; Mitchell, R. Chemotactic and growth responses of marine bacteria to algal extracellular products. *Biol. Bull.* **1972**, *143*, 265–277.

68. MacArthur, R. Fluctuations of animal populations and a measure of community stability. *Ecology* **1955**, *36*, 533–536.

69. Gardner, M.; Ashby, W. Connectance of large dynamic (cybernetic) systems: Critical values for stability. *Nature* **1970**, *228*, 784.

70. Pimm, S. The complexity and stability of ecosystems. *Nature* **1984**, *307*, 321–326.

71. Elton, C. *The Ecology of Invasions by Animals and Plants*; Springer: Dordrecht, The Netherlands, 1958.

72. McCann, K.S. The diversity-stability debate. *Nature* **2000**, *405*, 228–233.

73. Ivanova, E.P.; Flavier, S.; Christen, R. Phylogenetic relationships among marine Alteromonas-like proteobacteria: Emended description of the family Alteromonadaceae and proposal of Pseudoalteromonadaceae fam. nov., Colwelliaceae fam. nov., Shewanellaceae fam. nov., Moritellaceae fam. nov., Ferri. *Int. J. Syst. Evolut. Microbiol.* **2004**, *54*, 1773–1788.

74. Rosenberg, E.; DeLong, E.F.; Lory, S.; Stackebrandt, E.; Thompson, F. *The Prokaryotes*; Springer: Berlin/Heidelberg, Germany, 2014.

75. Lee, S.O.; Kato, J.; Takiguchi, N.; Kuroda, A.; Ikeda, T. Involvement of an Extracellular Protease in Algicidal Activity of the Marine Bacterium *Pseudoalteromonas* sp. Strain A28. *Appl. Environ. Microbiol.* **2000**, *66*, 4334–4339.

76. Venkateswaran, K.; Dohmoto, N. *Pseudoalteromonas peptidolytica* sp. nov., a novel marine mussel-thread-degrading bacterium isolated from the Sea of Japan. *Int. J. Syst. Evolut. Microbiol.* **2000**, *50*, 565–574.

77. Chen, X.; Xie, B.; Lu, J.; He, H.; Zhang, Y. A novel type of subtilase from the psychrotolerant bacterium *Pseudoalteromonas* sp. SM9913: Catalytic and structural properties of deseasin MCP-01. *Microbiology* **2007**, *153*, 2116–2125.

78. Khudary, R.A.; Venkatachalam, R.; Katzer, M. A cold-adapted esterase of a novel marine isolate, Pseudoalteromonas arctica: Gene cloning, enzyme purification and characterization. *Extremophiles* **2010**, *14*, 273–285.

79. Lu, M.; Wang, S.; Fang, Y.; Li, H.; Liu, S.; Liu, H. Cloning, expression, purification, and characterization of cold-adapted α-amylase from Pseudoalteromonas arctica GS230. *Protein J.* **2010**, *29*, 591–597.

80. Albino, A.; Marco, S.; Maro, A.D. Characterization of a cold-adapted glutathione synthetase from the psychrophile Pseudoalteromonas haloplanktis. *Mol. Biosyst.* **2012**, *8*, 2405–2414.

81. He, H.; Guo, J.; Chen, X.; Xie, B.; Zhang, X. Structural and functional characterization of mature forms of metalloprotease E495 from Arctic sea-ice bacterium *Pseudoalteromonas* sp. SM495. *PLoS ONE* **2012**, *7*, e35442.

82. Holmström, C.; Kjelleberg, S. Marine Pseudoalteromonas species are associated with higher organisms and produce biologically active extracellular agents. *FEMS Microbiol. Ecol.* **1999**, *30*, 285–293.

83. Ivanova, E.P.; Mikhaïlov, V.V. A new family of Alteromonadaceae fam. nov., including the marine proteobacteria species Alteromonas, Pseudoalteromonas, Idiomarina i Colwellia. *Mikrobiologiia* **2001**, *70*, 15–23.

84. Reid, R.; Butler, A. Investigation of the mechanism of iron acquisition by the marine bacterium Alteromonas luteoviolaceus: Characterization of siderophore production. *Limnol. Oceanogr.* **1991**, *36*, 1783–1792.

85. Holt, P.D.; Reid, R.R.; Lewis, B.L.; Luther, G.W.; Butler, A. Iron(III) coordination chemistry of alterobactin A: A siderophore from the marine bacterium Alteromonas luteoviolacea. *Inorg. Chem.* **2005**, *44*, 7671–7677.

86. Vraspir, J.M.; Butler, A. Chemistry of marine ligands and siderophores. *Annu. Rev. Mar. Sci.* **2009**, *1*, 43–63.

87. Bruland, K.W.; Donat, J.R.; Hutchins, D.a. Interactive influences of bioactive trace metals on biological production in oceanic waters. *Limnol. Oceanogr.* **1991**, *36*, 1555–1577.

88. Morel, F.M.M.; Price, N.M. The biogeochemical cycles of trace metals in the oceans. *Science* **2003**, *300*, 944–947.

89. Yoon, J.; Jo, Y.; Kim, G.J.; Choi, H. Gramella lutea sp. nov., a Novel Species of the Family Flavobacteriaceae Isolated from Marine Sediment. *Curr. Microbiol.* **2015**, *71*, 252–258.

90. Glöckner, F.O.; Fuchs, B.M.; Amann, R. Bacterioplankton compositions of lakes and oceans: A first comparison based on fluorescence in situ hybridization. *Appl. Environ. Microbiol.* **1999**, *65*, 3721–3726.

91. Abell, G.; Bowman, J. Ecological and biogeographic relationships of class Flavobacteria in the Southern Ocean. *FEMS Microbiol. Ecol.* **2005**, *51*, 265–277.

92. DeLong, E.F.; Preston, C.M.; Mincer, T.; Rich, V.; Hallam, S.J.; Frigaard, N.U.; Martinez, A.; Sullivan, M.B.; Edwards, R.; Brito, B.R.; et al. Community genomics among stratified microbial assemblages in the ocean's interior. *Science* **2006**, *311*, 496–503.

93. Manz, W.; Amann, R.; Ludwig, W.; Vancanneyt, M.; Schleifer, K.H. Application of a suite of 16S rRNA-specific oligonucleotide probes designed to investigate bacteria of the phylum cytophaga-flavobacter-bacteroides in the natural environment. *Microbiology* **1996**, *142*, 1097–1106.

94. Kirchman, D.L. The ecology of Cytophaga-Flavobacteria in aquatic environments. *FEMS Microbiol. Ecol.* **2002**, *39*, 91–100.

95. Cottrell, M.T.; Kirchman, D.L. Natural assemblages of marine proteobacteria and members of the Cytophaga-Flavobacter cluster consuming low- and high-molecular-weight dissolved organic matter. *Appl. Environ. Microbiol.* **2000**, *66*, 1692–1697.

96. Pinhassi, J.; Sala, M.M.; Havskum, H.; Peters, F.; Guadayol, Ò.; Malits, A.; Marrasé, C. Changes in bacterioplankton composition under different phytoplankton regimens. *Appl. Environ. Microbiol.* **2004**, *70*, 6753–6766.

97. Smayda, T. Harmful algal blooms: Their ecophysiology and general relevance to phytoplankton blooms in the sea. *Limnol. Oceanogr.* **1997**, *42*, 1137–1153.

98. Starr, M.P.; Stolp, H.; Trüper, H.G.; Balows, A.; Schlegel, H.G. *The Prokaryotes*; Springer: Berlin/Heidelberg, Germany, 1981.

99. Anzai, Y.; Kim, H.; Park, J.Y.; Wakabayashi, H.; Oyaizu, H. Phylogenetic affiliation of the pseudomonads based on 16S rRNA sequence. *Int. J. Syst. Evolut. Microbiol.* **2000**, *50*, 1563–1589.

100. Isnansetyo, A.; Kamei, Y. Bioactive substances produced by marine isolates of *Pseudomonas*. *J. Ind. Microbiol. Biotechnol.* **2009**, *36*, 1239–1248.

101. Zelezniak, A.; Andrejev, S.; Ponomarova, O.; Mende, D.R.; Bork, P.; Patil, K.R. Metabolic dependencies drive species co-occurrence in diverse microbial communities. *Proc. Natl. Acad. Sci. USA* **2015**, *112*, 6449–6454.

102. Ponomarova, O.; Patil, K.R. Metabolic interactions in microbial communities: Untangling the Gordian knot. *Curr. Opin. Microbiol.* **2015**, *27*, 37–44.

103. Tao, H.; Bausch, C.; Richmond, C.; Blattner, F.R.; Conway, T. Functional genomics: Expression analysis of Escherichia coli growing on minimal and rich media. *J. Bacteriol.* **1999**, *181*, 6425–6440.

104. Beg, Q.K.; Zampieri, M.; Klitgord, N.; Collins, S.B.; Altafini, C.; Serres, M.H.; Segrè, D. Detection of transcriptional triggers in the dynamics of microbial growth: Application to the respiratorily versatile bacterium Shewanella oneidensis. *Nucleic Acids Res.* **2012**, *40*, 7132–7149.

105. Bertrand, E.M.; Allen, A.E.; Dupont, C.L.; Norden-Krichmar, T.M.; Bai, J.; Valas, R.E.; Saito, M.A. Influence of cobalamin scarcity on diatom molecular physiology and identification of a cobalamin acquisition protein. *Proc. Natl. Acad. Sci. USA* **2012**, *109*, E1762–E1771.

106. Horgan, R.P.; Kenny, L.C. 'Omic' technologies: Genomics, transcriptomics, proteomics and metabolomics. *Obstet. Gynaecol.* **2011**, *13*, 189–195.

107. Lozupone, C.; Knight, R. UniFrac: A New Phylogenetic Method for Comparing Microbial Communities. *Appl. Environ. Microbiol.* **2005**, *71*, 8228–8235.

Permissions

All chapters in this book were first published in MDPI; hereby published with permission under the Creative Commons Attribution License or equivalent. Every chapter published in this book has been scrutinized by our experts. Their significance has been extensively debated. The topics covered herein carry significant findings which will fuel the growth of the discipline. They may even be implemented as practical applications or may be referred to as a beginning point for another development.

The contributors of this book come from diverse backgrounds, making this book a truly international effort. This book will bring forth new frontiers with its revolutionizing research information and detailed analysis of the nascent developments around the world.

We would like to thank all the contributing authors for lending their expertise to make the book truly unique. They have played a crucial role in the development of this book. Without their invaluable contributions this book wouldn't have been possible. They have made vital efforts to compile up to date information on the varied aspects of this subject to make this book a valuable addition to the collection of many professionals and students.

This book was conceptualized with the vision of imparting up-to-date information and advanced data in this field. To ensure the same, a matchless editorial board was set up. Every individual on the board went through rigorous rounds of assessment to prove their worth. After which they invested a large part of their time researching and compiling the most relevant data for our readers.

The editorial board has been involved in producing this book since its inception. They have spent rigorous hours researching and exploring the diverse topics which have resulted in the successful publishing of this book. They have passed on their knowledge of decades through this book. To expedite this challenging task, the publisher supported the team at every step. A small team of assistant editors was also appointed to further simplify the editing procedure and attain best results for the readers.

Apart from the editorial board, the designing team has also invested a significant amount of their time in understanding the subject and creating the most relevant covers. They scrutinized every image to scout for the most suitable representation of the subject and create an appropriate cover for the book.

The publishing team has been an ardent support to the editorial, designing and production team. Their endless efforts to recruit the best for this project, has resulted in the accomplishment of this book. They are a veteran in the field of academics and their pool of knowledge is as vast as their experience in printing. Their expertise and guidance has proved useful at every step. Their uncompromising quality standards have made this book an exceptional effort. Their encouragement from time to time has been an inspiration for everyone.

The publisher and the editorial board hope that this book will prove to be a valuable piece of knowledge for researchers, students, practitioners and scholars across the globe.

List of Contributors

Dongyang Yang
Division of Biostatistics, Dalla Lana School of Public Health, University of Toronto, Toronto, ON M5T 3M7, Canada

Wei Xu
Division of Biostatistics, Dalla Lana School of Public Health, University of Toronto, Toronto, ON M5T 3M7, Canada
Department of Biostatistics, Princess Margaret Cancer Centre, Toronto, ON M5G 2M9, Canada

Daniel Craig Zielinski and Arjun Patel
Department of Bioengineering, University of California, San Diego, San Diego, CA 92093, USA

Bernhard O. Palsson
Department of Bioengineering, University of California, San Diego, San Diego, CA 92093, USA
Novo Nordisk Foundation Center for Biosustainability, Technical University of Denmark, 2800 Lyngby, Denmark

Fadoua El Moustaid
Department of Biological Sciences, Virginia Tech University, Blacksburg, VA 24061, USA

Federica Villa
Department of Food, Environmental and Nutritional Sciences, Università degli Studi di Milano, 20133 Milano, Italy

Isaac Klapper
Department of Mathematics, Temple University, Philadelphia, PA 19122, USA

Andrea G. Capodaglio and Daniele Cecconet
DICAr, University of Pavia, 27100 Pavia, Italy

Daniele Molognoni
DICAr, University of Pavia, 27100 Pavia, Italy
LEITAT Technological Centre, 08225 Terrassa, Barcelona, Spain

Ashley E. Beck
Microbiology and Immunology, Center for Biofilm Engineering, Montana State University, Bozeman, MT 59717, USA

Hans C. Bernstein
Biological Sciences Division, Pacific Northwest National Laboratory, Richland, WA 99352, USA

Ross P. Carlson
Chemical and Biological Engineering, Center for Biofilm Engineering, Montana State University, Bozeman, MT 59717, USA

St. Elmo Wilken, Mohan Saxena and Michelle A. O'Malley
Department of Chemical Engineering, University of California, Santa Barbara, CA 93106, USA

Linda R. Petzold
Department of Computer Science, University of California, Santa Barbara, CA 93106, USA

Emily B. Graham and James C. Stegen
Pacific Northwest National Laboratory, Richland, WA 99352, USA

Marco Zaccaria, Sandra Dedrick and Babak Momeni
Department of Biology, Boston College, Chestnut Hill, MA 02467, USA

Aisling J. Daly, Jan M. Baetens and Bernard De Baets
KERMIT, Department of Data Analysis and Mathematical Modelling, Ghent University, Coupure links 653, B-9000 Ghent, Belgium

Johanna Vandermaesen and Dirk Springael
Division of Soil and Water Management, KU Leuven, Kasteelpark Arenberg 20 bus 2459, B-3001 Heverlee, Belgium

Nico Boon
Center for Microbial Ecology and Technology (CMET), Ghent University, Coupure links 653, B-9000 Ghent, Belgium

Tianyu Zhang
Department of Mathematical Sciences, Montana State University, Bozeman, MT 59717, USA

Fiona Wanjiku Moejes and Julie Maguire
Bantry Marine Research Station, Gearhies, Bantry P75 AX07, Co. Cork, Ireland

Antonella Succurro
Cluster of Excellence on Plant Sciences (CEPLAS), Heinrich-Heine University, Universitätsstrasse 1, 40225 Düsseldorf, Germany
Botanical Institute, University of Cologne, Zülpicher Strasse 47b, 50674 Cologne, Germany

Ovidiu Popa and Oliver Ebenhöh
Cluster of Excellence on Plant Sciences (CEPLAS), Heinrich-Heine University, Universitätsstrasse 1, 40225 Düsseldorf, Germany
Institute of Quantitative and Theoretical Biology, Heinrich-Heine University, Universitätsstrasse 1, 40225 Düsseldorf, Germany

Index

Printed in the USA
CPSIA information can be obtained
at www.ICGtesting.com
JSHW051431221024
72173JS00006B/1434